AN INTRODUCTION TO Theory and Applications of Quantum Mechanics

AN INTRODUCTION TO

Theory and Applications of Quantum Mechanics

AMNON YARIV
California Institute of Technology

1807 1982
175 YEARS OF PUBLISHING

JOHN WILEY & SONS

New York
Chichester
Brisbane
Toronto
Singapore

Library of Congress Cataloging in Publication Data:

Yariv, Amnon.
An introduction to theory and applications of
quantum mechanics.

Bibliography: p. 293
Includes index.
1. Quantum theory. I. Title.
QC174.12.Y37 530.1'2 81-16007
ISBN 0-471-06053-4 AACR2

Printed in the United States of America

10 9 8 7 6 5 4 3 2 1

PREFACE

At the California Institute of Technology we have witnessed over the last ten years or so a trend that by now has become an established new direction in our undergraduate and graduate education. We refer to this new direction as applied physics and have even created new undergraduate and graduate degree options bearing this name.

The applied physics student is by training a physics student who has chosen to apply his or her knowledge to solving technological problems rather than unlocking and elucidating the secrets of nature. Up to now most of our graduate applied physics students have received traditional physics undergraduate education. Their undergraduate introduction to physics was often heavily weighted in the direction of atomic scattering theory, elementary particles, and relativity, reflecting the specialization of the departments teaching the course. This was done at the expense of such topics as atomic and molecular spectroscopy, interaction of radiation with matter, and the electronic properties of crystals. These topics form the backbone of new applied disciplines such as lasers, optical communication, and semiconductor electronics, which have come to play important roles in our new communication-oriented technologies.

This text is based on notes accumulated in the process of teaching a new course at the California Institute of Technology that is taken by juniors in applied physics. The course is meant to serve as an introduction to formal quantum mechanics. The intent, however, is to treat in the same course not only the basic formalism and related phenomena, but to take the student a step further to a consideration of generic and important applications.

The two "practical" topics highlighted in this book are the semiconductor transistor and the laser. The transistor is emphasized because it is the basic building block of our electronic technology, including the computer. We examine the laser because of its revolutionary impact on so many aspects of our science and technology, including fusion research, atomic and molecular spectroscopy, and optical communication via silica fibers.

The typical student will follow this course with a one-, or possibly two-, year course of formal quantum mechanics. The hope is that this early introduction to the application of quantum mechanics will lead to a quiet simmering that will pay handsome dividends when, and if, the student chooses to make a career of applied physics.

I am grateful to Ruth Stratton and my wife, Frances, for expert typing of the manuscript and editing and to John Stephen Smith for critical and creative proofreading.

Amnon Yariv
Pasadena, California

CONTENTS

AN INTRODUCTION TO Theory and Applications of Quantum Mechanics

CHAPTER ONE

Why Quantum Mechanics?

In the late 1800s and early 1900s it was becoming clear that the science of physics was due for a major revision. An increasing number of phenomena and observations failed to be adequately, or even approximately, described by the laws of physics as they were then formulated. Problems arose especially in attempts to provide an explanation for phenomena involving "small" particles such as electrons and atoms and their interaction with electromagnetic fields.

At first these deficiencies in physics were patched by ad hoc hypotheses and postulates. However, as their number grew it became clear that physics needed a complete reformulation, especially the physics of small systems. The result was *quantum mechanics*—one of the towering intellectual achievements of humankind.

In this first chapter we will place the formal development of quantum mechanics in context by outlining some of the basic results of classical physics. We will then recount some of the phenomena that, historically, defied explanation by classical physics.

1.1 NEWTONIAN MECHANICS AND CLASSICAL ELECTROMAGNETISM

Newtonian Mechanics

In classical nonrelativistic physics particles are assumed to move under the influence of forces. The law describing the motion is

$$\mathbf{F} = m\frac{d\mathbf{v}}{dt} \tag{1.1}$$

where m is the mass of the particle, $\mathbf{F}$ the force, and $\mathbf{v}$ the velocity. This law, together with the law of gravitational attraction, for example, proved adequate for describing the motion of heavenly bodies and for predicting accurately the orbits of artillery shells. One important aspect of Newtonian mechanics was its determinism. Once the position and velocity of a particle were specified at some instant of time, and if the forces acting on it were known, then its behavior at all other times was exactly determined.

Electromagnetism

The electric and magnetic fields, $\mathbf{E}(\mathbf{r}, t)$ and $\mathbf{B}(\mathbf{r}, t)$, are described in classical electromagnetism by Maxwell's equations, which in free space and in MKS units take the form[1]

$$\nabla \times \mathbf{E} = -\frac{\partial \mathbf{B}}{\partial t} \tag{1.2a}$$

$$\nabla \times \mathbf{B} = \frac{1}{c^2}\frac{\partial \mathbf{E}}{\partial t} \tag{1.2b}$$

where c = velocity of light. Using the vector identity $\nabla \times (\nabla \times \mathbf{A}) = -\nabla^2 \mathbf{A} + \nabla \nabla \cdot \mathbf{A}$ plus the fact that in free space $\nabla \cdot \mathbf{E} = 0$, we obtain from (1.2$a$) and (1.2$b$)

$$\nabla^2 \mathbf{E} - \frac{1}{c^2}\frac{\partial^2 \mathbf{E}}{\partial t^2} = 0 \tag{1.3}$$

and a similar equation for $\mathbf{B}$. Equation (1.3) admits solutions of the form

$$\mathbf{E} = \mathrm{Re}\left(\mathbf{E}_0 e^{-i(\omega t - \mathbf{k}\cdot\mathbf{r})}\right) \tag{1.4}$$

provided

$$k = \omega / c \tag{1.5}$$

The field (1.4) describes a plane wave propagating with a velocity c along $\mathbf{k}$. A stationary observer will observe the field to oscillate at a frequency $\nu = \omega / 2\pi$. (The unit frequency is 1 Hz = 1 cycle/sec.) The wavelength is given by

$$\lambda = \frac{2\pi}{k} = \frac{c}{\nu} \tag{1.6}$$

Classical physics thus provides two formalisms with which to describe natural phenomena. The first—mechanics—deals with particles; the second—electromagnetic theory—deals with radiation waves. The two classes of phenomena were assumed to be distinct but coupled through the Lorentz force equation

$$\mathbf{F} = e[\mathbf{E} + \mathbf{v} \times \mathbf{B}]$$

which gives the force $\mathbf{F}$ exerted on a particle of charge e moving with velocity $\mathbf{v}$ in fields $\mathbf{E}, \mathbf{B}$.

---

[1] $c^2 = (\mu_0 \varepsilon_0)^{-1}$, where μ_0 and ε_0 are the magnetic permeability and dielectric constant of free space, respectively.

1.2 BLACK BODY RADIATION

One of the major unsolved problems occupying physicists around the late 1800s and the early 1900s was that of black body radiation. An idealized "black body" is a material that absorbs perfectly at all wavelengths. Many common materials—lampblack, for example—are excellent absorbers over large spectral regions. General thermodynamic arguments have shown that the spectral intensity (watts/m^2 per unit frequency interval) $I(\nu)$ of emitted radiation should be the same for all black bodies at a given temperature. This indeed was found to hold true, and experimental measurements of $I(\nu)$ yielded the curves shown in Fig. 1.1. The intensity reaches a maximum at some frequency ν_m while dropping to zero on either side of ν_m. The frequency ν_m, as well as the height of the peak, increase with temperature.

Theoretical attempts to predict the behavior of the black body spectral intensity from the then known first principles were unsuccessful until 1900. The application of statistical thermodynamics and the ordinary laws of mechanics and electromagnetic theory led to the so-called Rayleigh–Jeans formula

$$I(\nu)=\frac{2\pi}{c^2}kT\nu^2 \tag{1.7}$$

where $k=1.3807\times10^{-23}$ joule/k is the Boltzmann constant. This law is plotted in Fig. 1.1 and, except for very low frequencies, is in total disagreement with experimental results. The Rayleigh–Jeans law predicts, as may be

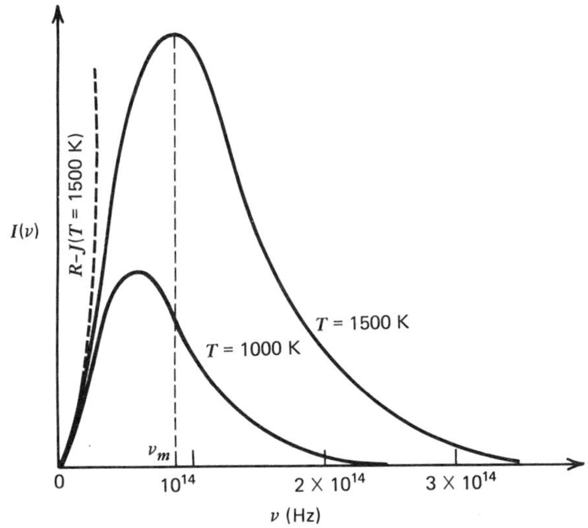

Figure 1.1 The spectral intensity $I(\nu)$ (power per unit area per unit frequency) of blackbody radiation as a function of frequency at $T=1000$ K and $T=1500$ K. The dashed curve is a plot of the Rayleigh–Jeans law, Eq. (1.7).

verified by integrating over all frequencies, an infinite amount of radiated intensity. Actually, the total radiated intensity is finite.

Max Planck resolved this controversy by postulating in 1900[2] that the exchange of energy between atoms and radiation involves *discrete* amounts of energy. At a given frequency ν, the smallest amount of energy that can be exchanged is equal to

$$\mathcal{E} = h\nu \tag{1.8}$$

where h is some constant. *Only multiples* of $h\nu$ are involved in the interaction. Applying this postulate to the problem of black body radiation theory, Planck obtained

$$I(\nu) = \frac{2\pi\nu^2}{c^2}\frac{h\nu}{e^{h\nu/kT}-1} \tag{1.9}$$

which yielded excellent agreement with the observation, provided the value of h was taken as

$$h = 6.62377 \times 10^{-34} \text{ joule-sec}$$

The constant h, whose quoted numerical value reflects the most recent determination, is known as Planck's constant.

The notion that field energy is quantized rather than a continuous quantity was a new and profound addition to physics. Under certain circumstances the behavior of waves was best described in terms of particles—*photons*—moving with the velocity of light c and having an energy $h\nu$.

1.3 THE HEAT CAPACITY OF SOLIDS AND THE PHOTOELECTRIC EFFECT

Another experimental observation that could not be explained by ordinary physics was that of the heat capacity C of solids. Experiments yielded the behavior shown in Fig. 1.2. At low temperature $C \propto T^3$. At higher temperatures C tends to a constant value. The application of ordinary statistical thermodynamics considers each atom in a crystal as an oscillator with an average thermal excitation energy of $kT/2$ per degree of freedom. The total energy per unit volume is thus

$$\mathcal{E} = 3NkT$$

where N is the density (atoms/m^3). The heat capacity $C_v = \delta\mathcal{E}/\partial T$ is then

$$C = 3Nk \tag{1.10}$$

and is independent of temperature. This disagreement between theory and experiment was resolved by Einstein[3] and Debye[4] who applied the Planck

[2] M. Planck, *Ann. Phys.* **4**, 553 (1901).

[3] A. Einstein, *Ann. Phys.* **22**, 180 (1907).

[4] P. Debye, *Ann. Phys.* **39**, 789 (1912).

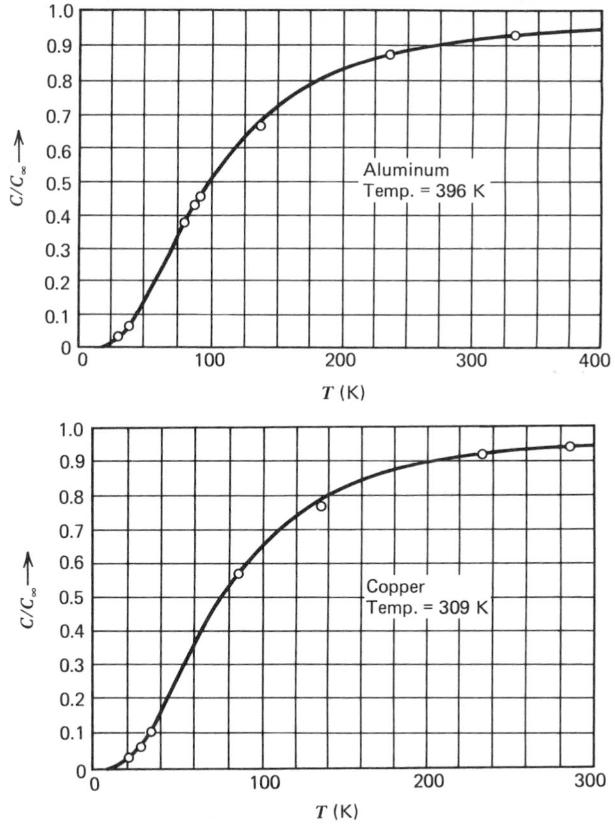

Figure 1.2 Observed values (dots) and theoretical plots using the Debye model of the heat capacity of aluminum and copper (after P. Debye, *Ann. Phys.*, **39**, 789 (1912)).

quantization condition (1.8) to the mechanical oscillations of the lattice. The most profound effect of applying Planck's postulate to an oscillator is that its average thermal equilibrium excitation energy is no longer equal to kT (the classical value) but to

$$\mathcal{E} = \frac{h\nu}{e^{h\nu/kT} - 1} \tag{1.11}$$

Note that $\mathcal{E} \to kT$ for $kT \gg h\nu$. The application of (1.11) to the heat capacity problem resulted in excellent agreement with experiment, as demonstrated in Fig. 1.2.

The Photoelectric Effect

A direct confirmation of the energy quantization of electromagnetic fields was provided by the phenomenon of photoelectric emission. When light of frequency ν is incident on a solid, electrons are emitted from the solid surface only when $h\nu > \phi$, where ϕ is a constant characteristic of the material (and often the nature of the surface). When $h\nu > \phi$, the electrons are emitted with a kinetic energy T, where

$$h\nu = \phi + T$$

These observations are independent of the intensity of the incident radiation. The latter determines merely the number (per second) of the emitted electrons. The explanation was provided by Einstein in 1905 and invoked the electromagnetic field particles, photons, each carrying an energy $h\nu$. Electrons in the material are held back from the vacuum by an energy barrier of height ϕ. The impinging photon can transmit its energy $h\nu$ to an electron near the surface. If $h\nu < \phi$, this energy is insufficient to surmount the barrier and no electrons are ejected. If $h\nu > \phi$, the excess $T = h\nu - \phi$ is the kinetic energy of the electron leaving the surface.

1.4 PHOTON MOMENTUM AND COMPTON SCATTERING

The spectacular success of Planck's postulate in explaining black body radiation and the heat capacity of solids shows that we may think of electromagnetic radiation as composed of particles—photons—with energy $\hbar\omega$ $(h\nu = \hbar\omega)$.[5] Further confirmation of the particle nature of radiation was provided by photoelectric experiments. It was thus natural to inquire whether photons possess momentum as well as energy. The classical (relativistic) relation between energy E and momentum p of a particle is

$$\mathcal{E} = (p^2c^2 + m^2c^4)^{1/2} = \frac{mc^2}{\sqrt{1-(v^2/c^2)}} \tag{1.12}$$

where m is the rest mass, c the velocity of light in free space, and v the velocity of the particle. If (1.12) is applied to the photon, then $v = c$. Since its energy $\mathcal{E}$ is finite ($\mathcal{E} = h\nu$), its rest mass m must be zero. It follows that

$$\mathcal{E} = pc \tag{1.13}$$

Using the relation $\mathcal{E} = \hbar\omega$ and $\omega = kc$, we have

$$p = \frac{\mathcal{E}}{c} = \hbar k = \frac{h}{\lambda} = \frac{h\nu}{c} \tag{1.14}$$

The wave expression (1.4)

$$E = E_0 e^{-i(\omega t - \mathbf{k}\cdot\mathbf{r})}$$

can then be rewritten in terms of the photon energy and momentum as

$$E = E_0 e^{-i(\mathcal{E}t - \mathbf{p}\cdot\mathbf{r})/\hbar} \tag{1.15}$$

---

[5]The constant $\hbar \equiv h/2\pi$ is often used.

The association of a particle momentum p with the carriers of the electromagnetic field—that is, the photons—must of course be checked experimentally. The most direct and convincing demonstration of photon momentum is provided by the scattering of short wavelength radiation (usually x-rays) from electrons—*Compton scattering.*

In a Compton scattering experiment radiation of a wavelength λ_1 ($\lambda_1 = c/\nu_1$) is incident on an assembly of electrons. A spectrometer measures the wavelength λ_2 of the radiation scattered by the electrons along some direction making an angle θ with that of the incident radiation. We represent the basic scattering as that of a photon of energy p_1c and momentum p_1 from an electron initially at rest, as shown in Fig. 1.3.

Equating the total energy (electron plus photon) before and after the scattering and taking the scattered electron momentum as $\mathbf{p}_e$ gives

$$p_1c + mc^2 = p_2c + \left(p_e^2 + m^2c^2\right)^{1/2}c \tag{1.16}$$

while conservation of momentum leads to

$$\mathbf{p}_1 = \mathbf{p}_e + \mathbf{p}_2 \tag{1.17}$$

so that after applying the law of cosines to the triangle defined by (1.17),

$$p_e^2 = p_1^2 + p_2^2 - 2p_1p_2\cos\theta$$

A simple manipulation of (1.16) and the last equation gives

$$\left(\frac{1}{p_2} - \frac{1}{p_1}\right)mc = 2\sin^2\frac{\theta}{2}$$

Recalling that the photon momentum and field wavelength are related by

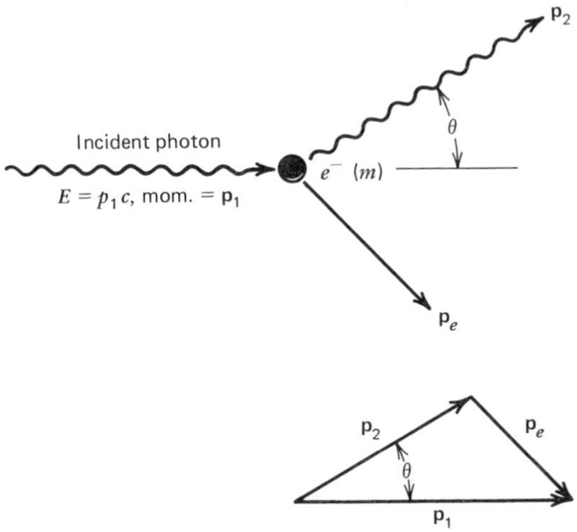

Figure 1.3 Compton scattering of electromagnetic radiation from electrons.

$p = h/\lambda$ enables us to rewrite the last relation as

$$\lambda_2 - \lambda_1 = \frac{4\pi\hbar}{mc}\sin^2\frac{\theta}{2} = 4\pi\lambda_e \sin^2\frac{\theta}{2} \tag{1.18}$$

where the Compton wavelength of the electron is defined as

$$\lambda_e \equiv \hbar / mc \simeq 4 \times 10^{-13} \text{ m}$$

Experiments reveal excellent agreement with the prediction of (1.18). Especially noteworthy is the fact that the wavelength shift $\Delta\lambda$ is independent of the wavelength of the incident radiation. Since the derivation of (1.18) was based on the premise that the photon possesses a momentum that is related to its energy by $p = \mathcal{E}/c$, the experimental verification of (1.18) can be viewed as a confirmation of that premise.

Numerical Example

Assume incident radiation with a wavelength $\lambda = 1$ μm (1 μm$=10^{-6}$ m). Taking $\theta = \pi$ we obtain from (1.19)

$$\Delta\lambda/\lambda \sim 5 \times 10^{-6}$$

This is a very small fractional change and one not easily resolved. If the experiment is repeated with X-ray radiation with a wavelength of $\lambda = 1$ Å, the fractional shift in the wavelength of the scattered radiation is

$$\Delta\lambda/\lambda \sim 5 \times 10^{-2}$$

which is easily detected.

1.5 WAVE ASPECTS OF PARTICLES

In the preceding sections it was shown how the awareness of the particle aspect of radiation was forced on physics by the weight of experimental evidence. It was thus highly satisfying to find out that particles, as well, led a dual life and under certain circumstances behaved as *waves*.

In 1924 Louis de Broglie proposed that an electron of momentum p has an associated wavelength λ_e, where

$$\lambda_e = h/p \tag{1.19}$$

The first experiment to most clearly demonstrate the particle-wave duality was performed by Davisson and Germer.[6] In their experiment they bombarded a crystal of nickel with a monoenergetic beam of electrons. They found that instead of a more or less uniform scattering in all directions, electrons were scattered only in well-specified directions. These could be predicted using the theory of Bragg diffraction of electromagnetic waves, say X rays, provided one associated with an electron of kinetic energy $\mathcal{E}$ and

---

[6] C. Davisson and L. Germer, *Nature* **119**, 558 (1927).

momentum **p** a *wave*:

$$\psi(\mathbf{r}, t) \propto e^{-i(\mathcal{E}t - \mathbf{p}\cdot\mathbf{r})/\hbar} \tag{1.20}$$

The wavelength λ_e of the electron was indeed that predicted by de Broglie:

$$\lambda_e = \frac{h}{p} = \frac{h}{\sqrt{2m\mathcal{E}}} \tag{1.21}$$

where we used $\hbar \equiv h/2\pi$, $\mathcal{E} = p^2/2m$ (non relativistic).

The diffraction is observed provided the incident electron beam satisfies the Bragg condition

$$d\sin\theta = l(\lambda_e/2) \quad (l=1,2,3,\ldots) \tag{1.22}$$

where d is the separation of two adjacent atomic planes and θ is the angle of incidence as shown in Fig. 1.4. The Bragg condition (1.22) guarantees that the portions of the electron wave reflected from neighboring atomic planes add up in phase (see Fig. 1.4*b*) along θ. What is profoundly different in this point of view is that a *single* electron is envisaged as reflected *simultaneously* from all the atoms in the crystal. This is certainly different from the "billiard ball" picture of particle behavior that prevailed at that time.

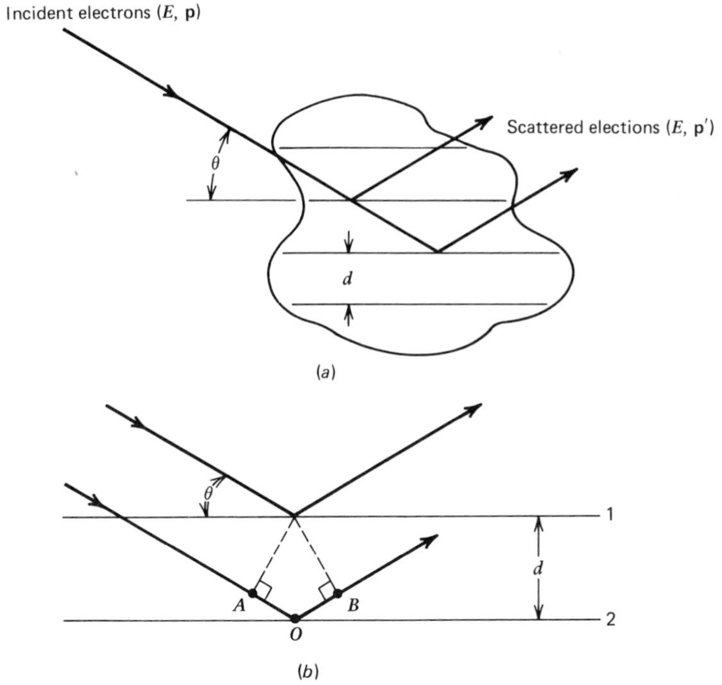

Figure 1.4 (*a*) Scattering of electrons from a crystal. (*b*) Strong scattering occurs when the path delay AOB for scattering from two adjacent planes is equal to an integer times the electron wavelength λ_e. Since $AOB = 2d\sin\theta$, this interference requirement leads to the Bragg condition (1.22).

Numerical Example

For electron waves we obtain from (1.21)

$$\lambda_e\,(\text{Å}) = \frac{12.26}{\sqrt{\mathcal{E}_{\text{eV}}}} \tag{1.23}$$

so that energies of a few electron volts (1 eV$=1.602\times10^{-19}$ joules is the energy of an electron that has been accelerated from rest through a potential difference of 1 volt) yield values of λ_e of a few angstroms that are appropriate for diffraction in crystals where typical values of interatomic distances are $d\sim3$–5 Å.

If we were to use neutrons instead of electrons, we would have

$$\lambda_n\,(\text{Å}) = \frac{2.86\times10^{-1}}{\sqrt{\mathcal{E}_{\text{eV}}}} \tag{1.24}$$

so that very "slow" neutrons with energies of say $\frac{1}{40}$ eV may be used in diffraction studies of crystals. Slow neutrons are indeed used in solid state crystal structure studies.[7]

1.6 THE HYDROGEN ATOM AND THE BOHR MODEL

As the last example of the failure of classical physics to account for observed phenomena, we consider the case of the hydrogen atom. According to the Rutherford model[8] the hydrogen atom consists of a single electron orbiting around a positively charged heavy nucleus (proton). This model failed to explain two main observational features of the hydrogen atom: (a) its stability and (b) the spectrum of its radiation. Let us consider these one at a time.

An electron in a curved orbit is accelerated and hence must radiate. As it radiates its energy away, the radius of its orbit must decrease until eventually it collapses into the nucleus. A simple classical calculation (see Problem 2 at the end of this chapter) shows that a typical lifetime of an electron in a hydrogenlike orbit is $\sim10^{-10}$ sec. Experience, however, shows that the hydrogen atom is remarkably stable.

The second discrepancy involves the observed radiation spectrum. The frequency of the radiated energy should be the same as the orbiting frequency. As the electron orbit collapses, its orbiting frequency increases continuously. We might thus expect the spectrum of radiation emitted by excited hydrogen atoms to be continuous. In contrast, the experimentally observed spectrum consists of families of discrete lines. The frequencies of one such group, called

---

[7] C. Kittel, *Introduction to Solid State Physics*, 5th ed. (J. Wiley & Sons, New York, 1974).

[8] E. Rutherford, *Phil. Mag.* **21**, 669 (1911).

the Balmer series, is found empirically to be described by

$$\hbar\omega(\text{eV}) = 13.64\left(\frac{1}{2^2} - \frac{1}{n^2}\right) \quad (n=3,4,5) \tag{1.25}$$

Bohr provided an explanation for both the spectral discreteness and the observed stability. He proposed in 1913[9] that in solving for the orbital motion of the electron in its hydrogenic orbit one should impose an added condition: *The angular momentum of the electron must be equal to some integer multiple of* $\hbar$.

$$l = n\hbar \quad (n=1,2,3) \tag{1.26}$$

To demonstrate the consequences of Bohr's condition, let us consider the problem of an electron in a circular orbit subject to the angular momentum quantization as expressed by (1.26). The electron orbits with a radius a about a nucleus with an electric charge $+e$.

Relating the Coulomb attraction to the centripetal acceleration gives

$$\frac{e^2}{4\pi\varepsilon_0 a^2} = \frac{mv^2}{a} = m\omega^2 a \quad \left(\omega = \frac{v}{a}\right) \tag{1.27}$$

or

$$\frac{e_M^2}{a^2} = m\omega^2 a \quad \left(e_M^2 \equiv \frac{e^2}{4\pi\varepsilon_0}\right)$$

Bohr's condition (1.26) takes the form

$$m\omega a^2 = n\hbar \tag{1.28}$$

where we used $l = m\omega a^2$.

Combining (1.27) and (1.28) gives

$$a_n = \frac{\hbar^2}{e_M^2/2} n^2 = a_0 n^2 \quad (n=1,2,\ldots) \tag{1.29}$$

where a_n is the radius a corresponding to the integer n. $a_0 = \hbar^2/e_m^2 = 0.53 \times 10^{-10}$ m is called the Bohr radius of the electron. The set a_n gives the admissible radii of the electron orbits. The radian rotation frequency associated with the orbit n is obtained from (1.29) and (1.28) as

$$\omega_n = \frac{n\hbar}{ma_n^2} = \frac{me_M^4}{\hbar^3 n^3} \tag{1.30}$$

The total energy in orbit n is the sum of the kinetic and potential (Coulomb) energies:

$$E_n = \frac{m}{2}(\omega_n a_n)^2 - \frac{e_M^2}{a_n} \tag{1.31}$$

---

[9] N. Bohr, *Phil. Mag.* **26**, 476 (1913).

Using (1.29) and (1.30) the expression for the energy can be written as

$$E_n = -\frac{me_M^4}{2\hbar^2 n^2} = -E_0\left(\frac{1}{n^2}\right) \tag{1.32}$$

where $E_0 = me_M^4/2\hbar^2 = 13.64$ eV.

The lowest energy state is that with $n=1$. It corresponds, according to (1.29), to the smallest orbit. The highest energy $E=0$ results when $n=\infty$—that is, when the electron and proton are separated by an infinite distance. The ionization energy of the hydrogen atom—that is, the least amount of energy needed to remove an electron from the $n=1$ orbit to infinity—is equal to E_0 (13.64 eV). A plot of E_n for different n is shown in Fig. 1.5.

The introduction of the Bohr model constituted a major advance in atomic physics. The discreteness of the energy levels helped explain the discreteness of the observed radiation from excited hydrogen atoms, since radiation was emitted when the atom executed a transition from some excited state to a lower state. The Balmer series, for example, is seen as the result of a transition from excited states $n=3,4,5,\ldots$ to the state $n=2$. By requiring that the energy of the emitted photon equal that lost by the atom, we find that the frequency ω resulting from a transition from the state n to the state $n=2$ may be given according to (1.32) by

$$\hbar\omega(\text{eV}) = 13.64\left(\frac{1}{2^2} - \frac{1}{n^2}\right) \quad (n=3,4,5,\ldots)$$

in agreement with the observed data as described by Eq. (1.25).

The Bohr model also helps us understand the observed stability of the hydrogen atom. Since the lowest allowed state is $n=1$, an atom in the state $n=1$ cannot lose energy by further downward transitions and is consequently stable.

The problem with the Bohr model was its ad hoc nature. The angular momentum condition (1.26) was not the consequence of a more general

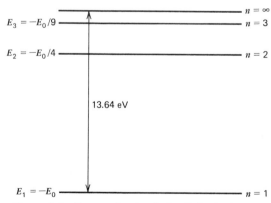

Figure 1.5 Energy levels of the Bohr hydrogenic model.

theory or the result of some fundamental insight. It was invented for the specific case of the hydrogen atom. The general dissatisfaction with this model helped pave the way to quantum mechanics.

In 1924 de Broglie proposed the relation [see (1.19)] $p = h/\lambda$ between the wavelength λ and the momentum p of a particle. Using this relation he showed that Bohr's angular momentum quantization condition $l = n\hbar$ [see (1.26)] was equivalent to requiring that

$$L_n = n\lambda_n \quad (n = 1, 2, 3, \ldots)$$

where L_n is the length of the nth Bohr orbit. Similar conditions occur often in problems involving wave propagation in spherical and cylindrical geometries. They reflect the required single-valued nature of the physically acceptable wave solutions. These solutions must return to their initial value when the azimuthal (coordinate) angle varies by some integral multiple of 2π.

It now became clear that such diverse phenomena as electron diffraction in crystals, Compton scattering, and the emission spectrum of hydrogen involved the wave aspect of the electron.

PROBLEMS

1. Show that in the hydrogen atom

$$\text{kinetic energy} = -\text{total energy} = -\tfrac{1}{2}(\text{potential energy})$$

2. Estimate the lifetime of a classical particle of mass m and charge $-e$ in orbit of radius r around a massive nucleus with charge e.

Hint: Define the lifetime as

$$\tau = \frac{\text{total energy}}{\text{radiated power}}$$

You will need to look up the expression for the power radiated by an accelerated charge.

3. Show that the angular momentum quantization condition (1.26) is equivalent to demanding that the electron orbit contain an integral (n) number of wavelengths.

4. Show that in the limit $n \gg 1$ the frequency

$$\nu = \frac{E_{n+1} - E_n}{h}$$

associated with the transition $n+1 \to n$ approaches that of the orbiting motion.

5. Using any standard text on thermal radiation, study and reproduce the derivation of Eq. (1.9).

6. The work function of aluminum is 4.25 eV. Calculate the maximum kinetic energy of the electrons ejected from an aluminum surface that is irradiated by an ultraviolet beam with $\lambda = 2100$ Å.

7. What is the maximum energy loss of a photon with energy $\hbar\omega = 10$ keV that collides with an electron at rest? What is the velocity of the electron after the collision?

8. In microscopy the smallest feature that can be revealed is of the order of the wavelength used. What is the theoretical resolution limit of an electron microscope using electrons accelerated through 150 keV? How does it compare to that of a microscope using visible light (4000 Å $< \lambda <$ 7000 Å)?

9. Referring to Figure 1.4, show that the Bragg condition can be written as

$$\mathbf{p}' - \mathbf{p} = \boldsymbol{\zeta} l \frac{2\pi\hbar}{d} \tag{1.33}$$

where $\boldsymbol{\zeta}$ is a unit vector normal to the planes. We may thus think of $2\pi/d$ as a unit of "crystal momentum" and of (1.33) as a law of conservation of momentum of electrons in crystals.

CHAPTER TWO

Operators

Mathematical operators play a central role in quantum mechanics, so that a study of their mathematical properties is essential. This will be done in the present chapter.

2.1 MATHEMATICAL PROPERTIES OF OPERATORS

Definitions

An operator is, loosely speaking, some instruction that, when applied to a function, changes it into another function. As an example, the operator

$$\hat{A} = 1 + \frac{\partial}{\partial x}$$

has the property that

$$\hat{A}f(x) = f(x) + \frac{\partial f}{\partial x} \tag{2.1}$$

We shall designate an operator by the small circumflex ("hat" symbol, ˆ) over the letter. The operators we encounter in this book will, in general, involve $\mathbf{r}$, ∇, that is, the spatial coordinates, and their derivatives. As a second example, consider the operator

$$A = \frac{\partial}{\partial x} x \tag{2.2}$$

so that

$$\begin{aligned} \hat{A}f(x) &= \frac{\partial}{\partial x}[xf(x)] = x\frac{\partial f}{\partial x} + f(x) \\ &= \left(1 + x\frac{\partial}{\partial x}\right)f(x) \end{aligned} \tag{2.3}$$

This result may be summarized by the equation

$$\frac{\partial}{\partial x}x = 1 + x\frac{\partial}{\partial x} \tag{2.4}$$

An operator equation such as (2.4) signifies that operating on any *arbitrary* function $f(x)$ with the operator on the left side of the equality sign yields the same function as that which results from an operation by the right side. We will restrict ourselves in this book to linear operators so that for any functions $f(x)$ and $g(x)$ and an arbitrary constant c,

$$\hat{A}cf(x) = c\hat{A}f(x)$$

and

$$\hat{A}[f(x) + g(x)] = \hat{A}f + \hat{A}g$$

The sum $\hat{A} + \hat{B}$ of two operators is defined by

$$(\hat{A} + \hat{B})f(x) = \hat{A}f(x) + \hat{B}f(x)$$

2.2 THE EIGENFUNCTIONS AND EIGENVALUES OF OPERATORS

With each linear operator $\hat{A}(x, \partial/\partial x)$ we associate a set of numbers a_n and a set of functions $u_n(x)$ such that

$$\hat{A}\left(x, \frac{\partial}{\partial x}\right) u_n(x) = a_n u_n(x) \tag{2.5a}$$

The numbers a_n are called the *eigenvalues* of the operator $\hat{A}$, and the functions $u_n(x)$ are called the *eigenfunctions*. The set a_n may be discrete or continuous. As an example, consider the operator $\hat{p}_x = -i\hbar\,\partial/\partial x$. To find its eigenvalues and eigenfunctions we need to solve the equation

$$-i\hbar \frac{\partial}{\partial x} u_p(x) = p u_p(x) \tag{2.5b}$$

The solution is

$$u_p(x) = e^{ipx/\hbar} \tag{2.6}$$

The eigenvalue p can be *any number* (including complex values). However, if we impose the boundary condition that $u_p(x)$ be periodic in some distance L, we obtain

$$\frac{p}{\hbar} = n\frac{2\pi}{L} \qquad (n = 0, \pm 1, \pm 2) \tag{2.7a}$$

and

$$u_n(x) = e^{in(2\pi/L)x} \tag{2.7b}$$

so that the eigenvalues are discrete and real. This is an important observation. It shows that the eigenvalues and eigenfunctions depend not only on the nature of the operator, but also on the boundary conditions. In an obvious extension of (2.5a) to three dimensions, we substitute

$$-i\frac{\partial}{\partial x} \to -i\nabla \equiv -i\left(\mathbf{i}\frac{\partial}{\partial x} + \mathbf{j}\frac{\partial}{\partial y} + \mathbf{k}\frac{\partial}{\partial z}\right) \tag{2.8}$$

where **i**, **j**, and **k** are unit vectors in the x, y, and z directions, respectively. The eigenvalue equation (2.5a) becomes

$$-i\hbar\nabla u_{lmn}(\mathbf{r})=\mathbf{a}_{lmn}u_{lmn}(\mathbf{r}) \tag{2.9}$$

or

$$-i\hbar\left(\mathbf{i}\frac{\partial}{\partial x}+\mathbf{j}\frac{\partial}{\partial y}+\mathbf{k}\frac{\partial}{\partial z}\right)u_{lmn}(\mathbf{r})=\mathbf{a}_{lmn}u_{lmn}(\mathbf{r}) \tag{2.10}$$

We postulate that u_{lmn} can be written as a product

$$u_{lmn}(x,y,z)=f_l(x)g_m(y)p_n(z) \tag{2.11}$$

which, when substituted in (2.10), leads to three separate equations of the form

$$\frac{\partial f_l}{\partial x}=ia_l f_l(x), \qquad \frac{\partial g_m}{\partial y}=ia_m g_m, \qquad \frac{\partial p_n}{\partial z}=ia_n p_n \tag{2.12}$$

where a_l, a_m, and a_n are constants. A possible solution is

$$u_{lmn}=e^{i(a_l x+a_m y+a_n z)} \tag{2.13}$$

The eigenvalue equation (2.10) becomes

$$-i\hbar\nabla u_{lmn}(x,y,z)=(\mathbf{i}a_l+\mathbf{j}a_m+\mathbf{k}a_n)\hbar u_{lmn} \tag{2.14}$$

The eigenvalue is thus a vector with arbitrary Cartesian components, that is,

$$\mathbf{a}_{lmn}=(\mathbf{i}a_l+\mathbf{j}a_m+\mathbf{k}a_n)\hbar \tag{2.15}$$

If a boundary condition is imposed that u_{lmn} be periodic along x, y, and z, with periods of A, B, and C, respectively, then in analogy with (2.7)

$$\mathbf{a}_{lmn}=\left(\mathbf{i}\frac{l2\pi}{A}+\mathbf{j}\frac{m2\pi}{B}+\mathbf{k}\frac{n2\pi}{C}\right)\hbar \quad (l,m,n=0,\pm,\pm 2,\ldots) \tag{2.16}$$

so that the eigenvalues are vectors with real discrete components.

2.3 HERMITIAN OPERATORS

Any operator corresponding to a physically observable property (as do most of the operators that we encounter in this book) possesses the property

$$\int f^*(\hat{A}g)\,d^3\mathbf{r}=\int(\hat{A}f)^*g\,d^3\mathbf{r} \tag{2.17}$$

where $f(r)$ and $g(r)$ are two arbitrary functions.[1,2] An operator satisfying relation (2.17) is called *Hermitian*. It is easy to show, as an example, that the operator $-i\partial/\partial x$ is Hermitian (see Problem 1 at the end of this chapter), so

---

[1] $f(\mathbf{r})$ and $g(\mathbf{r})$ must both approach zero as $\mathbf{r}\to\infty$ in a manner that is satisfied by the functions of physical interest.

[2] An integral without specified limits will be understood in this book to extend over the domain of the function.

that for any f and g such that $fg \to 0$ as $x \to \infty$,

$$\int f^* \left(i \frac{\partial g}{\partial x} \right) dx = \int \left(i \frac{\partial f}{\partial x} \right)^* g \, dx \tag{2.18}$$

2.4 ORTHOGONALITY OF THE EIGENFUNCTIONS OF A HERMITIAN OPERATOR

Theorem: Any two eigenfunctions u_n and u_m of a Hermitian operator $\hat{A}$ satisfy

$$\int u_n^*(\mathbf{r}) u_m(\mathbf{r}) \, d^3\mathbf{r} = 0 \quad (m \neq n) \tag{2.19}$$

provided their eigenvalues a_n and a_m are not equal. If u_n and u_m satisfy (2.19), they are said to be *orthogonal* to each other.

Proof: Since $u_n(\mathbf{r})$ and $u_m(\mathbf{r})$ are eigenfunctions of $\hat{A}$, they satisfy

$$\hat{A} u_m = a_m u_m$$

$$\hat{A} u_n = a_n u_n$$

$$\int u_m^* \hat{A} u_n \, d^3\mathbf{r} = a_n \int u_m^* u_n \, d^3\mathbf{r} \tag{2.20}$$

Similarly,

$$\int u_n^* \hat{A} u_m \, d^3\mathbf{r} = a_m \int u_n^* u_m \, d^3\mathbf{r}$$

Since $\hat{A}$ is Hermitian, we may use (2.17) and rewrite the last equality as

$$\int (\hat{A} u_n)^* u_m \, d^3\mathbf{r} = a_m \int u_n^* u_m \, d^3\mathbf{r}$$

and upon taking the complex conjugate of the last equation

$$\int u_m^* \hat{A} u_n \, d^3\mathbf{r} = a_m^* \int u_m^* u_n \, d^3\mathbf{r} \tag{2.21}$$

Subtract (2.21) from (2.20)

$$(a_n - a_m^*) \int u_m^* u_n \, d^3\mathbf{r} = 0 \tag{2.22}$$

By taking the case $n = m$, we obtain

$$(a_n - a_n^*) \int |u_n|^2 \, d^3\mathbf{r} = 0$$

Since the integral is nonzero, it follows that $a_n = a_n^*$; that is, *the eigenvalues of Hermitian operators are real numbers.*

If $n \neq m$, then the integral on the right side of (2.22) is zero, provided $a_n \neq a_m$. This is the orthogonality statement of Eq. (2.19). In the case of two (or more) eigenfunctions with the same eigenvalue, $a_n = a_m$, the orthogonality proof given above does not apply. In this case it is possible to choose linear combinations of u_n and u_m that are orthogonal to each other.

2.5 NORMALIZATION OF EIGENFUNCTIONS

Since the eigenvalue equations such as (2.5) involve linear operators, the eigenfunctions are specified only to within an arbitrary constant. This constant is chosen so that

$$\int u_n^* u_n \, d^3\mathbf{r} = 1 \tag{2.23}$$

Condition (2.23) is called the normalization condition. We can express the orthogonality condition (2.19) and the normalization condition (2.23) simultaneously by writing

$$\int u_n^* u_m \, d^3\mathbf{r} = \delta_{nm} \tag{2.24}$$

where δ_{nm} is the Kronecker delta symbol.

2.6 COMPLETENESS OF EIGENFUNCTIONS

It is a property of the set of eigenfunctions $u_n(\mathbf{r})$ of a Hermitian operator $\hat{A}$ $(\hat{A}u_n = a_n u_n)$ that they can be used to expand an arbitrary function $\psi(\mathbf{r})$, that is,

$$\psi(\mathbf{r}) = \sum_n b_n u_n(\mathbf{r}) \tag{2.25}$$

There are some restrictions on the type of function $\psi(\mathbf{r})$ and on its behavior at infinity. We will avoid this problem, however, by stating that all the functions $\psi(\mathbf{r})$ we encounter in quantum mechanics are "well behaved" and obey these restrictions. The expansion coefficient b_i (in general, a complex number) is obtained by multiplying (2.25) by $u_i^*(r)$ and integrating

$$\int u_i^*(\mathbf{r})\psi(\mathbf{r}) \, d^3r = \sum_n b_n \int u_i^*(\mathbf{r}) u_n(\mathbf{r}) \, d^3\mathbf{r}$$

Using (2.24) we obtain

$$b_i = \int u_i^*(\mathbf{r})\psi(\mathbf{r}) \, d^3\mathbf{r} \tag{2.26}$$

It is convenient to think of b_i in an abstract sense as the "projection" of $\psi(\mathbf{r})$ on $u_i(\mathbf{r})$. The set $u_n(\mathbf{r})$ is thus equivalent to the unit vectors in an infinite-dimension space. The function $\psi(\mathbf{r})$ can be thought of as an arbitrary vector in this space, and (2.25) expresses the expansion of that vector in terms of the unit vectors $u_n(\mathbf{r})$. It is formally equivalent to expanding some real vector $\mathbf{T}$ as

$$\mathbf{T}=\sum \mathbf{a}_i T_i \tag{2.27}$$

with $\mathbf{a}_i$ being a unit vector along the ith Cartesian coordinate, and $T_i=\mathbf{T}\cdot\mathbf{a}_i$ is the projection of $\mathbf{T}$ along $\mathbf{a}_i$.

We may thus think of the integral (2.26) as the scalar product of $u_i(\mathbf{r})$ and $\psi(\mathbf{r})$. The orthonormality condition of the unit Cartesian vector

$$\mathbf{a}_i\cdot\mathbf{a}_j=\delta_{ij} \tag{2.28}$$

is thus equivalent to (2.24).

2.7 DIRAC NOTATION

A very useful notation of the integrals occurring in quantum mechanics is due to Dirac. This notation greatly simplifies the manipulation involved in obtaining such integrals and is useful in formal proofs.

Consider as an example the integral

$$\int f^* g \, d^3\mathbf{r}$$

where f and g are two-state functions. In the Dirac notation we represent this integral by a *bracket*:

$$\int f^* g \, d^3\mathbf{r} \equiv \langle f | g \rangle \tag{2.29}$$

We can represent the individual functions f and g in (2.29) by a *ket* vector

$$g \rightarrow | g \rangle$$

and a *bra*

$$f^* \rightarrow \langle f |$$

When a bra and ket are joined together as in (2.29), the result is, *by convention*, the integral (2.29).

The result of operating on g with $\hat{A}$ is thus represented by

$$\hat{A}g \rightarrow \hat{A}| g \rangle$$

and

$$\int f^* \hat{A} g \, d^3\mathbf{r} \rightarrow \langle f | \hat{A} | g \rangle$$

The orthonormality condition (2.24)

$$\int u_n^* u_m \, d^3\mathbf{r} = \delta_{nm}$$

is written in the Dirac notation as

$$\langle u_n | u_m \rangle \equiv \langle n | m \rangle = \delta_{nm} \tag{2.30}$$

while the "projection" (2.26) of $\psi(\mathbf{r})$ on $u_i(\mathbf{r})$ is

$$b_i = \langle u_i | \psi \rangle \tag{2.31}$$

The expansion (2.25) of an arbitrary state function $|\psi\rangle$ is thus expressed as

$$\begin{aligned} |\psi\rangle &= \sum_n [\langle n|\psi\rangle]|n\rangle \\ &= \sum_n |n\rangle\langle n|\psi\rangle \end{aligned} \tag{2.32a}$$

A formal interpretation of (2.32a) is that the result of operating on the arbitrary state function $|\psi\rangle$ with the operator $\Sigma_n |n\rangle\langle n|$ is to obtain back $|\psi\rangle$. It follows that

$$\sum_n |n\rangle\langle n| = \hat{I} \tag{2.32b}$$

where $\hat{I}$ is the identity operator.[3] The set $|n\rangle$, we recall, is any orthonormal complete set. We will make frequent use of (2.32b).

EXAMPLE: Expansion in the eigenfunction of an operator

According to (2.25) we may expand an arbitrary function $\psi(\mathbf{r})$ in terms of the eigenfunctions of any Hermitian operator $\hat{A}$. Consider, an an example, the operator $-i\,\partial/\partial x$, whose eigenfunctions, limited to the interval $-L/2 < x < L/2$, are given by (2.7b) as

$$u_n = \frac{1}{\sqrt{L}} e^{in(2\pi/L)x} \tag{2.33}$$

[The $L^{-1/2}$ factor was added to satisfy the normalization condition (2.23).] The members of the set u_n are clearly orthogonal to each other in the sense of (2.19) with the limits of integration being $-L/2$ and $L/2$. We may thus use them to expand any function $\psi(x)$, which is defined in the interval $-L/2 < x < L/2$, that is,

$$\begin{aligned} \psi(x) &= \frac{1}{\sqrt{L}} \sum_{n=-\infty}^{+\infty} a_n e^{in(2\pi/L)x} \\ a_n &= \frac{1}{\sqrt{L}} \int_{-L/2}^{L/2} e^{-in(2\pi/L)x} \psi(x)\, dx \end{aligned} \tag{2.34}$$

Equation (2.34) is recognized as the Fourier series expansion of $\psi(x)$.

---

[3]The identity operator is the operator that leaves the functions on which it operates unaltered—that is, $\mathbf{I}f = f$ for f, an arbitrary function.

Hermitian Adjoint Operators

Given a linear operator $\hat{A}$, there exists an operator $\hat{B}$—the Hermitian adjoint of $\hat{A}$—such that

$$\int f^*(\mathbf{r})\hat{A}g(\mathbf{r})\,d^3\mathbf{r}=\int\left[\hat{B}f(\mathbf{r})\right]^*g(\mathbf{r})\,d^3\mathbf{r} \tag{2.35}$$

where f and g are two arbitrary state functions. We note this by writing

$$\hat{B}=\hat{A}^\dagger \qquad \text{(pronounced "}A\text{-dagger")}$$

so that

$$\int f^*(\mathbf{r})\hat{A}g(\mathbf{r})\,d^3\mathbf{r}=\int\left[\hat{A}^\dagger f(\mathbf{r})\right]^*g(\mathbf{r})\,d^3\mathbf{r} \tag{2.36}$$

It follows by comparing (2.36) to (2.17) that an operator $\hat{A}$ is Hermitian when it is its own Hermitian adjoint; that is, for $\hat{A}$ to be Hermitian,

$$\hat{A}^\dagger=\hat{A} \tag{2.37}$$

To gain some practice, and for future reference, we express the above relations in the Dirac notation. Using the definition of Section 2.6, we write (2.35) as

$$\langle f\,|\hat{A}g\rangle\equiv\langle f\,|\hat{A}|\,g\rangle=\langle\hat{B}f\,|\,g\rangle \tag{2.38}$$

and (2.36) as

$$\langle f\,|\hat{A}|\,g\rangle=\langle\hat{A}^\dagger f\,|\,g\rangle \tag{2.39}$$

The elegance of this notation becomes apparent if we show, using it, that

$$\text{if}\quad \hat{A}=\hat{B}^\dagger,\quad \text{then}\quad \hat{B}=\hat{A}^\dagger \tag{2.40}$$

so that

$$(\hat{A}^\dagger)^\dagger=\hat{A} \tag{2.41}$$

This task is assigned as a problem.

PROBLEMS

1. Show by integration by parts that $-i\,\partial/\partial x$ is a Hermitian operator.
2. Show that operators corresponding to physical observables must be Hermitian.
3. Supply the proof of Eqs. (2.40) and (2.41) using the Dirac notation.
4. Find the eigenfunctions in three dimensions of the operator $i\,\nabla$. Show how you can expand an arbitrary function $f(\mathbf{r})$ in terms of these eigenfunctions.
5. Show that $i(\hat{A}^\dagger-\hat{A})$ is a Hermitian operator for any $\hat{A}$.

CHAPTER THREE

The Basic Postulates of Quantum Mechanics

In classical mechanics it is implicit that the motion of a particle can be measured and described in an exact manner. It is possible to measure both the velocity and the coordinate of a particle at some time t without disturbing it.

According to quantum mechanics, the act of measurement interferes with the system and modifies it. The resulting perturbation is negligible in "large" (classical) systems, but assumes major importance in small systems such as atoms, electrons, and nucleons.

To illustrate this point, consider the experimental task of photographing an electron in a hydrogenic orbit. We know [see (1.29)] that the typical radius of the electron orbit is ~1 Å. We thus need a photographic method capable of resolving to better than 1 Å. It is a basic result of electromagnetic theory (specifically, of the theory of diffraction) that the spatial limit of resolution is approximately equal to the wavelength of the electromagnetic radiation used. We thus need to employ frequencies higher than

$$\nu = \frac{c}{\lambda} = \frac{3\times 10^{8}}{10^{-10}} = 3\times 10^{18}\ \text{Hz}$$

The corresponding photon energy is

$$h\nu = 1.95\times 10^{-15}\ \text{j} = 1.23\times 10^{4}\ \text{eV}$$

We thus find that the process of measuring the position of a hydrogenic electron whose binding energy is [see (1.32)] ~10 eV involves bombarding it with projectiles (photons) whose energy is $\sim 10^4$ eV. The act of measuring the position of the electron will thus perturb its velocity. The finer the position accuracy that is required, the shorter the wavelength λ that must be used, which leads to a larger uncertainty in the final velocity of the measured electron.

At first glance the situation appears hopeless. If "small" systems are so delicate and the means of observation so gross, is there any hope of learning anything about them at all? The answer is yes. Although there are questions we may not ask, certain useful information can be deduced and certain measurements can be performed even on small systems. The formalism of quantum mechanics concerns itself with this problem. It is based on a small number of postulates—not proofs—that we accept. Some of the implications of these postulates are still a topic of philosophical debate.[1] In this book we adopt the point of view to which we alluded earlier. We justify our use of quantum mechanics not because of some inherent belief in its rightness, but solely because it is simple, self-consistent, and successful. We shall continue to do so until a more successful theory becomes available.

3.1 THE BASIC POSTULATES OF QUANTUM MECHANICS

(1) Each type of physical observable (momentum, position, energy, angular momentum, etc.) is associated with a Hermitian operator. A measurement of an observable must result in a number that is a member of a set a_n (discrete or continuous). This set corresponds to the eigenvalues of the associated operator. The possible results of a measurement are thus determined by solving an eigenvalue equation such as (2.5). A definite value of the observable a_n is thus associated with the eigenfunction $u_n(\mathbf{r})$.

Let us associate the operator $\hat{A}$ with the measurement of the physical variable A, and $\hat{B}$ with the measurement of B. We will denote the process of measuring A first, followed by that of B, by $\hat{B}\hat{A}$. Measuring the observables in the reverse order is represented by $\hat{A}\hat{B}$. Since each measurement disturbs the system, the two procedures may not lead to the same result.

According to postulate 1 above, the measurements of A and B interfere with each other (so that the result of measuring A and B depends on the order in which they are executed) when

$$\hat{A}\hat{B} - \hat{B}\hat{A} \neq 0 \tag{3.1}$$

We refer to the expression $\hat{A}\hat{B} - \hat{B}\hat{A}$ as the commutator of the operators $\hat{A}$ and $\hat{B}$, and represent it by

$$[\hat{A}, \hat{B}] \equiv \hat{A}\hat{B} - \hat{B}\hat{A} \tag{3.2}$$

Since we have already found in Chapter 1 that the ad hoc remedies to classical physics involve the constant $\hbar$, it will not be surprising to learn, as we soon shall, that the interference between A and B, which is described by $[\hat{A}, \hat{B}]$, is of the order of magnitude of $\hbar$.

---

[1]See, for example, D. Bohm, *Quantum Theory* (Prentice-Hall, Englewood Cliffs, N.J., 1955).

(2) A physical system is characterized by a wavefunction $\psi(\mathbf{r}, t)$ that contains all the knowable information about the system. We shall choose ψ to be normalized according to

$$\int \psi^*(\mathbf{r}, t)\psi(\mathbf{r}, t)\, d^3\mathbf{r} = 1 \tag{3.3}$$

If $\psi(\mathbf{r}, t_0) = u_n(\mathbf{r})$, where $u_n(\mathbf{r})$ is an eigenfunction of $\hat{A}$ ($\hat{A}u_n = a_n u_n$), then a measurement at A_0 of the observable A is certain to yield the value a_n. If[2] $\psi(\mathbf{r})$ is not equal to $u_n(\mathbf{r})$, then the probability that the measurement of A will yield the value a_n is

$$P(a_n) = \left| \int u_n^*(\mathbf{r})\psi(\mathbf{r})\, d^3\mathbf{r} \right|^2 \tag{3.4}$$

that is, is equal to the squared magnitude of the "projection" of $\psi(\mathbf{r})$ along $u_n^*(\mathbf{r})$ [see the discussion following (2.26)]. The plausibility of (3.4) may be appreciated by expanding $\psi(\mathbf{r})$ in the eigenfunctions $u_n(\mathbf{r})$ of $\hat{A}$:

$$\psi(\mathbf{r}) = \sum_i b_i u_i(\mathbf{r}) \tag{3.5}$$

Multiplying both sides by $u_n^*(\mathbf{r})$, integrating over all space, and using (2.24) gives

$$b_n = \int u_n^*(\mathbf{r})\psi(\mathbf{r})\, d^3\mathbf{r} \tag{3.6}$$

We can thus rewrite (3.4) as

$$P_\psi(a_n) = |b_n|^2 \tag{3.7}$$

so that the probability of obtaining a_n when measuring A is equal to $|b_n|^2$, which can be viewed as the squared amplitude of finding the system in the state u_n.

It follows from (3.4) directly that if $\psi(\mathbf{r}) = u_n(\mathbf{r})$, then the probability $P_\psi(a_i)$ of measuring a_i is unity for $i = n$ and zero otherwise. Expressed in words: When the wavefunction $\psi(\mathbf{r}, t_0)$ is equal to one of the eigenfunctions of $\hat{A}$, say $u_n(\mathbf{r})$, all measurements of A are certain to yield the value a_n.

3.2 THE AVERAGE VALUE OF AN OBSERVABLE

Consider a very large number of *identical* systems each described by the *same* wavefunction $\psi(\mathbf{r})$. The probability that any single measurement of A will yield the value a_n was postulated in Section 3.1 to be given by

$$\begin{aligned} P_\psi(a_n) &= |b_n|^2 \\ &= \left| \int u_n^*(\mathbf{r})\psi(\mathbf{r})\, d^3\mathbf{r} \right|^2 \end{aligned} \tag{3.8}$$

[2] We will ignore in the remainder of this chapter the time dependence of $\psi(\mathbf{r}, t)$ so that all the developments that follow are assumed to apply at some specific instant, say $t = t_0$. $\psi(\mathbf{r})$ in (3.4) thus stands for $\psi(\mathbf{r}, t_0)$. t_0 can be considered as one of the parameters characterizing $\psi(\mathbf{r})$. The specific dependence of ψ on t will be introduced in Chapter 10.

The average value that results from measuring A in each individual system is obtained by adding the allowed values a_n, each multiplied by the probability $P_\psi(a_n)$ of its occurrence:

$$\langle A\rangle = \sum_n P_\psi(a_n)a_n = \sum_n |b_n|^2 a_n \tag{3.9}$$

It can also be written as

$$\langle A\rangle = \int \psi^*(\mathbf{r})\hat{A}\psi(\mathbf{r})\, d^3\mathbf{r} \tag{3.10}$$

Proof of (3.10): We use (3.5) to rewrite (3.10) as

$$\langle A\rangle = \int \sum_n \sum_i b_n^* b_i u_n^* \hat{A} u_i \, d^3\mathbf{r} \tag{3.11}$$

Using $\hat{A}u_i = a_i u_i$ as well as the orthonormality condition (2.24) yields, after interchanging the order of the summation and integration,

$$\begin{aligned}\langle A\rangle &= \sum_n \sum_i b_n^* b_i a_i \delta_{ni} \\ &= \sum_n |b_n|^2 a_n \end{aligned} \tag{3.12}$$

which is the same as (3.9). We will refer to $\langle A\rangle$ as the ensemble average of the observable A, or as the (quantum mechanical) *expectation value* of the variable A.

3.3 THE FORM OF QUANTUM MECHANICAL OPERATORS

In Section 3.1 we advanced a number of basic postulates that establish a correspondence between physical observables and operators, and between the eigenvalues of these operators and the possible results of physical measurements. We have not, however, laid down any rules as to the manner by which such operators are to be constructed. We need to insure that in the limit of large ("classical") systems operator equations go smoothly into their classical observables counterparts. This requirement is satisfied if we insist that the defining relations of quantum mechanical operators are the same as their classical (observable) counterparts.

For example, the relation

$$\mathbf{l} = \mathbf{r} \times \mathbf{p} \tag{3.13}$$

relating angular momentum $\mathbf{l}$ to the linear momentum $\mathbf{p}$ is to become

$$\begin{gathered}\hat{\mathbf{l}} = \hat{\mathbf{r}} \times \hat{\mathbf{p}} \\ \left(\hat{l}_x = \hat{y}\hat{p}_z - \hat{z}\hat{p}_y, \text{ etc.}\right)\end{gathered} \tag{3.14}$$

Clearly, the simplest way to construct the position operator $\hat{x}$ is to represent it by a multiplication by x, that is,

$$\hat{x}\psi(\mathbf{r}) = x\psi(\mathbf{r}) \tag{3.15}$$

The plausibility of such an assignment is considered in Section 3.4. The operator representing, as an example, the total energy of a one-dimensional harmonic oscillator with a spring constant K and mass M is thus

$$\hat{\mathcal{H}} = \frac{\hat{p}_x^2}{2M} + \tfrac{1}{2}Kx^2 \tag{3.16}$$

where x is the deviation of the point mass from its equilibrium position.

There still remains the problem of determining the form of the momentum operator $\hat{\mathbf{p}}$. In Section 1.5 we have shown that the observed wave aspects of particles can be explained if we associate with a particle having momentum p_x a wave

$$\psi(x,t) = e^{-i(Et - p_x x)/\hbar} \tag{3.17}$$

[Here we rewrite (1.20) for the case of particles moving along the x coordinate.] If $\psi(\mathbf{r},t)$ is to correspond to a state with a well-defined momentum p_x, then it must, according to Postulate 1 in Section 3.1, be the eigenfunction of the operator $\hat{p}_x$ with an eigenvalue p_x.

Suppressing the time dependence (which is irrelevant in the present argument), the eigenvalue equation for the momentum operator is

$$\hat{p}_x e^{ip_x x/\hbar} = p_x e^{ip_x x/\hbar} \tag{3.18}$$

From which it follows that the operator $\hat{p}_x$ must have the form

$$\hat{p}_x \to -i\hbar\frac{\partial}{\partial x} \tag{3.19}$$

and in three dimensions

$$\hat{\mathbf{p}} \to -i\hbar\nabla \tag{3.20}$$

3.4 THE COMMUTATION RELATION FOR THE MOMENTUM AND POSITION OPERATORS. COMMUTING OPERATORS AND THEIR EIGENFUNCTIONS

Since the position and momentum are the key observables characterizing a particle in classical mechanics, it is important to know if the corresponding quantum mechanical operators commute or not. Using (3.16) and (3.20) we obtain

$$\begin{aligned}[\hat{x},\hat{p}_x]f(x) &= -i\hbar\left(x\frac{\partial f}{\partial x} - \frac{\partial}{\partial x}(xf)\right) \\ &= i\hbar f\end{aligned} \tag{3.21}$$

Since $f(x)$ is an arbitrary function we can write

$$[\hat{x},\hat{p}_x] = i\hbar$$

or in general

$$[\hat{r}_i, \hat{p}_j] = i\hbar\,\delta_{ij} \tag{3.22}$$

where $\hat{r}_x = x$, and so on. We thus find that the momentum and position operators of a particle *do not commute.* The corresponding measurements interfere and influence each other so these observables *cannot be measured simultaneously with arbitrary accuracy.*

Commuting Operators

If two operators—say $\hat{A}$ and $\hat{B}$—commute, they possess common eigenfunctions; that is, the eigenfunctions of $\hat{A}$ are eigenfunctions of $\hat{B}$ and vice versa. The proof of the above statement proceeds as follows: Since $\hat{B}\hat{A} = \hat{A}\hat{B}$, it follows that $\hat{A}\hat{B}u_B = \hat{B}\hat{A}u_B = \hat{A}bu_B = b\hat{A}u_B$, where u_B is some eigenfunction of $\hat{B}$ with an eigenvalue b. The function $\hat{A}u_B$ is thus an eigenfunction of $\hat{B}$ with an eigenvalue b. If there is only one eigenfunction of $\hat{B}$ associated with the eigenvalue b, then $\hat{A}u_B = cu_B$, where c is some constant, that is, u_B is also an eigenfunction of $\hat{A}$. If there is more than one eigenfunction with an eigenvalue b, then the above proof shows that $\hat{A}u_b$ is a linear combination of these functions. It is then possible to construct a linear combination of this degenerate eigenfunction of $\hat{B}$, that is, those having the same eigenvalue, which is also an eigenfunction of $\hat{A}$.

The proof that if two hermitian operators have common eigenfunctions, they commute, is left as an exercise.

We can now understand better why the observables corresponding to two commuting operators can be determined with arbitrary accuracy without the effect of one measurement influencing the other. Let us measure first the observable B corresponding to the operator $\hat{B}$. The result according to the basic postulate of quantum mechanics (Section 3.1) must be one of the eigenvalues, say b, of $\hat{B}$. The particle is thus "left" in the eigenstate u_B corresponding to b. Since $\hat{A}$ and $\hat{B}$ have common eigenfunctions, a subsequent measurement of the observable A will yield, with certainty, the value c, which is the eigenvalue of $\hat{A}$ corresponding to the eigenfunction u_B, that is, $\hat{A}u_B = cu_B$. We can go back and measure B again and obtain the same value b. We can thus make a simultaneous determination of A and B.

We now can better understand why the observables corresponding to noncommuting operators cannot be measured simultaneously with arbitrary accuracy. Consider, for example, two (noncommuting) operators $\hat{A}$ and $\hat{B}$. The measurement of $\hat{A}$ leaves the system in an eigenstate of the operator $\hat{A}$, say $\phi_a^{(A)}$ (the eigenstate that corresponds to the eigenvalue a measured). Since $\hat{A}$ and $\hat{B}$ do not commute, this eigenfunction is generally not an *eigenfunction* of $\hat{B}$. We can, however, expand $\phi_a^{(A)}$ in terms of the eigenfunctions $\phi_b^{(B)}$ of $\hat{B}$:

$$\phi_a^{(A)}(x) = \sum_b c_b \phi_b^{(B)}(x), \qquad \sum_b |c_b|^2 = 1$$

A measurement of the observable B can result in any member of the set b with a corresponding probability $|c_b|^2$. In other words, the process of measuring A has introduced an uncertainty into the measurement of B, since the result of the measurement of B can only be predicted probabilistically.

If $\hat{B}$ and $\hat{A}$ commute, they possess common eigenfunctions so that the last expansion will consist of a single term, say b'. The measurement of B, following that of A, will then lead to a single number, since $|c_{b'}|^2 = 1$.

3.5 THE SIGNIFICANCE OF $\psi(\mathbf{r})$

According to (3.12), the expectation value $\langle A \rangle$ of some observable A is obtained by performing the integral

$$\langle A \rangle = \int \psi^*(\mathbf{r}) \hat{A} \psi(\mathbf{r})\, d^3\mathbf{r} \tag{3.23}$$

where ψ is the state function.

Let us apply (3.23) to calculate the expectation value of the position vector $\mathbf{r}$. Since $\hat{\mathbf{r}} \to \mathbf{r}$, we obtain

$$\langle \mathbf{r} \rangle = \int |\psi(\mathbf{r})|^2 \mathbf{r}\, d^3\mathbf{r} \tag{3.24}$$

Using the classical definition for the average value of some vector $\mathbf{S}(\mathbf{r})$, given a probability distribution function $P(\mathbf{r})$ $[\int P(\mathbf{r})\, d^3\mathbf{r} = 1]$,

$$\mathbf{S}_{\text{av}} \equiv \int P(\mathbf{r}) \mathbf{S}(\mathbf{r})\, d^3\mathbf{r} \tag{3.25}$$

We associate $|\psi(\mathbf{r})|^2\, d^3\mathbf{r}$ with the *probability of finding the particle inside the differential volume* $d^3\mathbf{r}$.

3.6 THE EIGENFUNCTIONS OF THE ENERGY OPERATOR—THE SCHRÖDINGER EQUATION

In quantum mechanics the eigenfunctions and eigenvalues of the energy operator play a very important role. This is due to the fact that physical measurements often involve a determination of the energy (or radiation frequency) emitted or absorbed as the system makes a transition from one energy eigenstate to another. The energy operator—the Hamiltonian—of a single particle is

$$\hat{\mathcal{H}} = \frac{\hat{p}^2}{2m} + \hat{V}(r)$$

where $V(r)$ is the potential energy function of the particle. The eigenfunction and eigenvalue are obtained as a solution of the equation

$$\left(-\frac{\hbar^2}{2m} \nabla^2 + V(\mathbf{r}) \right) u_E(\mathbf{r}) = E u_E(\mathbf{r}) \tag{3.26}$$

where we use $\hat{\mathbf{p}} \to -i\hbar\nabla$. Equation (3.26) is called the time-independent Schrödinger equation. The function $u_E(\mathbf{r})$ is the wavefunction describing the system in a state with energy E.

3.7 THE UNCERTAINTY PRINCIPLE

We established in Section 3.3 that the momentum and position of a particle cannot be measured simultaneously with arbitrary accuracy. This was due to the fact that the operators $\hat{p}_x$ and $\hat{x}$ do not commute. To quantify the interference between these two measurements, consider a particle that is confined in one dimension to a typical distance, dimension Δx. Mathematically, we can accomplish this confinement by choosing

$$\psi(x) = e^{-x^2/2(\Delta x)^2} \tag{3.27}$$

[$\psi(x)$ as given by (3.27) is not normalized. The normalization constant, however, is immaterial to the development that follows.] The probability distribution for finding the particle at x is thus proportional to $P(x) = |\psi(x)|^2 = \exp[-x^2/(\Delta x)^2]$ and decreases rapidly for $x > \Delta x$. It is plotted in Fig. 3.1a. The probability for finding the particle with a momentum p is obtained

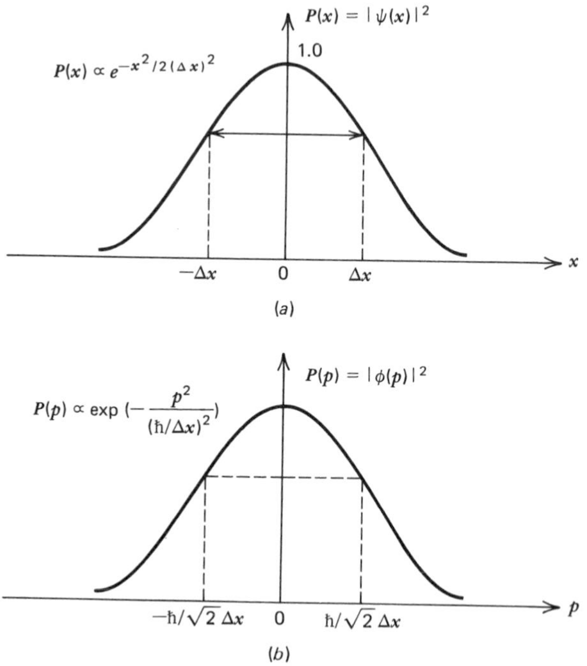

Figure 3.1 (a) A Gaussian probability distribution function $P(x)$ for the coordinate x of a particle. (b) The momentum probability distribution $P(p)$ of the same particle obtained using Eq. (3.29).

using (3.4) as

$$P(p)=|\phi(p)|^2=\left|\int \psi(x)u_p^*(x)\,dx\right|^2 \tag{3.28}$$

where $u_p(x)$, the eigenfunction of the momentum operator, is given by (3.18):

$$u_p(x)=e^{ipx/\hbar}$$

(Here again we ignore the normalization constant.) The integral in (3.28) becomes

$$\phi(p)\equiv\int \psi(x)u_p^*(x)\,dx=\int_{-\infty}^{\infty} e^{-x^2/2(\Delta x)^2}e^{-ipx/\hbar}\,dx \tag{3.29}$$

$$=\exp\left(\frac{-p^2(\Delta x)^2}{2\hbar^2}\right)\int_{-\infty}^{\infty}\exp\left\{-\frac{1}{2}\left[\frac{x}{\Delta x}+i\left(\frac{p\Delta x}{\hbar}\right)\right]^2\right\}dx \tag{3.30}$$

The last integral is a constant independent of p, so that

$$P(p)=|\phi(p)|^2=\left(\frac{2(\Delta x)^2}{\pi\hbar^2}\right)^{1/4}\exp\left(\frac{-p^2}{(\hbar/\Delta x)^2}\right) \tag{3.31}$$

A plot of $P(p)$, the momentum distribution function, is shown in Fig. 3.1*b*. According to (3.31) the range of momenta of the particle is limited to an interval

$$\Delta p\sim\hbar/\Delta x \tag{3.32}$$

If we use Δx as a measure of the uncertainty (fuzziness) in position, and Δp as a measure of the momentum spread, we have

$$\Delta p\Delta x\sim\hbar \tag{3.33}$$

An attempt to localize the particle more precisely—that is, to make Δx small—thus results in an increase of Δp, the momentum uncertainty, and vice versa.

A certain duality of describing the particle in coordinate (x) or momentum (p) space is manifested. We note that $\phi(p)$ plays, in momentum space, a role similar to that of $\psi(x)$ in coordinate space. We may thus regard $\phi(p)$ as the *wave function in momentum space*. According to (3.29) $\phi(p)$ is given by

$$\phi(p)=\int_{-\infty}^{\infty}\psi(x)e^{-ipx/\hbar}\,dx \tag{3.34}$$

so that $\phi(p/\hbar)$ is the Fourier transform of $\psi(x)$. Since Fourier transform pairs contain the same information, it is possible to conduct our quantum mechanical business either in the momentum representation or, as we have, in coordinate space.

Another important uncertainty relationship involves the energy of a particle and the time duration Δt used to measure it. Let us assume that the velocity of a particle is measured by performing the measurement over some distance L. The duration of the measurement is thus

$$\Delta t \simeq L/v$$

We also have

$$\Delta v = \Delta p/m$$

so that

$$\Delta E = \Delta(p^2/2m) = (p/m)\Delta p$$

$$\Delta E \Delta t \sim \Delta p L = \Delta p \Delta x$$

so that using (3.33)

$$\Delta E \Delta t \sim \hbar \tag{3.35}$$

The direct implication of (3.35) is that the higher the precision of measurement of the energy of a system, the longer the time needed for the measurement. It is gratifying to find that the basic uncertainty measure $\hbar$ is the same constant advanced by Planck to explain the observation of black body radiation.

3.8 THE UNCERTAINTY PRINCIPLE APPLIED TO ELECTROMAGNETIC FIELDS

Although we derived the uncertainty principle using a free electron as an example, it holds for any physical system. The implication of applying this principle to photons, for example, manifests itself in interesting ways. Let us consider for example the problem of designing an antenna for launching an information-carrying microwave beam from the earth to a communication satellite as shown in Fig. 3.2. The beam-spread angle θ can be ascribed to the fact that the photons (considered here as particles) are confined initially to a distance

$$\Delta x \sim 2R$$

corresponding to the dish diameter. Their transverse momentum spread is, according to (3.32),

$$\Delta p \sim \hbar/2R$$

The spreading angle θ can be taken as

$$\theta \sim \frac{\Delta p}{p} \sim \frac{\hbar}{2R(2\pi\hbar/\lambda)} = \frac{\lambda}{4\pi R} \tag{3.36}$$

where we used $p = 2\pi\hbar/\lambda$ (1.14).

This is, of course, a well-known result. Approached as a problem in electromagnetic theory, (3.36) reflects the Fourier transform relationship that exists between the field distribution in the antenna plane and the far-field angular distribution of the radiated field. This Fourier transform relationship

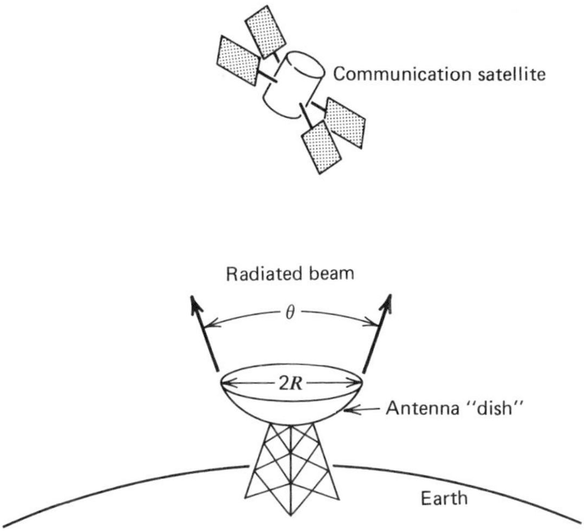

Figure 3.2 A microwave (or optical) antenna of diameter $2R$ beams a wave toward a communication satellite. The beam spread angle is θ.

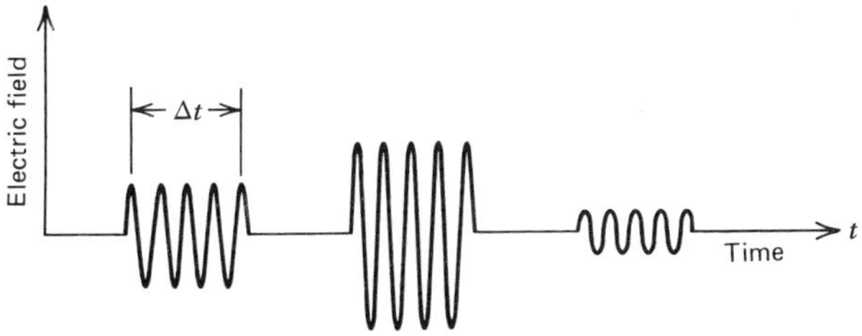

Figure 3.3 A sequence of radiation pulses with the information coded into the amplitudes.

is also the basis for the uncertainty relations in quantum mechanics, as shown in the discussion surrounding equation (3.34).

As a second example, we consider the basic quantum mechanical limitation on the rate at which information may be sent via electromagnetic fields. To be specific, consider the information carried by a sequence of electromagnetic pulses as shown in Fig. 3.3. The average power is taken as $\bar{p}$, so that the average energy per pulse is $\bar{p}\,\Delta t$. The transmitted information is coded into the amplitude of the pulses so that each predetermined value of pulse energy is associated with a message. According to information theory[3] the information content, in bits, of each pulse is given by the logarithm (to the base 2) of the number of distinguishable values (in this case energies) it may possess. The

[3]C. E. Shannon and W. Weaver, *The Mathematical Theory of Communication* (University of Illinois Press, Urbana, Ill., 1949), p. 67.

average information content per pulse is thus

$$\bar{I}_{\text{pulse}} = \log_2\left(\frac{\bar{p}\,\Delta t}{\Delta E}\right) \tag{3.37}$$

where the energy resolution ΔE is the smallest pulse energy increment that can be measured. Since the time available for measuring the energy of a given pulse is Δt, the energy resolution ΔE is given according to the uncertainty principle (3.35) by $\Delta E \sim \hbar/\Delta t$. Substitution in (3.37) gives

$$\bar{I}_{\text{pulse}} \sim \log_2\left(\frac{\bar{p}(\Delta t)^2}{\hbar}\right)$$

The average information transmission rate is equal to the average information content per pulse multiplied by the number of pulses per second $1/\Delta t$:

$$C \sim \log_2\left(\frac{\bar{p}(\Delta t)^2}{\hbar}\right)\left(\frac{1}{\Delta t}\right) \tag{3.38}$$

It is clear that C increases as Δt is made smaller. The limit for this procedure is reached when Δt becomes comparable to the oscillation period f^{-1}. At this point (3.38) becomes

$$C \sim \left[\log_2\left(\frac{\bar{p}}{\hbar\omega B}\right)\right] B \tag{3.39a}$$

where the transmission bandwidth is taken as $B = (\Delta t)^{-1}$. Shannon's classical result[4] is,

$$C = \left[\log_2(1 + \bar{p}/\text{noise power})\right] B \tag{3.39b}$$

Equation (3.39*a*) describes a fundamental limit to the rate at which information can be transmitted. The quantity $\hbar\omega B$ has the dimension of power and by analogy to (3.39*b*) can be thought of as a noise "contaminating" the signal power $\bar{p}$. Since it results from the application of the uncertainty principle, we view the noise power $\hbar\omega B$ to be nonclassical and of quantum mechanical origin.

PROBLEMS

1. If the behavior of $\psi(r, t)$ as $r \to \infty$ is dominated by r^{-n}, what values can n assume if the integral

$$\int_A (\psi^* \nabla \psi - \psi \nabla \psi^*) \cdot \mathbf{n}\, da$$

taken over the surface at infinity is to vanish.

2. Prove that hermitian operators $\hat{A}$ and $\hat{B}$ commute if they have the same eigenfunctions.

---

[4]Shannon and Weaver, p. 67.

3. Prove that if $\hat{\mathcal{H}} = \hat{p}^2/2m + V(r)$,

$$\frac{\hbar^2}{m}\int_{-\infty}^{\infty} u_i^* \frac{\partial u_i}{\partial x}\, dv = (E_i - E_j)\int_{-\infty}^{\infty} u_j^* x u_i \, dv$$

where $\hat{\mathcal{H}} u_i = E_i u_i$.

Hint: Calculate first the commutator $[\hat{\mathcal{H}}, x]$.

4. Show that $[x, p_x^n] = i\hbar n p_x^{n-1}$.

Hint: What is $(\partial^n/\partial x^n)(xu)$?

5. In which step of the derivation leading to the uncertainty relation (3.33) did we use (implicitly) the noncommutation of the momentum and coordinate operators?

CHAPTER FOUR

One-Dimensional Energy Eigenvalue Problems

Among eigenvalue problems, the problem of finding the eigenfunctions and eigenvalues of the energy operator (the Hamiltonian) plays a special role. The total energy of a system is a constant, and is thus an important characteristic. Another reason is that the energy levels (i.e., energy eigenvalues) of a system often determine properties such as chemical bonding, crystal structure, electrical and optical properties, and rates of chemical reactions. The nature of the energy levels thus has a direct bearing on the way we perceive materials (color, as an example) and utilize them.

In this chapter we consider some simple one-electron, one-dimensional energy eigenvalue problems. In each case we solve the time-independent Schrödinger equation

$$\left(\frac{\hat{p}^2}{2m}+V(x)\right)u_E(x)=Eu_E(x) \tag{4.1}$$

where the energy operator (Hamiltonian) $\hat{p}^2/2m+V(x)$ is the sum of the kinetic and potential energies of a particle. The solution yields the energy eigenvalues E and the eigenfunctions $u_E(x)$. $V(x)$ is the potential energy function.

4.1 INFINITE POTENTIAL WELL

Consider a particle of mass m moving in a potential one-dimensional well that is zero over the interval $-a<x<a$ and infinite elsewhere, as shown in Fig.

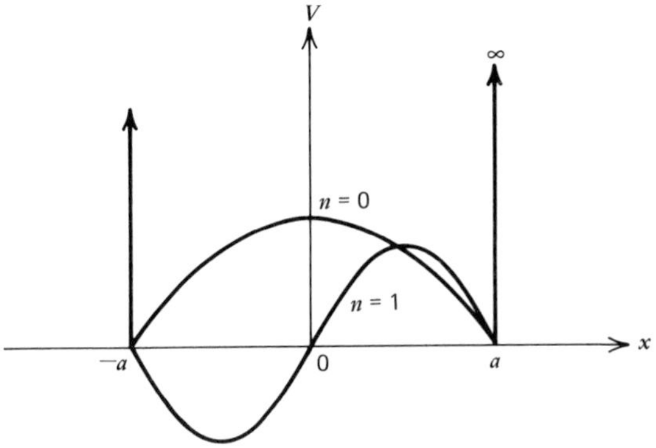

Figure 4.1 The infinite potential well and the first two ($n=0,1$) eigenfunctions of the bound particle.

4.1. The energy eigenvalue equation is given in (4.1), where, following (3.20), we set $\hat{p} \to -i\hbar\,\partial/\partial x$. The result is

$$\left(-\frac{\hbar^2}{2m}\frac{\partial^2}{\partial x^2}+V(x)\right)u_E(x)=Eu_E(x) \tag{4.2}$$

The solution in the region $-a<x<a$, where $V(x)=0$, is of the form

$$u_E=\cos kx \quad \text{or} \quad u_E=\sin kx \tag{4.3}$$

where

$$k=\sqrt{2mE/\hbar^2} \tag{4.4}$$

Outside the region $-a<x<a$, $u_E(x)$ must be zero. This follows from (4.2), since the potential $V(x)$ is infinite. In order that $u_E(x)$ vanish at $x=\pm a$, we must choose

$$k=(2n+1)\frac{\pi}{2a} \quad (n=0,1,2,3,\ldots) \tag{4.5}$$

in the case of the even (cosine) solution, and

$$k=2n\left(\frac{\pi}{2a}\right) \quad (n=1,2,3) \tag{4.6}$$

for the odd (sine) functions.

The even and odd solutions thus assume, respectively, the form

$$u_E(x)=\frac{1}{\sqrt{2a}}\cos\left[\left(n+\frac{1}{2}\right)\frac{\pi x}{a}\right] \tag{4.7}$$

$$u_E(x)=\frac{1}{\sqrt{2a}}\sin\left(n\frac{\pi x}{a}\right) \tag{4.8}$$

where the normalization factor $(2a)^{-1/2}$ was added so that $\int_{-a}^{a}u_E^2(x)\,dx=1$.

The energy expression

$$E = \frac{\hbar^2 k^2}{2m} \tag{4.9}$$

becomes, combining (4.5) and (4.6),

$$E_l = l^2 \frac{\pi^2 \hbar^2}{8ma^2} \quad (l=1,2,3,\ldots) \tag{4.10}$$

The odd values of l belong to even solutions and vice versa. The lowest energy is associated with the even solution (cosine) with $n=0$ (or $l=1$). The next highest energy is that of the odd solution with $n=1$ ($l=2$). It follows from (4.5) and (4.6) that the solutions alternate between those of odd and even symmetry as the energy increases.

4.2 FINITE POTENTIAL WELL

Next consider the motion of an electron in a potential well with finite barriers of height V, as shown in Fig. 4.2. Since the potential well is symmetric $[V(x)=V(-x)]$, the solutions of the Schrödinger equation

$$\left(-\frac{\hbar^2}{2m}\frac{d^2}{dx^2} + V(x)\right) u_E(x) = E u_E(x) \tag{4.11}$$

must possess odd symmetry $[u_E(-x) = -u_E(x)]$ or even symmetry[1] $[u_E(x) = u_E(-x)]$. It is convenient to consider two different cases.

Case 1: $E<V$

In the internal region $|x| \le a$ the solution of (4.11) is of the form

$$u_E(x) = \cos k_0 x, \sin k_0 x \tag{4.12}$$

$$k_0 = \sqrt{2mE/\hbar^2} \tag{4.13}$$

The solution for $x > a$ is of the form[2]

$$u_E(x) = Ce^{-\kappa|x|} \quad (|x|>a) \tag{4.14}$$

$$\kappa = \sqrt{2m(V-E)/\hbar^2} \tag{4.15}$$

The alternate solution $\exp(\kappa x)$ has been eliminated on physical grounds, since $\int |u_E(x)|^2\, dx$ must be finite. Since $V(x)$, and with it $u_E''(x)$, are finite everywhere, $u_E'(x)$ and $u_E(x)$ must be continuous everywhere including $x = \pm a$. We apply the continuity condition to the *even* solution and its derivative at

---

[1]The relation between the symmetry of the eigenfunction and that of $V(x)$ is discussed more fully in Chapter 5.

[2]The solution for $x < -a$ can be obtained directly from the symmetry condition.

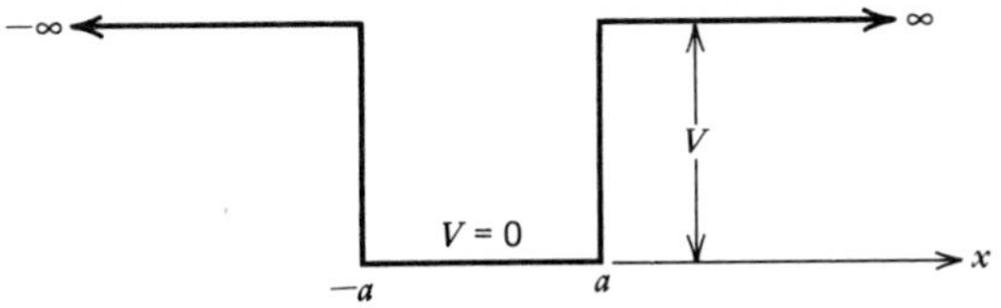

Figure 4.2 The finite potential well.

$x = a$ obtaining

$$\cos k_0 a = Ce^{-\kappa a} \tag{4.16}$$

and

$$-k_0 \sin k_0 a = -\kappa Ce^{-ka}$$

so that

$$k_0 a \tan k_0 a = \kappa a \tag{4.17a}$$

Applying the boundary conditions to the odd solutions results in

$$k_0 a \cot k_0 a = -\kappa a \tag{4.17b}$$

In either case, the constants k_0 and κ must satisfy (4.17) while simultaneously being related through (4.13) and (4.15). The last two mentioned equations can be combined as

$$\kappa^2 a^2 + k_0^2 a^2 = \frac{2mV}{\hbar^2} a^2 \tag{4.18}$$

which is the equation of a circle in the κa, $k_0 a$ plane with a radius $\sqrt{2mVa^2/\hbar^2}$. We thus need to solve (4.17) and (4.18) for k_0 and κ. A graphical procedure for obtaining k_0 and k is shown in Fig. 4.3.

The solutions correspond to the intersections in the upper half-plane ($\kappa > 0$) of the circle (4.18) with a plot of (4.17*a*). The first three solutions are designated as $n = 0, 2, 4$. We refer to the corresponding eigenstates as u_0, u_2, u_4.

The constants κ and k_0 of the odd solutions are obtained in a similar manner as the coordinates of the intersections of the circle (4.18) with a plot of (4.17*b*). These are designated as $n = 1, 3, \ldots$. The corresponding eigenstates are $u_1, u_3, \ldots$.

It is evident from the graphical construction of Fig. 4.3 that the number of bound states—that is, the number of intersections—increases with V. If the value of V is such that

$$s\frac{\pi}{2} < \sqrt{\frac{2mVa^2}{\hbar^2}} < (s+1)\frac{\pi}{2} \tag{4.19}$$

there exist exactly s+1 bound states. The state index n is thus equal to the number of zero crossings of $u_n(x)$. The first three eigenfunctions $n = 0, 1, 2$ are shown in Fig. 4.4. We note from Fig. 4.3 that the higher the state index n, the

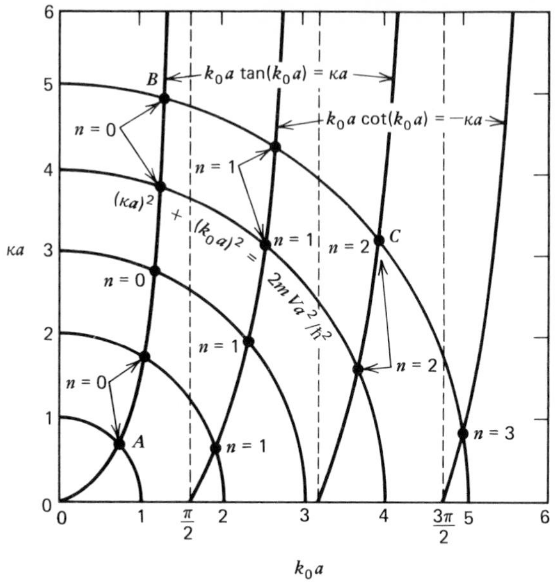

Figure 4.3 The graphical solution for obtaining the κ and k_0 of the eigenfunction of a particle in a one-dimensional rectangular potential well. These are determined by the intersections (black dots) of the plots of (4.17*a*) (for even modes) and (4.17*b*) (for odd modes) with the circles (4.18). The radius of the circles is $\sqrt{2mV}a/\hbar$. (A given problem involves only one circle)

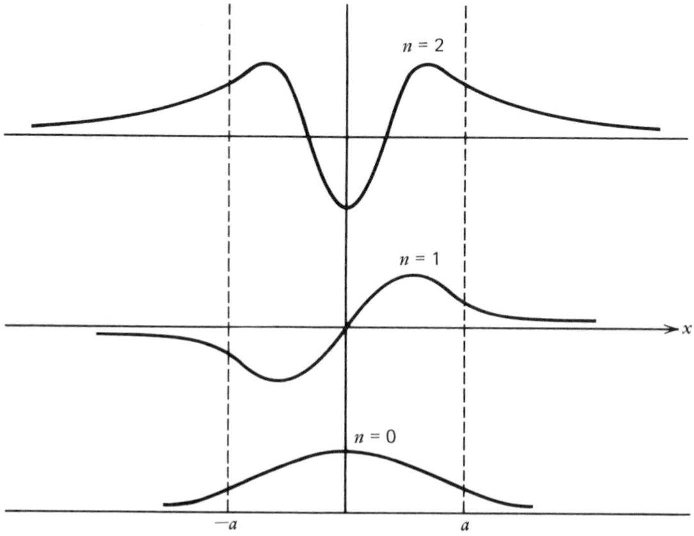

Figure 4.4 The first three bound states of a particle in a finite potential well.

smaller the value of κ, hence the deeper the penetration of the corresponding $u_n(x)$ into the $|x|>a$ region.

The energy of the eigenstate $u_n(x)$ is, according to (4.13),

$$E_n = \frac{\hbar^2 k_{0n}^2}{2m}$$

where k_{0n} is the value of k_0 of the mode u_n.

The solutions discussed in this section were all subject to the constraint $E<V$. Classical particles with energies $E<V$ cannot penetrate beyond the barrier to $|x|>a$. Yet in quantum mechanics we find that the probability distribution $|u_E(x)|^2$ is finite even for $|x|>a$. The quantum mechanical particle rushes in where its classical counterpart fears to tread.

Case 2: $E>V$

In this regime the particle energy exceeds that of the barrier and $\kappa=\sqrt{2m(V-E)}/\hbar$ is imaginary so that the solution for (4.12)–(4.14) for $u_E(x)$ is sinusoidal everywhere. Consequently, the probability density is distributed over all space and the particle *is not bound*. Here it makes physical sense to consider the case of a particle approaching the well from one direction, and to inquire about the probability of either its passing across the barrier or its reflection from it. This subject is considered in the next section.

4.3 FINITE POTENTIAL BARRIER

Here we consider a particle incident from the left on a barrier of height V and width a, as shown in Fig. 4.5a. Our results will also apply to the potential well of Fig. 4.2 by replacing V with $-V$. We first consider the case when $E<V$, so that classically the particle would be turned back at $x=0$. The solution of the Schrödinger equation (4.2) is satisfied by taking

$$u_E(x)=\begin{cases} e^{ikx}+Ae^{-ikx} & (x<0) \\ Be^{-\kappa x}+Ce^{\kappa x} & (0<x<a) \\ De^{ik(x-a)} & (x>a) \end{cases} \tag{4.20}$$

$$k=\sqrt{2mE}/\hbar, \qquad \kappa=\sqrt{2m(V-E)}/\hbar \tag{4.21}$$

In (4.20) we take the incident wave as $\exp(ikx)$ with a unity amplitude. The wave $A\exp(-ikx)$ corresponds to a reflected wave (its momentum eigenvalue $-\hbar k$ is negative). At $x>a$, $u_E(x)$ is in the form of a particle wave traveling to the right. The form of $|u_E(x)|^2$ for $E<V$ is sketched in Fig. 4.5b. By imposing the continuity condition on $u_E(x)$ and $u'_E(x)$ at the boundaries $x=0$, $x=a$, we obtain four linear equations with the unknown coefficients A, B, C, D. Their solution is

$$D=\frac{2i\kappa k}{2i\kappa k\cosh\kappa a+(k^2-\kappa^2)\sinh\kappa a} \tag{4.22}$$

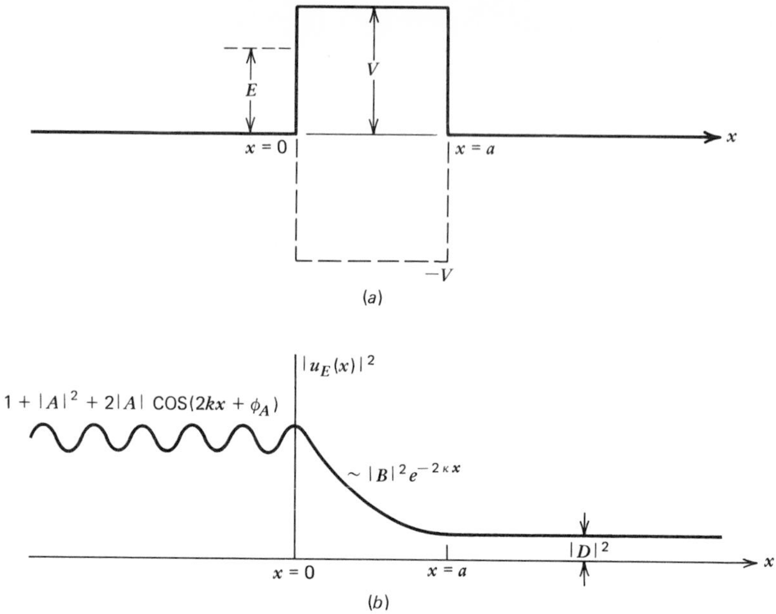

Figure 4.5 (*a*) The potential profile $V(x)$. (*b*) A plot of a typical probability distribution $|u_E(x)|^2$ for $E<V$. The plot inside the well, $0<x<a$, is for the case $\kappa a \gg 1$, where it approaches an exponential decay $|u_E(x)|^2 \propto \exp(-2\kappa x)$.

The tunneling probability; that is, the probability that an incident particle penetrates beyond the barrier ($x>a$), is given by $|D|^2$:

$$T(E<V) \equiv |D|^2 = \left(1 + \frac{\sinh^2\left(\sqrt{\frac{2m(V-E)}{\hbar^2}}\, a \right)}{4(E/V)(1-E/V)} \right)^{-1} \tag{4.23}$$

We note that when $E<V$ the transmissivity $T(E)$ is always less than unity. When the incident particle energy exceeds that of the barrier ($E>V$), the constant κ becomes imaginary, and instead of (4.23) we have

$$T(E>V) = \left(1 + \frac{\sin^2\left(\sqrt{\frac{2m(E-V)}{\hbar^2}}\, a \right)}{4(E/V)(E/V-1)} \right)^{-1} \tag{4.24}$$

The behavior of $T(E)$ for both regimes, $E<V$ and $E>V$, is sketched in Fig. 4.6. Also plotted is the reflectivity $R=|A|^2$. It will be left as an exercise to show that

$$R+T=1 \tag{4.25}$$

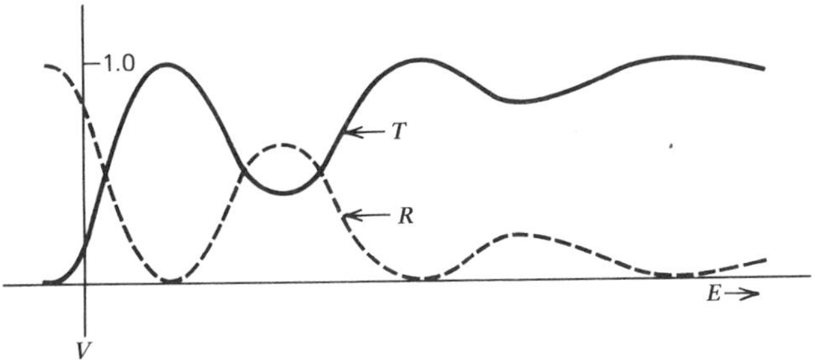

Figure 4.6 Transmission and reflection coefficients of a rectangular potential barrier.

Two important features of (4.23) and (4.24) stand out: (1) When the incident particle energy $E<V$, a region where classically the particle is turned back at $x=0$, there still exists a finite probability for penetrating—tunneling—through the barrier. This probability is given by (4.23). In the limit of small tunneling probability, $[2m(V-E)]^{1/2}(a/\hbar)\gg 1$, we may approximate (4.23) by

$$T(E)\simeq 4\left(\frac{E}{V}\right)\left(1-\frac{E}{V}\right)\exp\left(-\sqrt{\frac{8m(V-E)}{\hbar^2}}\,a\right) \tag{4.26}$$

(2) At $E>V$ the transmission $T(E)$ goes through a series of unity maxima. Two such neighboring peaks, say E_2 and E_1 in Fig. 4.6, are characterized by having the electronic round trip "phase shift" $\sqrt{8m(E-V)}\,a/\hbar$ differ by 2π. At each such peak the reflections from the barriers at $x=0$ and $x=a$ interfere destructively so that the reflection is zero. In this regime the behavior of the barrier is reminiscent of that of the optical Fabry–Perot interferometer (etalon).[3] This is an optical device that in its simplest embodiment consists of a slab of transparent solid with flat and parallel end faces. The transmission as a function of frequency of light incident on the etalon is described by a function similar to (4.24) and the mathematics involved in treating it is identical to that used in this section.

In conclusion we may note that the case of the inverted barrier, shown as a dashed curve in Fig. 4.5, is obtained by merely replacing V with $-|V|$ in (4.24). The potential well in this case is identical to the one considered in Section 4.2, except that here $E>V$ and, consequently, the solutions for E larger than 0 are unbound (sinusoidal everywhere), while in Section 4.2 we considered bound states only. In the unbound regime the energy E may take on any value. In other words, we can obtain a solution of the Schrödinger equation in the form (4.20) satisfying the boundary conditions for any E. In the bound regime discussed in Section 4.2, only a finite number of discrete eigenvalues E exist. We thus find that the complete spectrum of eigenvalues E of the potential well is part discrete and part continuous.

[3] M. Born and E. Wolf, *Principles of Optics*, 3rd ed. (Pergamon, New York, 1965), Chapter 7.

4.4 PHYSICAL MANIFESTATION OF PARTICLE TUNNELING

α Decay of Nuclei

The decay of nuclei by emission of an α particle can be viewed as a tunneling process. The nucleus before a decay event can be thought of as an α particle (a helium nucleus) trapped in a spherical potential well, which represents its interaction with the rest of the nucleus. The potential energy of the α particle as a function of its distance r from the rest of the nuclear mass is sketched in Fig. 4.7. The behavior at large r is due to Coulomb repulsion between the like charges, while at $r < R$ ($R \sim 10^{-15}$ cm) it is dominated by the nuclear forces and is attractive. The particle of energy E less than the maximum in the potential well may be thought of as bouncing between the two sides of the well with a small probability of tunneling through and escaping upon each incidence.

The tunneling probability per unit time (the inverse of the decay lifetime) is equal to the number of bounces per unit time multiplied by the tunneling probability per incidence. The bouncing rate is approximately

$$\omega \sim \frac{v}{R} \simeq \frac{p}{mR} \simeq \frac{\hbar k}{mR} \simeq \frac{\hbar \pi}{mR^2}$$

where v is the velocity of the particle and we used (4.5) to write $k \sim \pi/R$. The

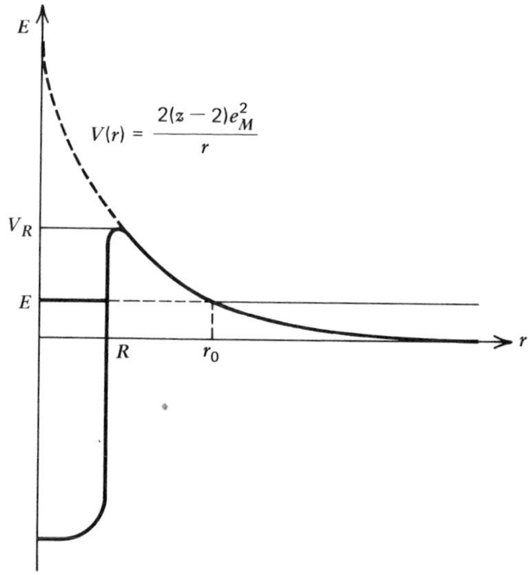

Figure 4.7 Energy diagram of the mutual potential between an α-particle ($z=2$) and "daughter" nucleus whose charge $=(z-2)e$. For large distances ($r \gg 10^{-14}$ m) it is simply the Coulomb repulsion. At short distances it is dominated by the nuclear attraction.

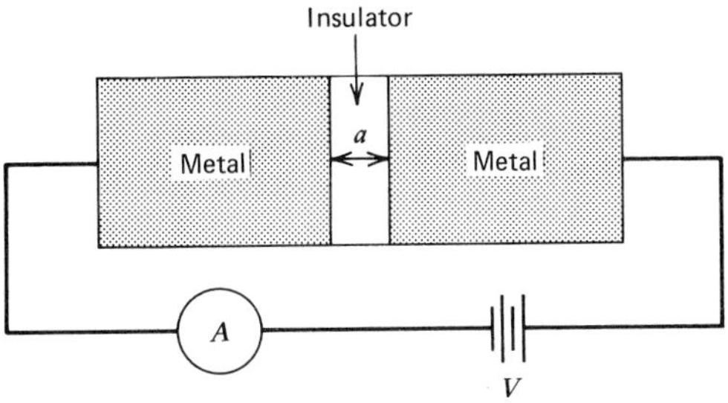

Figure 4.8 A metal-insulator/metal-sandwich. When a voltage is applied across the structure, a current flow is established. The electrons cross the insulating material by tunneling.

tunneling probability per incidence is given by (4.26) as

$$T(E) \sim 4\left(\frac{E}{V}\right)\left(1-\frac{E}{V}\right)\exp\left(-\frac{\sqrt{8m(V-E)}}{\hbar^2}(r_0-R)\right)$$

The decay rate is thus determined predominantly by the exponential factor in $T(E)$. Its actual value is very sensitive to the exact shape of the potential curve and can vary by many orders of magnitude from nucleus to nucleus.

Tunneling in Solids

Another manifestation of tunneling occurs in solid state physics. Consider two conductors (Fig. 4.8) (this may include superconductors and semiconductors) that are separated by a thin (~ 10 Å) layer of an insulator. When a voltage is impressed across the "sandwich" a current will be observed to flow. This current is due to electrons crossing from one metal to another by tunneling through the potential barrier presented by the insulator.[4] The tunneling nature of this current is established by noting its exponential dependence on the insulator thickness a in accordance with (4.26).

For the student of electromagnetic theory we may point out the exact formal similarity of electron tunneling and propagation of electromagnetic modes in waveguides below cutoff.[5]

PROBLEMS

1. Obtain the solution $u_E(x)$ for a potential well

$$V(x)=\infty \ (x<0);\ V(x)=0\ (0<x<a);\ V(x)=V_0\ (x>a).$$

---

[4]This potential barrier is due to the fact that when dissimilar materials are brought into contact their chemical potentials are equalized through electric charge transfer so that potential gradients are set up.

[5]See, for example: S. Ramo, J. R. Whinnery, and T. Van Duzer, *Fields and Waves in Communication Electronics* (John Wiley and Sons, New York, 1965), p. 422.

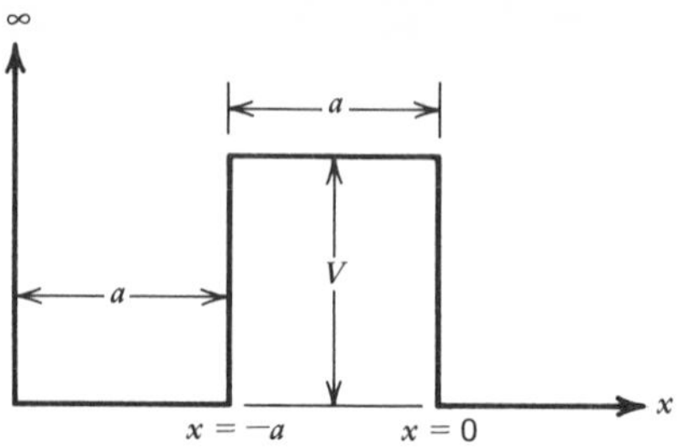

Figure 4.9

2. (a) Complete the algebra and derive D in (4.22).
   (b) Prove (4.25).
3. Using any convenient references, describe the formal analogy between the evanescent phenomena of electron tunneling and that of propagation of electromagnetic waveguide modes at frequencies below cutoff.
4. Formulate the one-dimensional problem of the transmission and reflection of an incident electron from two potential barriers, each of the form of Fig. 4.5*a*, which are separated by a distance d.

   Hint: It will prove profitable to develop a matrix formalism to describe the effect of any single well on the eigenfunction. The matrix should relate the incident and reflected waves at one plane to those at some other plane.

5. Extend the matrix technique of Problem 4 to describe the propagation of an electron through an arbitrary sequence of rectangular barriers.
6. Estimate the lifetime of a particle of mass m trapped in the potential well shown in Fig. 4.9.

   Hint: The system shown does not possess trapped particle eigenstates corresponding to a trapped particle, since such a particle will "leak" away through tunneling. For the purpose of the approximate estimate of the lifetime, we may assume that $V=\infty$ when deriving the eigenfunctions of the trapped particle. The latter may be assumed to be incident on the boundary $x=0$ with a velocity $v\simeq \hbar k/m$ and a frequency v/a.

CHAPTER FIVE

The Harmonic Oscillator

In this chapter we consider the eigenvalue problem of the harmonic oscillator. The idealized harmonic oscillator is taken as a point mass connected to the end of a frictionless idealized spring (i.e., a spring in which the restoring force is proportional to its elongation). A number of very important problems—including the quantum treatment of electromagnetic modes, lattice vibrations, and even the electrical engineer's RLC "tank" circuit—can be modeled as a harmonic oscillator. The study of the harmonic oscillator is thus of fundamental importance in quantum mechanics. The mathematical techniques employed are very elegant and are crucial to consideration of quantum optics, fluctuation theory, noise, and coherence. Obviously, this is one topic we cannot avoid.

5.1 PARITY

Before delving into the problem of the harmonic oscillator we introduce the concept of parity to which we already alluded in Section 4.1.

Consider the time-independent Schrödinger equation of a particle moving in a potential field $V(\mathbf{r})$:

$$-\frac{\hbar^2}{2m}\nabla^2 u_E(\mathbf{r})+V(\mathbf{r})u_E(\mathbf{r})=Eu_E(\mathbf{r}) \tag{5.1}$$

Let the potential function $V(\mathbf{r})$ possess inversion symmetry, that is,

$$V(-\mathbf{r})=V(\mathbf{r}) \tag{5.2}$$

It follows that

$$-\frac{\hbar^2}{2m}\nabla^2 u_E(-\mathbf{r})+V(\mathbf{r})u_E(-\mathbf{r})=Eu_E(-\mathbf{r}) \tag{5.3}$$

so that $u_E(-\mathbf{r})$ is an eigenfunction of the same Hamiltonian as $u_E(\mathbf{r})$ with the same eigenvalue. If there exists only one eigenfunction with the eigenvalue E (i.e., the set E is nondegenerate), then $u_E(\mathbf{r})$ and $u_E(-\mathbf{r})$ are not independent; we may write

$$u_E(-\mathbf{r}) = \lambda u_E(\mathbf{r}) \tag{5.4}$$

where λ is some constant. Let a parity operator $\hat{P}$ be defined by

$$\hat{P}f(\mathbf{r}) = f(-\mathbf{r}) \tag{5.5}$$

where $f(\mathbf{r})$ is any function. We may now restate (5.4) as

$$\hat{P}u_E(\mathbf{r}) = u_E(-\mathbf{r}) = \lambda u_E(\mathbf{r}) \tag{5.5a}$$

so that

$$\hat{P}^2 u_E(\mathbf{r}) = \lambda^2 u_E(\mathbf{r}) = u_E(\mathbf{r}) \tag{5.5b}$$

(operating twice with $\hat{P}$ must return the function u_E to its original form). From (5.5*b*)

$$\lambda^2 = 1 \quad (\lambda = \pm 1) \tag{5.6}$$

Let us first take up the case $\lambda = 1$. From (5.4),

$$u_E(-\mathbf{r}) = u_E(\mathbf{r})$$

The function $u_E(\mathbf{r})$ remains invariant under inversion and is said to possess *even parity.* The second root, $\lambda = -1$, leads to

$$u_E(-\mathbf{r}) = -u_E(\mathbf{r}) \tag{5.7}$$

and the function possesses *odd parity.* In a centrosymmetric potential field the eigenfunction must possess odd or even parity. The above proof does not work when the set u_E is degenerate (i.e., more than one eigenfunction has the same energy). In that case we may construct linear combinations that possess definite parity (i.e., either remain the same or change sign upon the inversion $\mathbf{r} \to -\mathbf{r}$).

We remind ourselves, however, that there exist many physical configurations where the potential function $V(\mathbf{r})$ is not centrosymmetric [i.e., $V(\mathbf{r}) \neq V(-\mathbf{r})$] so that the eigenfunctions do not possess parity.

5.2 THE HARMONIC OSCILLATOR

The wave equation of the harmonic oscillator in one dimension with a restoring force Kx is

$$-\frac{\hbar^2}{2m}\frac{d^2u}{dx^2} + \tfrac{1}{2}Kx^2 u = Eu \tag{5.8}$$

Using the following substitutions:

$$\xi = \alpha x, \quad \alpha^4 = \frac{mK}{\hbar^2} = \left(\frac{m\omega}{\hbar}\right)^2, \quad \omega^2 = \frac{K}{m}, \quad \lambda = \frac{2E}{\hbar\omega}$$

Equation (5.8) becomes

$$\frac{d^2u}{d\xi^2}+(\lambda-\xi^2)u=0 \tag{5.9}$$

For $\xi^2 \gg \lambda$ the behavior of u is dominated by $e^{-(1/2)\xi^2}$, so that it is natural to try a solution of the form

$$u(\xi)=H(\xi)e^{-(1/2)\xi^2} \tag{5.10}$$

where $H(\xi)$ is some function of ξ.

Substituting $u(\xi)$ from (5.10) in (5.9) leads to the equation

$$\frac{d^2H}{d\xi^2}-2\xi\frac{dH}{d\xi}+(\lambda-1)H=0 \tag{5.11}$$

We assume a power series expansion

$$H(\xi)=\xi^s(a_0+a_1\xi+a_2\xi^2+\cdots) \tag{5.12}$$

where $a_0\xi^s$ is the first nonzero term of H.

Substituting (5.12) in (5.11) and equating separately the coefficients of the various powers of ξ to zero yields

$$\begin{aligned} s(s-1)a_0&=0\\ (s+1)sa_1&=0\\ (s+2)(s+1)a_2-(2s+1-\lambda)a_0&=0\\ &\vdots\\ (s+v+2)(s+v+1)a_{v+2}-(2s+2v+1-\lambda)a_v&=0 \end{aligned} \tag{5.13}$$

Since $a_0 \neq 0$ it follows from the first of Eq. (5.13) that $s=0$ or $s=1$, from the second that $s=0$ or $a_1=0$, or both. The last equation shows how the general coefficient a_{v+2} can be determined from a_v.

The first case to be considered is that of $s=0$. Since $a_0 \neq 0$, the only way to terminate the sequence of a_v with v even is to have

$$\lambda=2v+1 \tag{5.14}$$

for some v. Since v is even, λ can take on the values $1,5,9,\ldots$. This choice of λ will not terminate, as is made clear from the last of (5.13), the odd a_v sequence. The only way to guarantee a finite number of terms in $H(\xi)$ is to put $a_1=0$. This prevents the odd a_v series from ever "getting off the ground." The same argument is repeated with $s=1$, again only even v terms are allowed and $a_1=0$. λ takes on the sequence of values

$$\lambda=2v+3=3,7,11,\ldots \tag{5.15}$$

and the resultant polynomial $H_n(\xi)$ is odd (since the even polynomial is now multiplied by ξ). Combining (5.14) and (5.15), the allowed values for λ

become

$$\lambda = 2n+1 \quad (n=1,2,3,\ldots)$$

which when using $\lambda = 2E/\hbar\omega$ gives

$$E_n = \hbar\omega\left(n+\tfrac{1}{2}\right) \tag{5.16}$$

for the energy of the nth eigenstate.

According to (5.16), the harmonic oscillator, even in its lowest energy state $n=0$, has a finite amount of energy, $\frac{1}{2}\hbar\omega$. The lowest energy of a classical harmonic oscillator is zero. This essential difference is a manifestation of the uncertainty principle. For the classical harmonic oscillator to have zero energy, both its momentum p_x and position x must be simultaneously zero. According to the uncertainty principle, this is impossible since $\Delta p \Delta x \geq \hbar$. The division of uncertainty between p and x which minimizes the total energy E_0 while satisfying the uncertainty principle, gives $E_0 \simeq \hbar\omega/2$. To show this we write off the oscillator in its ground state as the total energy as

$$E = \frac{1}{2}\left(K(\Delta x)^2 + \frac{(\Delta p)^2}{m} \right)$$

letting $\Delta p \Delta x = \hbar/2$:

$$E = \frac{1}{2}\left(\frac{K\hbar^2}{4(\Delta p)^2} + \frac{(\Delta p)^2}{m} \right)$$

which when minimized with respect to Δp gives

$$E_{\text{min}} = \tfrac{1}{2}\hbar\omega$$

The Hermite Polynomials

The solutions of Eq. (5.11) that correspond to the different values of $\lambda = 2n+1$ are seen to be polynomials of order n. The polynomials are even when n is an even integer or odd when n is odd. Putting $\lambda = 2n+1$, the differential equation for the polynomials $H_n(\xi)$, which are known as the Hermite polynomials, becomes

$$\frac{d^2H_n}{d\xi^2} - 2\xi\frac{dH_n}{d\xi} + 2nH_n = 0 \tag{5.17}$$

These polynomials are conveniently derived by means of the power series expansion of the function $e^{-s^2+2s\xi}$ according to

$$G(\xi, s) = e^{\xi^2-(s-\xi)^2} = e^{-s^2+2s\xi} = \sum_{n=0}^{\infty} \frac{H_n(\xi)}{n!} s^n \tag{5.18}$$

To show that the function $H_n(\xi)$ defined by the expansion (5.18) satisfies the differential equation (5.17), we first derive two important recursion relations. We start by differentiating both sides of (5.18) with respect to ξ:

$$2se^{-s^2+2s\xi}=2s\sum_{n=0}^{\infty}\frac{H_n(\xi)}{n!}s^n=\sum\frac{H_n'(\xi)}{n!}s^n$$

where the prime indicates differentiation with respect to the argument. Equating the coefficients of equal powers of s leads to

$$H_n'(\xi)=2nH_{n-1}(\xi) \tag{5.19}$$

A second relation results from differentiating (5.18) with respect to s:

$$e^{-s^2+2s\xi}(2\xi-2s)=\sum_{n=0}^{\infty}\frac{H_n(\xi)ns^{n-1}}{n!} \tag{5.20}$$

Using (5.18) to expand $\exp(-s^2+2s\xi)$ we obtain

$$\sum_{n=0}^{\infty}\left(-2\frac{H_n(\xi)}{n!}s^{n+1}+2\frac{H_n(\xi)}{n!}\xi s^n\right)=\sum_{n=0}^{\infty}\frac{H_n(\xi)ns^{n-1}}{n!}$$

From which

$$\xi H_n(\xi)=\tfrac{1}{2}H_{n+1}(\xi)+nH_{n-1}(\xi) \tag{5.21}$$

From (5.19) it follows that

$$H_n''=2nH_{n-1}' \tag{5.21a}$$

From (5.21)

$$\begin{aligned}2nH_{n-1}'&=\frac{d}{d\xi}(2\xi H_n)-H_{n+1}'\\&=2\xi H_n'+2H_n-H_{n+1}'\end{aligned}$$

and using (5.19), (5.21a) and in the last expression leaves:

$$H_n''(\xi)-2\xi H_n'(\xi)+2nH_n(\xi)=0 \tag{5.22}$$

which is the same as the differential equation (5.17). We may thus use either (5.17) or (5.18), or the set of recursion relations (5.19) and (5.21) as the basic relations defining the Hermite polynomials $H_n(\xi)$.

To generate the Hermite polynomials we notice that, according to (5.18),

$$\begin{aligned}H_n(\xi)&=\left(\frac{\partial^n}{\partial s^n}\left(e^{\xi^2-(s-\xi)^2}\right)\right)_{s=0}\\&=e^{\xi^2}(-1)^n\left(\frac{\partial^n}{\partial\xi^n}e^{-(s-\xi)^2}\right)_{s=0}\\&=e^{\xi^2}(-1)^n\frac{d^n}{d\xi^n}e^{-\xi^2}\end{aligned} \tag{5.23}$$

Applying (5.23) to generate, as an example, the first three H_n's gives

$$H_0(\xi)=1,\quad H_1(\xi)=2\xi,\quad H_2(\xi)=4\xi^2-2 \tag{5.24}$$

This particular sequence of $H_n(\xi)$ corresponds to choosing the coefficient of the highest power of ξ to be equal to 2^n.

Some Useful Integrals

According to (5.10), the harmonic oscillator wavefunction is

$$u_n(x) = N_n e^{-(1/2)\alpha^2 x^2} H_n(\alpha x) \tag{5.25}$$

The normalization constant N_n is determined by requiring that $\int_{-\infty}^{+\infty} u_n^* u_n \, dx = 1$:

$$\begin{aligned}\int_{-\infty}^{+\infty} u_n^* u_n \, dx &= N_n^2 \int_{-\infty}^{+\infty} e^{-\alpha^2 x^2} H_n^2(\alpha x)\, dx \\ &= \frac{N_n^2}{\alpha} \int_{-\infty}^{+\infty} e^{-\xi^2} H_n^2(\xi)\, d\xi \\ &= 1\end{aligned} \tag{5.26}$$

The integration (5.26) is most conveniently performed with the aid of the generating function (5.18). Consider the integral

$$\int_{-\infty}^{+\infty} e^{-s^2+2s\xi-t^2+2t\xi-\xi^2}\, d\xi = \sum_{n=0}^{\infty} \sum_{m=0}^{\infty} \frac{s^n t^m}{n!m!} \int_{-\infty}^{+\infty} H_n(\xi) H_m(\xi) e^{-\xi^2}\, d\xi$$

Replacing the definite integral on the left side with its value $\pi^{1/2} e^{2st}$ gives

$$\begin{aligned}\pi^{1/2} e^{2st} &= \pi^{1/2} \sum_{0} \frac{2^n s^n t^n}{n!} \\ &= \sum_n \sum_m \frac{s^n t^m}{n!m!} \int_{-\infty}^{+\infty} H_n(\xi) H_m(\xi) e^{-\xi^2}\, d\xi\end{aligned}$$

Equating the coefficients of equal powers of $s^n t^m$ on both sides results in

$$\int_{-\infty}^{+\infty} H_n(\xi) H_m(\xi) e^{-\xi^2}\, d\xi = \begin{cases} 0 & (m \neq n) \\ \pi^{1/2} n! 2^n & (m = n) \end{cases} \tag{5.27}$$

A comparison with (5.26) identifies N_n as

$$N_n = \left(\frac{\alpha}{\pi^{1/2} n! 2^n} \right)^{1/2} \tag{5.28}$$

The complete normalized harmonic oscillator wavefunction is thus

$$u_n(x) = \left(\frac{\alpha}{\pi^{1/2} n! 2^n} \right)^{1/2} H_n(\alpha x) e^{-(1/2)\alpha^2 x^2} \tag{5.29}$$

Plots of some of the low-order $u_n(x)$ are given in Fig. 5.1.

We will now use the generating function to evaluate the integral

$$\int_{-\infty}^{\infty} u_n \frac{du_m}{dx}\, dx$$

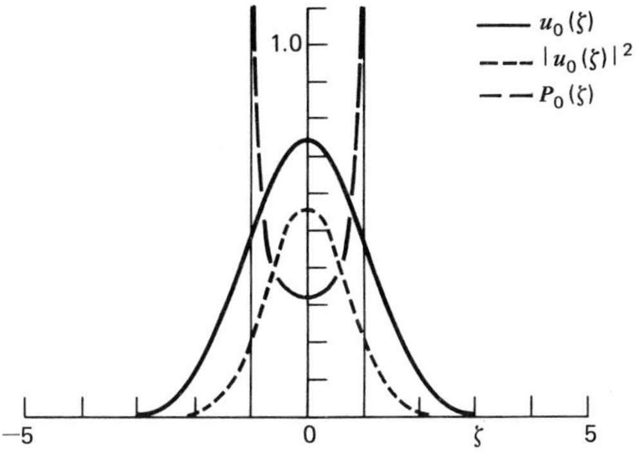

(a)

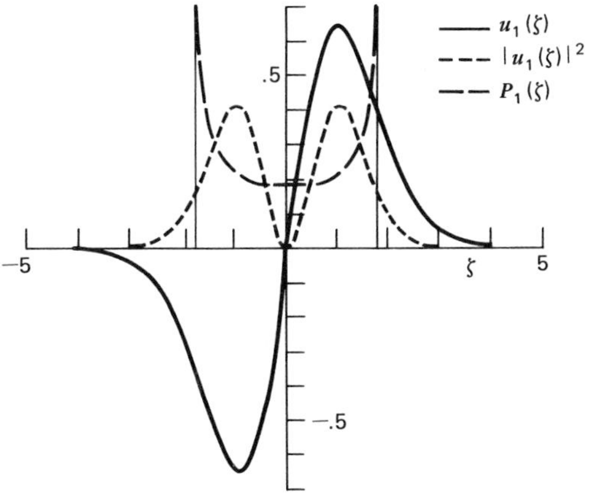

(b)

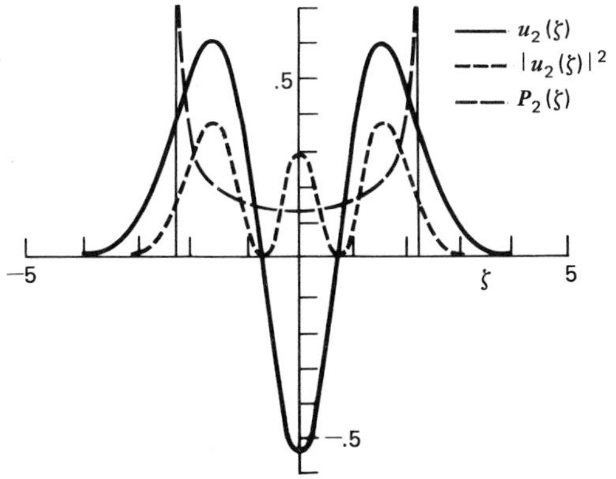

(c)

Figure 5.1 Harmonic-oscillator wavefunctions. The solid curves represent the functions $\alpha^{-1/2}u_n(\alpha x)$ with $\alpha x = \xi$ for $n = 0$, 1, 2, 3, and 10. The dotted curves represent $\alpha^{-1}u_n u_n^*$ for the same values of n. The dashed curves represent the probability distribution for a classical oscillator having the same energy as the corresponding quantum-mechanical oscillator. The vertical lines define the limits of the classical motion. [From R. B. Leighton, *Principles of Modern Physics* (McGraw-Hill, New York, 1959).]

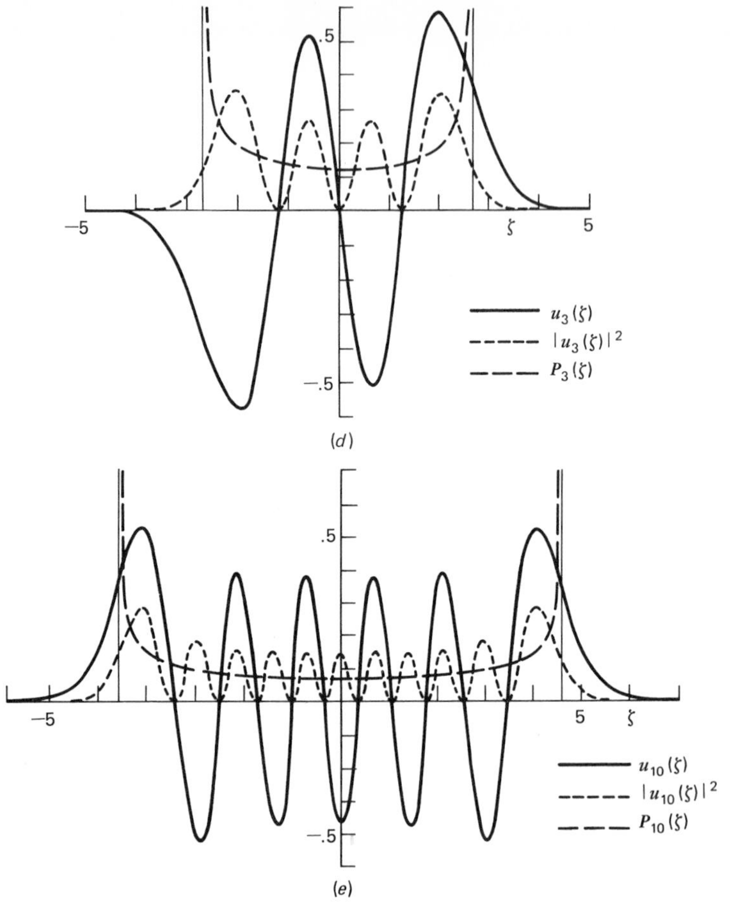

Figure 5.1 (*Continued*)

Consider first the integral

$$\int_{-\infty}^{+\infty} G(\xi, s) e^{-(\xi^2/2)} \frac{\partial}{\partial \xi}\left[G(\xi, t) e^{-(\xi^2/2)}\right] d\xi \tag{5.30}$$

which, using (5.18), becomes

$$\int_{-\infty}^{+\infty} e^{-s^2+2s\xi} e^{-\xi^2/2} \frac{\partial}{\partial \xi}\left(e^{-(\xi^2/2)+2t\xi-t^2}\right) d\xi$$

$$= \sum_n \sum_m \frac{s^n t^m}{n!m!} \int_{-\infty}^{+\infty} H_n(\xi) e^{-(\xi^2/2)} \frac{\partial}{\partial \xi}\left[H_m(\xi) e^{-(\xi^2/2)}\right] d\xi$$

$$= \int_{-\infty}^{+\infty} e^{-(s^2+t^2)} e^{+2\xi(s+t)} e^{-\xi^2}(-\xi+2t)\, d\xi \tag{5.31}$$

This last integral is next replaced by the value of the two definite integrals (in the same order) that compose it:

$$-\pi^{1/2}(s+t)e^{2st}+\pi^{1/2}2te^{2st}=\pi^{1/2}(t-s)e^{2st}$$

$$=\pi^{1/2}\sum_{n=0}^{\infty}2^{n}\frac{s^{n}t^{n+1}-s^{n+1}t^{n}}{n!}$$

Equating equal coefficients of $s^n t^m$ and using (5.29) gives

$$\int_{-\infty}^{+\infty}u_n\frac{du_m}{dx}dx=\begin{cases}\alpha\left(\dfrac{n+1}{2}\right)^{1/2} & (m=n+1)\\ -\alpha\left(\dfrac{n}{2}\right)^{1/2} & (m=n-1)\\ 0 & (\text{otherwise})\end{cases}\tag{5.32}$$

We can use (5.32) to obtain another useful integral involving u_n and u_m. We make use of the following relation that holds (see Problem 3, Chapter 3) for any pair of one-particle wavefunctions:

$$\int_{-\infty}^{+\infty}u_j^*(\mathbf{r})xu_i(r)\,d^3\mathbf{r}=\frac{\hbar^2}{m(E_i-E_j)}\int_{-\infty}^{+\infty}u_j^*\frac{\partial u_i}{\partial x}d^3\mathbf{r}\tag{5.33}$$

A direct substitution of (5.32) gives

$$\int_{-\infty}^{+\infty}u_n(x)xu_m(x)\,dx=\begin{cases}\dfrac{1}{\alpha}\left(\dfrac{n+1}{2}\right)^{1/2}=\sqrt{\dfrac{\hbar}{2m\omega}}\,(n+1)^{1/2} & (m=n+1)\\ \dfrac{1}{\alpha}\left(\dfrac{n}{2}\right)^{1/2}=\sqrt{\dfrac{\hbar}{2m\omega}}\,n^{1/2} & (m=n-1)\\ 0 & \text{otherwise}\end{cases}\tag{5.34}$$

where the relation $E_{n+1}-E_n=\hbar\omega$ was used. Another important integral, whose proof is assigned as a problem, is

$$\int_{-\infty}^{+\infty}u_n(x)x^2u_m(x)\,dx=\begin{cases}\dfrac{2n+1}{2\alpha^2} & (m=n)\\ \dfrac{\sqrt{(n+1)(n+2)}}{2\alpha^2} & (m=n+2)\\ 0 & (n\neq m\neq n\pm 2)\end{cases}\tag{5.35}$$

5.3 THE ANNIHILATION AND CREATION OPERATORS

Let us define two new operators $\hat{a}$ and $\hat{a}^\dagger$ by

$$\begin{aligned}\hat{a} &= \frac{\alpha}{\sqrt{2}}x + i\frac{1}{\sqrt{2}\,\hbar\alpha}\hat{p}_x \\ &= \frac{\alpha}{\sqrt{2}}x + \frac{1}{\sqrt{2}\,\alpha}\frac{\partial}{\partial x} \\ \hat{a}^\dagger &= \frac{\alpha}{\sqrt{2}}x - i\frac{1}{\sqrt{2}\,\hbar\alpha}\hat{p}_x \\ &= \frac{\alpha}{\sqrt{2}}x - \frac{1}{\sqrt{2}\,\alpha}\frac{\partial}{\partial x}\end{aligned} \tag{5.36}$$

where

$$\alpha^4 = \left(\frac{m\omega}{\hbar}\right)^2$$

In terms of the variable $\xi = \alpha x$,

$$\begin{aligned}\hat{a} &= \frac{1}{\sqrt{2}}\left(\xi + \frac{\partial}{\partial \xi}\right) \\ \hat{a}^\dagger &= \frac{1}{\sqrt{2}}\left(\xi - \frac{\partial}{\partial \xi}\right)\end{aligned} \tag{5.37}$$

Consider, next, the result of operating with $\hat{a}$ on the harmonic oscillator eigenfunction $u_n(\alpha x)$:

$$\begin{aligned}\hat{a}u_n(\alpha x) &= \frac{1}{\sqrt{2}}\left(\xi + \frac{\partial}{\partial \xi}\right)N_n H_n(\xi)e^{-(1/2)\xi^2} \\ &= \frac{N_n}{\sqrt{2}}(\xi H_n - \xi H_n + H_n')e^{-(1/2)\xi^2} \\ &= \frac{N_n}{\sqrt{2}}H_n'(\xi)e^{-(1/2)\xi^2}\end{aligned} \tag{5.38}$$

Using $N_n = (\alpha/\pi^{1/2}2^n n!)^{1/2}$ and (5.19) we obtain

$$\frac{N_n}{\sqrt{2}}H_n'(\xi)e^{-(1/2)\xi^2} = n^{1/2}N_{n-1}H_{n-1}(\xi)e^{-(1/2)\xi^2}$$

so that

$$\hat{a}u_n = n^{1/2}u_{n-1}$$

and, similarly,

$$\hat{a}^\dagger u_n = (n+1)^{1/2}u_{n+1} \tag{5.39}$$

The result of operating with $\hat{a}$ on u_n is to generate the next lowest eigenfunction u_{n-1}, which has one unit ($\hbar\omega$) of energy less than u_n. For this reason $\hat{a}$ is called the (harmonic oscillator) *annihilation operator*. $\hat{a}^\dagger$ is called the *creation operator*, since it transforms u_n into u_{n+1}, which has an extra quantum $\hbar\omega$ of energy. It follows directly from (5.36) that $\hat{a}$ and $\hat{a}^\dagger$ are not Hermitian. This is due to the fact that they are defined as a complex sum of the two Hermitian operators x and $\hat{p}_x$, and thus do not correspond to any physical observable.

The commutator of $\hat{a}$ and $\hat{a}^\dagger$ is

$$[\hat{a}, \hat{a}^\dagger] = \left[\left(\frac{\alpha}{\sqrt{2}}x + \frac{i}{\sqrt{2}\,\hbar\alpha}\hat{p}_x\right), \left(\frac{\alpha}{\sqrt{2}}x - \frac{i}{\sqrt{2}\,\hbar\alpha}\hat{p}_x\right)\right]$$

$$= -\frac{i}{2\hbar}[x, \hat{p}_x] + \frac{i}{2\hbar}[\hat{p}_x, x]$$

which, through the relation (3.22), becomes

$$[\hat{a}, \hat{a}^\dagger] = 1 \tag{5.40}$$

Using (5.36) we can express x and $\hat{p}_x$ in terms of $\hat{a}$ and $\hat{a}^\dagger$ as

$$x = \frac{1}{\sqrt{2}\,\alpha}(\hat{a}^\dagger + \hat{a})$$

$$\hat{p}_x = \frac{i\hbar\alpha}{\sqrt{2}}(\hat{a}^\dagger - \hat{a}) \tag{5.41}$$

Substituting (5.41) in the Hamiltonian,

$$\hat{\mathcal{H}} = \frac{\hat{p}_x^2}{2m} + \tfrac{1}{2}Kx^2$$

and using the relations $\alpha^4 = mK/\hbar^2$ and $\omega = \sqrt{K/m}$ yields

$$\hat{\mathcal{H}} = \frac{\hbar\omega}{2}(\hat{a}\hat{a}^\dagger + \hat{a}^\dagger\hat{a})$$

Using $[\hat{a}, \hat{a}^\dagger] = 1$ we obtain

$$\hat{\mathcal{H}} = \hbar\omega(\hat{a}^\dagger\hat{a} + \tfrac{1}{2}) \tag{5.42}$$

This is a most useful form of the harmonic oscillator Hamiltonian, and it will be encountered in several subsequent developments.

The operator $\hat{a}^\dagger\hat{a}$ commutes with $\hat{\mathcal{H}}$ and has the number of quanta n for its eigenvalue. To show this we use Eq. (5.39):

$$\hat{a}^\dagger\hat{a}u_n = \hat{a}^\dagger\sqrt{n}\,u_{n-1}$$

$$= nu_n$$

Before closing this section, it may be of interest to note that the solutions of the harmonic oscillator wavefunctions obtained above also characterize other types of physical phenomena. One of these involves the propagation of light in optical fibers whose index of refraction is a quadratic function of the distance from the axis

$$n(r) = n\left(1 - \frac{n_2}{2n} r^2\right)$$

The solutions[1] of the electric field of the propagating modes is formally analogous to that of the harmonic oscillator. Such fibers are playing an important role in optical communication.

PROBLEMS

1. Using the generating function method, solve the integral

$$\int_{-\infty}^{+\infty} u_m x^2 u_n \, dx$$

where $u_m(x)$ is the harmonic oscillator wavefunction.

2. Show that the operators $\hat{a}$ and $\hat{a}^\dagger$ are a Hermitian adjoint pair.

3. (a) Show that the commutation relation $[a, a^\dagger] = 1$ is consistent with the relation $a = d/da^\dagger$.

   (b) Show that

$$a^n f(a^\dagger) u_0 = \frac{d^n f(a^\dagger)}{d(a^\dagger)^n} u_0$$

where u_0 is the ground-state harmonic oscillator wavefunction and $f(a^\dagger)$ is any polynomial in $a^\dagger$.

4. Prove Eqs. (5.33), (5.34), and (5.35) using annihilation and creation operators and relation (5.39). Show that

$$u_n = \frac{(a^\dagger)^n}{\sqrt{n!}} u_0$$

---

[1]A. Yariv, *Introduction to Optical Electronics*, 2nd ed. (Holt Rinehart and Winston, New York, 1976), p. 40.

CHAPTER SIX

The Quantum Mechanics of Angular Momentum

The study of angular momentum operators and their eigenfunctions undertaken in this chapter is a prerequisite for treating the general problem of electrons in spherically symmetric potential fields $[V(\mathbf{r})=V(r)]$ and for the quantum mechanical analysis of molecular rotation.

6.1 THE ANGULAR MOMENTUM OPERATORS

It is convenient to express the angular momentum operators in terms of spherical coordinates. The coordinate system employed is shown in Fig. 6.1. The coordinates of a point (r, θ, ϕ) are related to (x, y, z) by

$$\begin{aligned} x &= r\sin\theta\cos\phi \\ y &= r\sin\theta\sin\phi \\ z &= r\cos\theta \end{aligned} \tag{6.1}$$

The classical definition

$$\mathbf{l}=\mathbf{r}\times\mathbf{p} \tag{6.2}$$

of the angular momentum vector $\mathbf{l}$ and the correspondence $\hat{\mathbf{p}} \rightarrow -i\hbar\nabla$

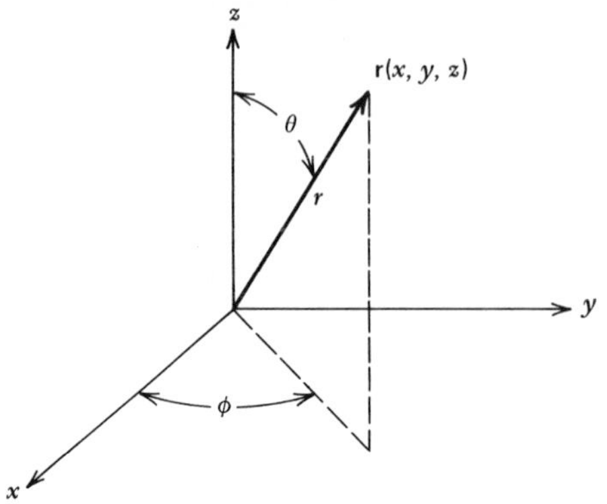

Figure 6.1 The relationship between the spherical coordinates (r, θ, ϕ) and the Cartesian coordinates (x, y, z) of a point $\mathbf{r}$.

enables us to write the Cartesian components of the operator $\hat{\mathbf{l}}$ as

$$\begin{aligned} \hat{l}_x &= y\hat{p}_z - z\hat{p}_y = -i\hbar\left(y\frac{\partial}{\partial z} - z\frac{\partial}{\partial y}\right) \\ \hat{l}_y &= -i\hbar\left(z\frac{\partial}{\partial x} - x\frac{\partial}{\partial z}\right) \\ \hat{l}_z &= -i\hbar\left(x\frac{\partial}{\partial y} - y\frac{\partial}{\partial x}\right) \end{aligned} \tag{6.3}$$

Our next task is to express $\hat{l}_x$, $\hat{l}_y$, and $\hat{l}_z$ in spherical coordinates. This is a somewhat tedious task. We will sketch the necessary steps in the case of

$$\hat{l}_x = -i\hbar\left(y\frac{\partial}{\partial z} - z\frac{\partial}{\partial y}\right)$$

as follows:

$$\frac{\partial}{\partial z} = \frac{\partial}{\partial r}\frac{\partial r}{\partial z} + \frac{\partial}{\partial \theta}\frac{\partial \theta}{\partial z} + \frac{\partial}{\partial \phi}\frac{\partial \phi}{\partial z} \tag{6.4}$$

where

$$\frac{\partial}{\partial r} = \left.\frac{\partial}{\partial r}\right|_{\theta,\phi}, \quad \frac{\partial}{\partial \phi} = \left.\frac{\partial}{\partial \phi}\right|_{r,\theta}, \quad \text{and} \quad \frac{\partial}{\partial \theta} = \left.\frac{\partial}{\partial \theta}\right|_{r,\phi}$$

From $r^2 = x^2 + y^2 + z^2$ we get

$$\frac{\partial r}{\partial z} = \frac{z}{r} \tag{6.5}$$

From the relation $\cos\theta = z/r$ one obtains

$$\frac{\partial \theta}{\partial z} = -\frac{\sin\theta}{r} \tag{6.6}$$

Lastly we have

$$\frac{\partial \phi}{\partial z}=0$$

Combining (6.4), (6.5), and (6.6) results in

$$\frac{\partial}{\partial z}=\cos\theta\frac{\partial}{\partial r}-\frac{\sin\theta}{r}\frac{\partial}{\partial\theta} \tag{6.7}$$

In a similar fashion one calculates

$$\begin{aligned}\frac{\partial}{\partial y}&=\frac{\partial}{\partial r}\frac{\partial r}{\partial y}+\frac{\partial}{\partial\theta}\frac{\partial\theta}{\partial y}+\frac{\partial}{\partial\phi}\frac{\partial\phi}{\partial y}\\&=\frac{y}{r}\frac{\partial}{\partial r}+\frac{\sin\phi\cos\theta}{r}\frac{\partial}{\partial\theta}+\frac{\cos\phi}{r\sin\theta}\frac{\partial}{\partial\phi}\end{aligned} \tag{6.8}$$

Substituting (6.7) and (6.8) in the first of (6.3) and using (6.1) gives after some algebra

$$\hat{l}_x=i\hbar\left(\sin\phi\frac{\partial}{\partial\theta}+\cot\theta\cos\phi\frac{\partial}{\partial\phi}\right) \tag{6.9}$$

By similar procedures we obtain

$$\hat{l}_y=i\hbar\left(-\cos\phi\frac{\partial}{\partial\theta}+\cot\theta\sin\phi\frac{\partial}{\partial\phi}\right) \tag{6.10}$$

$$\hat{l}_z=-i\hbar\frac{\partial}{\partial\phi} \tag{6.11}$$

and after some additional drudgery

$$\begin{aligned}\hat{\mathbf{l}}^2&\equiv\hat{\mathbf{l}}\cdot\hat{\mathbf{l}}=\hat{l}_x^2+\hat{l}_y^2+\hat{l}_z^2\\&=-\hbar^2\left[\frac{1}{\sin\theta}\frac{\partial}{\partial\theta}\left(\sin\theta\frac{\partial}{\partial\theta}\right)+\frac{1}{\sin^2\theta}\frac{\partial^2}{\partial\phi^2}\right]\end{aligned} \tag{6.12}$$

This completes our task.

6.2 THE EIGENFUNCTIONS AND EIGENVALUES OF $\hat{l}_z$

Since, according to (6.11),

$$\hat{l}_z=-i\hbar\frac{\partial}{\partial\phi} \tag{6.13}$$

we need to solve for $u(\phi)$ and l_z in the eigenvalue equation

$$-i\hbar\frac{\partial}{\partial\phi}u(\phi)=l_zu(\phi) \tag{6.14}$$

where l_z is the eigenvalue of $\hat{l}_z$. The solution of (6.14) is

$$u(\phi)=e^{i(l_z/\hbar)\phi} \tag{6.15}$$

Since $u(\phi)$ has to be a single-valued function of ϕ, we require that

$$u(\phi+2\pi)=u(\phi) \tag{6.16}$$

which restricts l_z to a *discrete* set

$$l_z = m\hbar \quad (m=0,\pm 1,\pm 2,\ldots) \tag{6.17}$$

Using the value of l_z given by (6.17), the eigenfunction (6.15) becomes

$$u_m(\phi)=(2\pi)^{-1/2}e^{im\phi} \tag{6.18}$$

We have already encountered the quantization condition (6.17) in the form of Bohr's postulate (1.26). It now appears as a simple and direct consequence of a formal theory rather than as a specialized ad hoc assumption.

6.3 THE EIGENFUNCTIONS AND EIGENVALUES OF THE SQUARED MAGNITUDE OF THE ANGULAR MOMENTUM

The eigenvalue equation for $\hat{\mathbf{l}}^2$ is given according to (6.12) by

$$-\hbar^2\left[\frac{1}{\sin\theta}\frac{\partial}{\partial\theta}\left(\sin\theta\frac{\partial}{\partial\theta}\right)+\frac{1}{\sin^2\theta}\frac{\partial^2}{\partial\phi^2}\right]Y_\beta(\theta,\phi)=\beta\hbar^2 Y_\beta(\theta,\phi) \tag{6.19}$$

where $\beta\hbar^2$ are the eigenvalues that need to be determined along with Y_β. The operators $\hat{\mathbf{l}}^2$ and $\hat{l}_z$ commute, $[\hat{\mathbf{l}}, l_z]=0$, since $l_z=-i\hbar(\partial/\partial\phi)$ commutes with $\partial^2/\partial\phi^2$. It follows that $Y(\theta,\phi)$ may be taken to be simultaneously an eigenfunction of both $\hat{\mathbf{l}}^2$ and $\hat{l}_z$. This can be satisfied by choosing[1]

$$Y_{\beta m}(\theta,\phi)=(2\pi)^{-1/2}P_{\beta m}(\theta)e^{im\phi} \tag{6.20}$$

where the subscript β denotes the dependence of $P_{\beta m}(\theta)$ on the eigenvalue of $\hat{\mathbf{l}}^2$. Using (6.11) we find

$$l_z Y_{\beta m}(\theta,\phi)=m\hbar Y_{\beta m}(\theta,\phi) \quad (m=0,\pm 1,\pm 2,\ldots) \tag{6.21}$$

We can eliminate the ϕ dependence from the eigenvalue equation (6.19) by substituting (6.20). The result is

$$\left[\frac{1}{\sin\theta}\frac{\partial}{\partial\theta}\left(\sin\theta\frac{\partial}{\partial\theta}\right)-\frac{m^2}{\sin^2\theta}\right]P_{\beta m}(\theta)=-\beta P_{\beta m}(\theta) \tag{6.22}$$

It is convenient to change variables to $w=\cos\theta$, which transforms (6.22) to

$$\frac{d}{dw}(1-w^2)\frac{dP}{dw}+\left(\beta-\frac{m^2}{1-w^2}\right)P=0 \tag{6.23}$$

---

[1]We can, alternatively, derive the form (6.20) by applying the method of the separation of variables to (6.19).

Equation (6.23) can also be written as

$$\frac{d^2P}{dw^2}-\frac{2w}{1-w^2}\frac{dP}{dw}+\left(\frac{\beta}{1-w^2}-\frac{m^2}{(1-w^2)^2}\right)P=0 \tag{6.24}$$

The function P has singularities at $w=\pm 1$. We start by exploring the behavior near $w=+1$. In this limit (6.24) becomes

$$\frac{d^2P}{dw^2}-\frac{1}{1-w}\frac{dP}{dw}-\frac{m^2}{4(1-w)^2}P=0 \tag{6.25}$$

In the spirit of (5.12) we assume a solution in the form

$$P(w\to 1)=(1-w)^{\alpha}\left[a_0+a_1(1-w)+a_2(1-w)^2+\cdots\right] \tag{6.26}$$

Taking, without loss of generality, $a_0\neq 0$, we substitute (6.26) into (6.25) and equate the coefficients of like powers of $(1-w)$. From the coefficient of $(1-w)^{\alpha-2}$ we obtain

$$a_0\left(\alpha(\alpha-1)+\alpha-\frac{m^2}{4}\right)=0$$

so that

$$\alpha=\pm m/2$$

There are thus two independent solutions of P near $w=1$; one resulting from $\alpha=-m/2$, tending to infinity, the other to zero as $w\to 1$. We denote the first as $P_{\infty}^{(1)}$ and the second as $P_0^{(1)}$. These solutions have the form

$$\begin{aligned}P_0^{(1)}&=(1-w)^{|m|/2}\left[a_0+a_1(1-w)+\cdots\right]\\ P_{\infty}^{(1)}&=(1-w)^{-|m|/2}\left[a_0'+a_1'(1-w)+\cdots\right]\end{aligned} \tag{6.27}$$

We repeat the procedure near $w=-1$. We assume a solution in the form

$$P(w\to -1)=(1+w)^{\alpha}\left[b_0+b_1(1+w)+\cdots\right] \tag{6.28}$$

obtaining again $\alpha=\pm m/2$ and, correspondingly, two independent solutions:

$$\begin{aligned}P_0^{(-1)}&=(1+w)^{|m|/2}\left[b_0+b_1(1+w)+\cdots\right]\\ P_{\infty}^{(-1)}&=(1+w)^{-|m|/2}\left[b_0'+b_1'(1+w)+\cdots\right]\end{aligned} \tag{6.29}$$

Since $w=1,-1$ correspond respectively to $\theta=0,\pi$, the solutions $P_{\infty}^{(1)}$ and $P_{\infty}^{(-1)}$, which diverge at $w=1,-1$, respectively, are not physically acceptable. We thus seek a solution of the full differential equation (6.24) that near $w=1$ approaches the form of $P_0^{(1)}(w)$ and near $w=-1$ that of $P_0^{(-1)}(w)$. This can be done by assuming the form

$$\begin{aligned}P_{\beta m}(w)&=(1-w)^{|m|/2}(1+w)^{|m|/2}Z_{\beta m}(w)\\ &=(1-w^2)^{|m|/2}Z_{\beta m}(w)\end{aligned} \tag{6.30}$$

The differential equation for $Z_{\beta m}(w)$ is obtained from (6.24):

$$(1-w^2)\frac{d^2Z}{dw^2}-2(|m|+1)w\frac{dZ}{dw}+[\beta-|m|(|m|+1)]Z$$
$$=0 \tag{6.31}$$

An assumed power series expansion

$$Z(w)=\sum_0^\infty a_k w^k \tag{6.32}$$

is substituted in (6.31). Equating the coefficients of like powers of w to zero results in

$$(k+2)(k+1)a_{k+2}=[(k+|m|)(k+|m|+1)-\beta]a_k \tag{6.33}$$

For $Z(w)$ to remain finite over the domain $1\leq w\leq -1$ (which corresponds to $0\leq\theta\leq\pi$) the series expansion (6.32) must terminate at some value of k. An examination of (6.33) shows that if the, so far unspecified, constant β is chosen so that

$$\beta=l(l+1) \tag{6.34a}$$

where

$$l=k+|m| \tag{6.34b}$$

then the series (6.32) will terminate with the term $a_k w^k$, since (6.33) guarantees that a_{k+2} is zero, provided the series contains either even or odd powers of w^k, but not both. We thus generate the even solutions by taking $a_0\neq 0$ and $a_1=0$, and the odd solutions with $a_0=0$, $a_1\neq 0$. In either case the highest power appearing in the expansion of $Z(w)$ is $k=l-|m|$. A solution of the differential equation (6.31) subject to the physical restriction of finiteness is thus characterized by an integer m, which can take on the set of values $m=0,\pm 1,\pm 2,\ldots$ and a positive integer l. It follows from (6.34b) that since k is a positive integer (or zero), then

$$|m|\leq l \tag{6.35}$$

The solution of (6.24) is denoted as $P_l^m(w)$ and called the associated Legendre function. For each l there correspond $2l+1$ solutions with $m=-l,-1+1,\ldots,0,\ldots,l$.

The complete solution of (6.19) is assembled according to (6.20) as

$$Y_l^m(\theta,\phi)=\frac{N_{lm}}{\sqrt{2\pi}}P_l^m(\theta)e^{im\phi} \tag{6.36}$$

where N_{lm} is a normalization constant to be determined and $P_l^m(w=\cos\theta)$ was re-expressed as $P_l^m(\theta)$.

The eigenvalue of the square of the angular momentum operator $\hat{\mathbf{l}}^2$ is equal to $\beta\hbar^2$, which, using (6.34), gives

$$\hat{\mathbf{l}}^2 Y_l^m(\theta,\phi)=l(l+1)\hbar^2 Y_l^m(\theta,\phi)$$

and (6.37)

$$\hat{l}_z Y_l^m(\theta,\phi) = m\hbar Y_l^m(\theta,\phi)$$

The absolute value of the angular momentum of a particle is thus restricted to a discrete set of numbers $\sqrt{l(l+1)}\,\hbar$. A particle in one of these states may, however, be found with the z component (l_z) of its angular momentum having a value of $m\hbar$, where m is one of the numbers $-l,\ldots,0,\ldots,+l$. Each value of m may be thought of classically as corresponding to a different direction of the axis of the angular momentum vector of rotation.

Using (6.9), (6.10), and (6.36) we find that

$$\langle \hat{l}_x \rangle = i\hbar \langle l, m | \sin\phi \frac{\partial}{\partial\theta} + \cot\theta \cos\phi \frac{\partial}{\partial\phi} | l, m \rangle$$

$$= 0$$

and, similarly,

$$\langle \hat{l}_y \rangle = 0$$

so that the eigenstate Y_l^m corresponding to definite values for $\langle \hat{l}_z \rangle$ and $\langle \hat{\mathbf{l}}^2 \rangle$ has zero values for the remaining two components of $\langle \hat{\mathbf{l}} \rangle$. This is just a manifestation of the noncommutation of the components of the angular momentum operator. The act of measuring $\langle \hat{l}_z \rangle$ destroys all information regarding $\langle \hat{l}_x \rangle$ and $\langle \hat{l}_y \rangle$. A graphical representation of ($\mathbf{l}$) is that of a vector of length $\sqrt{l(l+1)}\,\hbar$ rotating along the surface of a cone so that its projection along z is always $m\hbar$. This picture, shown in Fig. 6.2, is only meant to help us

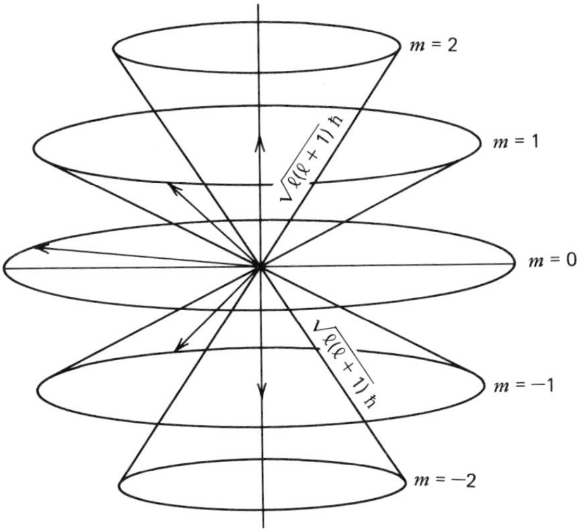

Figure 6.2 Geometrical representation of the angular momentum states for $l=2$.

remember the various relationships between the components of $\langle \mathbf{l} \rangle$ and is not to be taken literally.

The Normalization of $Y_l^m(\theta,\phi)$

The normalization condition for $Y_l^m(\theta,\phi)$ is

$$\int |Y_l^m(\theta,\phi)|^2\, d\Omega = 1 \tag{6.38}$$

where the integration extends over the unit sphere and $d\Omega = \sin\theta\, d\theta\, d\phi$ is the differential solid angle. Condition (6.38) follows from the fact that $|Y_l^m(\theta,\phi)|^2\, d\Omega$ is the probability of finding the particle inside the solid angle $d\Omega$ centered about the direction θ,ϕ. Using (6.36) the normalization is expressed as

$$\int_0^{\pi}\int_0^{2\pi} \frac{N_{lm}^2}{2\pi}\left[P_l^m(\theta)\right]^2 \sin\theta\, d\theta\, d\phi = 1 \tag{6.39}$$

or using $w = \cos\theta$,

$$|N_{lm}|^2 \int_{-1}^{1} \left[P_l^m(w)\right]^2 dw = 1 \tag{6.40}$$

Consultation of any standard mathematics textbook gives

$$\int_{-1}^{1} P_l^m(w) P_{l'}^m(w)\, dw = \begin{cases} \left(\dfrac{2}{2l+1}\right)\dfrac{(l+|m|)!}{(l-|m|)!} & (l=l') \\ 0 & (l \neq l') \end{cases} \tag{6.41}$$

which, when applied to (6.40), yields

$$N_{lm} = \left[\left(\frac{2l+1}{2}\right)\frac{(l-|m|)!}{(l+|m|)!}\right]^{1/2}$$

$$Y_l^m(\theta,\phi) = (-1)^m \sqrt{\frac{(2l+1)}{4\pi}\frac{(l-|m|)!}{(l+|m|)!}}\, P_l^m(\theta) e^{im\phi} \tag{6.42}$$

with

$$Y_l^{-m} = (-1)^m (Y_l^m)^*$$

The Parity of $Y_l^m(\theta,\phi)$

To determine the parity of $Y_l^m(\theta,\phi)$ we need to determine the effect of the parity operator on $Y_l^m(\theta,\phi)$. The parity (inversion) operator results in $\theta \rightarrow \pi - \theta$ and $\phi \rightarrow \phi + \pi$:

$$\hat{P} Y_l^m(\theta,\phi) = Y_l^m(\pi - \theta, \phi + \pi) \tag{6.43}$$

We recall that

$$Y_l^m(\theta,\phi) \propto P_l^m(w) e^{im\phi}$$

so that

$$Y_l^m(\pi-\theta,\phi+\pi)\propto P_l^m(-w)e^{im\phi}e^{im\pi} \tag{6.44}$$

Since $P_l^m(w)$ is equal to $(1-w^2)^{|m|/2}$ times a polynomial (consisting of even or odd powers) of order $l-|m|$, it follows that

$$P_l^m(-w)=P_l^m(w)(-1)^{l-|m|}$$

which, when applied to (6.44), gives

$$Y_l^m(\pi-\theta,\phi+\pi)=(-1)^l Y_l^m(\theta,\phi) \tag{6.45}$$

so that $Y_l^m(\theta,\phi)$ has even (odd) parity when l is even (odd).

It should also be stressed that the choice of the z axis was *completely arbitrary*, since there is nothing in the way the problem was stated that can distinguish any one direction from another. We thus need to interpret (6.17) as stating that the measurement of the component of the angular momentum along *any one direction* must lead to a value $m\hbar$. One cannot, however, measure the components of **l** along two orthogonal directions simultaneously, since the corresponding operators do not commute.

As an example, consider the commutator of $\hat{l}_x$ and $\hat{l}_y$. Using (6.3) and (3.22) we have

$$\begin{aligned}
[\hat{l}_x,\hat{l}_y]&=[(y\hat{p}_z-z\hat{p}_y),(z\hat{p}_x-x\hat{p}_z)]\\
&=[y\hat{p}_z,z\hat{p}_x]+[z\hat{p}_y,x\hat{p}_z]\\
&=i\hbar[\hat{p}_y x-\hat{p}_x y]\\
&=i\hbar\hat{l}_z
\end{aligned} \tag{6.46}$$

In a similar fashion we can also show that

$$[l_y,l_z]=i\hbar l_x \tag{6.47}$$

$$[l_z,l_x]=i\hbar l_y \tag{6.48}$$

A convenient mnemonic for remembering (6.46), (6.47), and (6.48) is the relation

$$\hat{\mathbf{l}}\times\hat{\mathbf{l}}=i\hbar\hat{\mathbf{l}} \tag{6.49}$$

Some low angular momentum eigenfunctions are

$$Y_0^0=\left(\frac{1}{4\pi}\right)^{1/2} \quad (s \text{ state})$$

$$\left.\begin{aligned}
Y_1^1&=-\left(\frac{3}{4\pi}\right)^{1/2}\sin\theta\frac{e^{i\phi}}{\sqrt{2}}\\
Y_1^0&=\left(\frac{3}{4\pi}\right)^{1/2}\cos\theta\\
Y_1^{-1}&=\left(\frac{3}{4\pi}\right)^{1/2}\sin\theta\frac{e^{-i\phi}}{\sqrt{2}}
\end{aligned}\right\}\quad (p \text{ states}) \tag{6.50}$$

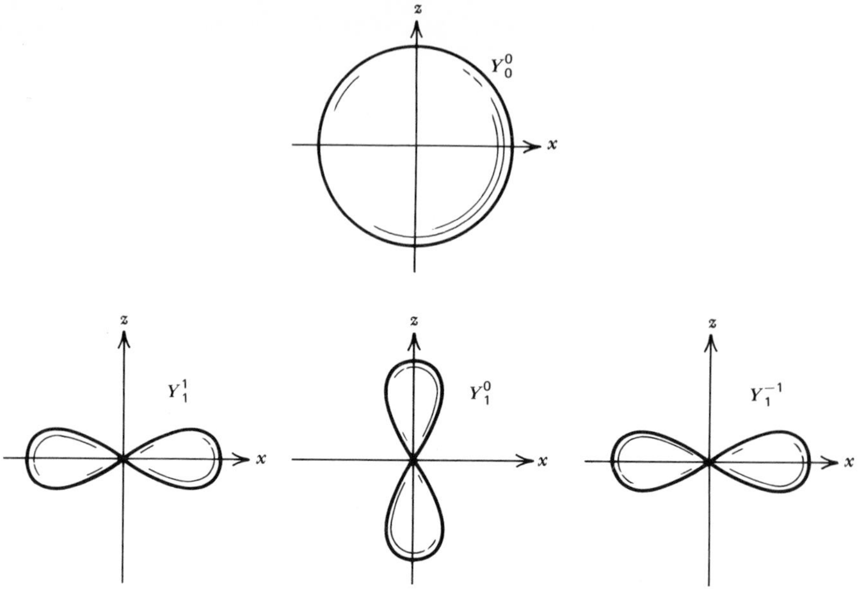

Figure 6.3 Polar plot representation of the angular dependence of the wavefunction "clouds" $|Y_l^m|^2$ for Y_0^0, Y_1^1, Y_1^0, and Y_1^{-1}. The distributions possess circular symmetry about the z axis.

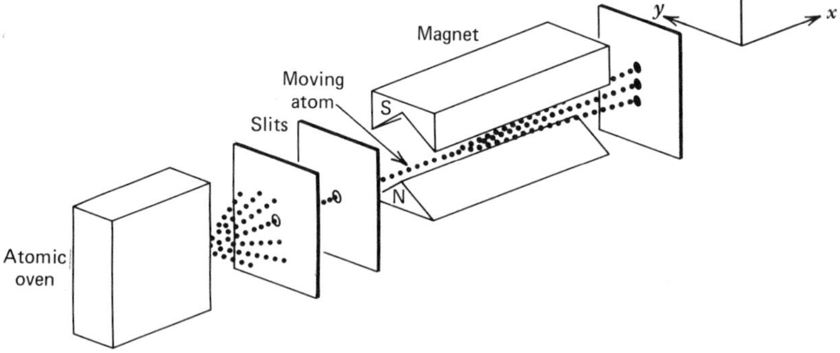

Figure 6.4 The Stern–Gerlach experiment.

Functions with l values of $0, 1, 2, 3$ are referred to by the letters s, p, d, f, respectively. The three states with $l=1$ are thus p states. The angular dependence of the four eigenfunctions $Y_0^0, Y_1^1, Y_1^0, Y_1^{-1}$ is shown schematically in the form of polar graphs in Fig. 6.3.

PROBLEMS

1. Complete the derivation of (6.9), (6.10), and (6.11).
2. Show that $[\hat{\mathbf{l}}^2, \hat{l}_x]$ and $[\hat{\mathbf{l}}^2, \hat{l}_y]=0$.
3. Assume that an atom with an orbital angular momentum $\mathbf{L}$ (we use $\mathbf{L}$ rather than $\mathbf{l}$ to denote the fact that the angular momentum may be due to the addition of momenta of more than one electron). Let there be a magnetic moment $\boldsymbol{\mu} = g\beta\mathbf{L}$ associated with $\mathbf{L}$ ($g\beta$ is a constant that is discussed in Chapter 18). A beam of atoms all with the same value of $\langle\hat{\mathbf{L}}^2\rangle = L(L+1)\hbar^2$ is moving in the x direction with a velocity v_0 through an inhomogeneous magnetic field $\mathbf{B}=\mathbf{a}_z(B_0+kz)$ (see Fig. 6.4). Taking the interaction energy of atoms with the field $\mathbf{B}$ ($\mathbf{B}=\mu_0\mathbf{H}$) as $w=-\boldsymbol{\mu}\cdot\mathbf{B}$ and the force as $\mathbf{F}=-\nabla w$, calculate the quantum mechanically allowed trajectories of the atoms in the magnetic field region. A similar experiment was performed by Stern and Gerlach[2,3] in 1921.

---

[2]O. Stern, A method of experimentally testing direction quantization in a magnetic field, *Z. Physik* **7**, 249 (1921); W. Gerlach and O. Stern, Experimental proof of direction quantization in a magnetic field, *Z. Physik* **9**, 349 (1922); On direction quantization in magnetic field, *Ann. Phys.* **74**, 673 (1924).

[3]For a lucid description of the experiment see F. K. Richtmeyer and E. H. Kennard, *Introduction to Modern Physics* (McGraw-Hill, New York, 1933), p. 40.

CHAPTER SEVEN

Particles in Spherical Symmetric Potential Fields and the Hydrogen Atom

In the first part of this chapter we explore some of the common features of the quantum mechanical analysis of particles moving under the influence of a spherical potential field—that is, a potential $V(r)$ whose strength at any point depends only on the distance r of the point from the origin. As a special case, we consider in the second part of the chapter the hydrogen atom, where $V(r)$ is due to the Coulomb attraction between the nucleus and the electron.

7.1 A PARTICLE IN A SPHERICALLY SYMMETRIC POTENTIAL FIELD

Consider a particle of mass m_e moving under the influence of a spherically symmetric potential field $V(r)$. The Schrödinger equation (3.25) for this particle is

$$\left(-\frac{\hbar^2}{2m_e}\nabla^2+V(r)\right)u_E(r,\theta,\phi)=Eu_E(r,\theta,\phi) \tag{7.1}$$

The Laplacian ∇^2 in spherical coordinates is

$$\nabla^2=\left[\frac{1}{r^2}\frac{\partial}{\partial r}\left(r^2\frac{\partial}{\partial r}\right)+\frac{1}{r^2\sin\theta}\frac{\partial}{\partial\theta}\left(\sin\theta\frac{\partial}{\partial\theta}\right)+\frac{1}{r^2\sin^2\theta}\frac{\partial^2}{\partial\phi^2}\right] \tag{7.2}$$

which, when incorporated into (7.1), results in

$$\left\{-\frac{\hbar^2}{2m_e}\left[\frac{1}{r^2}\frac{\partial}{\partial r}\left(r^2\frac{\partial}{\partial r}\right)+\frac{1}{r^2\sin\theta}\frac{\partial}{\partial\theta}\left(\sin\theta\frac{\partial}{\partial\theta}\right)+\frac{1}{r^2\sin^2\theta}\frac{\partial^2}{\partial\phi^2}\right]\right.$$
$$\left.+V(r)\right\}u_E(r,\theta,\phi)=Eu_E(r,\theta,\phi) \tag{7.3}$$

According to (6.11),

$$\mathbf{l}^2=-\hbar^2\left[\frac{1}{\sin\theta}\frac{\partial}{\partial\theta}\left(\sin\theta\frac{\partial}{\partial\theta}\right)+\frac{1}{\sin^2\theta}\frac{\partial^2}{\partial\phi^2}\right] \tag{7.4}$$

we can thus rewrite (7.3) as

$$\left[-\frac{\hbar^2}{2m_e}\left(\frac{1}{r^2}\right)\frac{\partial}{\partial r}\left(r^2\frac{\partial}{\partial r}\right)+\frac{\hat{l}^2(\theta,\phi)}{2m_er^2}+V(r)\right]u_E(r,\theta,\phi)=Eu_E(r,\theta,\phi) \tag{7.5}$$

It is evident from (7.5) that the Hamiltonian of a particle in a spherically symmetric field commutes with $\hat{l}^2$:

$$[\hat{\mathcal{H}},\hat{l}^2]=0 \tag{7.6}$$

We have shown in Section 6.3 that $\hat{l}_z$ commutes with $\hat{l}^2$:

$$[\hat{l}^2,l_z]=0 \tag{7.7}$$

It follows that the operators $\hat{l}^2$, $\hat{l}_z$, and $\hat{\mathcal{H}}$ have common eigenfunctions. The solution of the Schrödinger equation (7.5) can be written as

$$u_E(r,\theta,\phi)=R_{nl}(r)Y_l^m(\theta,\phi) \tag{7.8}$$

where, according to (6.37),

$$\hat{\mathbf{l}}^2Y_l^m(\theta,\phi)=l(l+1)\hbar^2Y_l^m(\theta,\phi) \tag{7.9}$$

$$\hat{l}_zY_l^m(\theta,\phi)=m\hbar Y_l^m(\theta,\phi) \tag{7.10}$$

$R_{nl}(r)$ is a function depending only on r. The constant n is the quantum number needed to specify the r dependence. If we now use (7.8) and (7.9) to replace the term $\hat{l}^2(\theta,\phi)u_E(r,\theta,\phi)$ in (7.5) with $l(l+1)\hbar^2Y_l^m(\theta,\phi)R_{nl}(r)$, and divide both sides of the resulting equation by $Y_l^m(\theta,\phi)$, we obtain

$$-\frac{\hbar^2}{2m_e}\frac{1}{r^2}\frac{\partial}{\partial r}\left(r^2\frac{\partial}{\partial r}\right)R_{nl}(r)+\left(\frac{l(l+1)\hbar^2}{2m_er^2}+V(r)\right)R_{nl}(r)=E_{nl}R_{nl}(r) \tag{7.11}$$

The constant n represents the new quantum number that results from the solution of (7.11).

To proceed further we need to specify the form of $V(r)$. We have established that the (θ,ϕ) dependence of the energy eigenfunction is described by the angular momentum eigenfunction $Y_l^m(\theta, \phi)$. It follows that the squared magnitude of the angular momentum as well as its projection along the z axis are constants of the motion of particles moving in a spherically symmetric potential field. Since in a spherically symmetric potential $V(r)$ the z axis has no special significance, it follows that the projection of $\mathbf{l}$ along any one direction, as well as l^2, may be determined with arbitrary accuracy.

7.2 THE HYDROGENIC ATOM

The hydrogen atom may be viewed as a special case of the spherically symmetric potential field. In this case $V(r)$ is due to the Coulomb attraction between the nucleus of charge Ze and an electron of charge $-e$. The potential energy is given by

$$V(r) = -\frac{Ze^2}{4\pi\varepsilon_0 r} = -\frac{Ze_M^2}{r}$$

$$e_M^2 = \frac{e^2}{4\pi\varepsilon_0} \tag{7.12}$$

and r is the distance from the electron to the nucleus. The expression (7.12) for $V(r)$ conforms to a (arbitrary) choice of the energy reference such that when the electron is infinitely far from the nucleus, the total potential energy of the electron-nucleus system is zero.

The Schrödinger equation (7.11) becomes

$$\left[-\frac{\hbar^2}{2m_e}\frac{1}{r^2}\frac{\partial}{\partial r}\left(r^2\frac{\partial}{\partial r}\right) - \frac{Ze_M^2}{r} + \frac{l(l+1)\hbar^2}{2m_e r^2}\right]R_{nl}(r) = E_{nl}R_{nl}(r) \tag{7.13}$$

The nuclear mass was assumed to be infinite so that the kinetic energy of the nucleus is zero. (The finite mass nucleus will be considered in Section 7.3.) In conformity with the above choice of the zero energy reference, bound states of the system possess negative energy, accordingly we put

$$E_{nl} = -|E_{nl}|$$

and define

$$\alpha_n^2 \equiv \frac{8m_e|E_{nl}|}{\hbar^2}$$

$$\rho \equiv \alpha_n r$$

$$\lambda \equiv \frac{Ze_M^2}{\hbar}\left(\frac{m_e}{2|E_{nl}|}\right)^{1/2} = \frac{2m_e Ze^2}{\alpha\hbar^2} \tag{7.14}$$

With these definitions (7.13) becomes

$$\left[\frac{1}{\rho^2}\frac{d}{d\rho}\left(\rho^2\frac{d}{d\rho}\right) + \frac{\lambda}{\rho} - \frac{1}{4} - \frac{l(l+1)}{\rho^2}\right]R_{\lambda l}(\rho) = 0 \tag{7.15}$$

We need to examine the form of the solution of (7.15) near the singularities at $\rho = 0$ and $\rho = \infty$. Near $\rho = \infty$ we may approximate (7.15) by

$$\frac{d^2R}{d\rho^2} - \tfrac{1}{4}R = 0$$

which has solutions

$$R = e^{\pm \rho/2}$$

The $e^{\rho/2}$ solution is not acceptable for an eigenfunction, since it increases without bound as $\rho \to \infty$.

Using arguments identical to those employed in connection with (5.12), we assume

$$R_{\lambda l}(\rho) = \rho^s L(\rho) e^{-\rho/2} \tag{7.16}$$

where $L(\rho)$ is a polynomial:

$$L(\rho) = a_0 + a_1\rho + a_2\rho^2 + \cdots + a_\nu \rho^\nu \tag{7.17}$$

with $a_0 \neq 0$ and s some positive number (if $s < 0$, then $R \to \infty$ as $\rho \to 0$).

Substituting (7.16) into (7.15) gives

$$\rho^2 \frac{d^2L}{d\rho^2} + \rho[2(s+1) - \rho]\frac{dL}{d\rho} + [\rho(\lambda - s - 1) + s(s+1) - l(l+1)]L = 0 \tag{7.18}$$

For (7.18) to remain valid at $\rho = 0$ we must have

$$s(s+1) - l(l+1) = 0$$

so that

$$s = l \tag{7.19}$$

or

$$s = -(l+1)$$

because $l \geqslant 0$, the root $s = -(l+1)$ is not acceptable, since its employment in (7.16) will lead to a solution $R(\rho)$, which diverges at $\rho = 0$. Using the remaining root $s = l$, we rewrite (7.18) as

$$\rho \frac{d^2L}{d\rho^2} + [2(l+1) - \rho]\frac{dL}{d\rho} + (\lambda - l - 1)L = 0 \tag{7.20}$$

Next we substitute (7.17) in (7.20) and equate the coefficients of the different powers of ρ to zero. The result is

$$a_1 = \frac{l+1-\lambda}{2l+2} a_0$$

$$a_2 = \frac{l+2-\lambda}{4l+6} a_1$$

and in general

$$a_{\nu+1}=\frac{\nu+l+1-\lambda}{(\nu+1)(\nu+2l+2)}a_\nu \tag{7.21}$$

The series (7.17) must terminate at some finite value of ν. Otherwise it follows from (7.21) that as $\nu\to\infty$, $a_{\nu+1}\to a_\nu/\nu$ so that as $\rho\to\infty$, $L(\rho)\to e^\rho$ will diverge. To ensure the termination of the series expansion (7.17) after, say, $\nu+1$ terms, we need, according to (7.21), to satisfy

$$\lambda=\nu+l+1\equiv n\ (=\text{positive integer}) \tag{7.22}$$

Since the lowest value that ν can assume is zero, it follows that

$$n\geq l+1 \tag{7.23}$$

The polynomial $L(\rho)$ in (7.17) generated by the above procedure will be denoted as $L_{nl}(\rho)$. The integer n is referred to as the *principal quantum number.* It follows from (7.16) and (7.19), and from the fact that $L(\rho)$ is a polynomial of degree $\nu=n-l-1$, that $R_{nl}(\rho)$ is equal to $\exp(-\rho/2)$ times a polynomial (in ρ) of order $(n-1)$.

Putting $\lambda=n$ in (7.20) leads to the differential equation,

$$\rho\frac{d^2L_{nl}}{d\rho^2}+[2(l+1)-\rho]\frac{dL_{nl}}{d\rho}+(n-l-1)L_{nl}=0 \tag{7.24}$$

Mathematicians, for reasons all their own, have chosen to deal with a function $L_q^p(\rho)$—the associated Laguerre polynomial—which satisfies

$$\rho\frac{d^2L_q^p}{d\rho^2}+(p+1-\rho)\frac{dL_q^p}{d\rho}+(q-p)L_q^p=0 \tag{7.25}$$

A comparison of (7.24) and (7.25) reveals that

$$L_{nl}(\rho)=L_{n+l}^{2l+1}(\rho) \tag{7.26}$$

We are now ready to assemble the solution for $R_{nl}(\rho)$. Using (7.16) and (7.26), and recalling that $s=l$, we obtain

$$\begin{aligned}R_{nl}(\rho)&=N_{nl}e^{-\rho/2}\rho^lL_{nl}(\rho)\\&=N_{nl}e^{-\rho/2}\rho^lL_{n+l}^{2l+1}(\rho)\end{aligned} \tag{7.27}$$

where N_{nl} is the normalization constant that will be determined later.

The Eigenvalues

To determine the eigenvalue—that is, the energy—of the state (n, l, m), we return to the last of (7.14), and putting $\lambda=n$ obtain

$$E_n=-|E_n|=-\frac{m_eZ^2e_M^4}{2\hbar^2n^2}=-\frac{Z^2e_M^2}{2a_0n^2} \tag{7.28}$$

where

$$a_0 \equiv \frac{\hbar^2}{m_e e_M^2} = 5.2917 \times 10^{-11}\text{m} = \text{“Bohr radius”}$$

$$e_M^2 \equiv \frac{e^2}{4\pi\varepsilon_0} = 2.3098 \times 10^{-28}\ (\text{Kg}-\text{m}^3-\text{sec}^{-2})$$

This is the same result obtained by Bohr [see Eq. (1.32)] after assuming that the angular momentum is quantized in multiples of $\hbar$.

Using (7.28) in expression (7.14) for α_n we get

$$\alpha_n = \frac{2Z}{na_0}$$

so that putting $\rho = \alpha_n r$ in (7.27) gives

$$R_{nl}(r) = N_{nl} e^{(-Z/na_0)r} \left(\frac{2Z}{na_0} r \right)^l L_{n+l}^{2l+1} \left(\frac{2Z}{na_0} r \right) \tag{7.29}$$

The Normalization Constant

The eigenfunction

$$u_{nlm}(r,\theta,\phi) = N_{nl} e^{-\rho/2} \rho^l L_{n+l}^{2l+1}(\rho) Y_l^m(\theta,\phi) \tag{7.30}$$

must be normalized so that

$$\int_0^\infty \int_0^\pi \int_0^{2\pi} |u_{nlm}(r,\theta,\phi)|^2 r^2 \sin\theta \, dr \, d\theta \, d\phi = 1 \tag{7.31}$$

We have already [see (6.38)] normalized $Y_l^m(\theta,\phi)$ so that

$$\int_0^\pi \int_0^{2\pi} |Y_l^m(\theta,\phi)|^2 \sin\theta \, d\theta \, d\phi = 1 \tag{7.32}$$

All that remains in order to satisfy (7.31) is to choose N_{nl} such that the indicated integral over the r coordinate is unity. In terms of $\rho = \alpha_n r$, this requirement translates to

$$\frac{N_{nl}^2}{\alpha_n^3} \int_0^\infty e^{-\rho} \rho^{2l} \left[L_{n+l}^{2l+1}(\rho) \right]^2 \rho^2 \, d\rho = 1 \tag{7.33}$$

Using the tabulated value of the definite integral

$$\int_0^\infty e^{-\rho} \rho^{2l} \left[L_{n+l}^{2l+1}(\rho) \right]^2 \rho^2 \, d\rho = \frac{2n[(n+l)!]^3}{(n-l-1)!} \tag{7.34}$$

we obtain

$$N_{nl} = \left[\left(\frac{2Z}{na_0} \right)^3 \frac{(n-l-1)!}{2n[(n+l)!]^3} \right]^{1/2} \tag{7.35}$$

where we used $\alpha_n \equiv 2Z/na_0$.

The complete hydrogenic wavefunction can finally be written, using (7.30), as

$$u_{nlm}(r,\theta,\phi)=\left[\left(\frac{2Z}{na_0}\right)^3\frac{(n-l-1)!}{2n[(n+l)!]^3}\right]^{1/2}e^{-Zr/na_0}\left(\frac{2Z}{na_0}r\right)^l$$

$$\times L_{n+l}^{2l+1}\left(\frac{2Z}{na_0}r\right)Y_l^m(\theta,\phi) \tag{7.36}$$

A plot of some low-order radial wavefunctions $R_{nl}(r)$ is shown in Fig. 7.1.

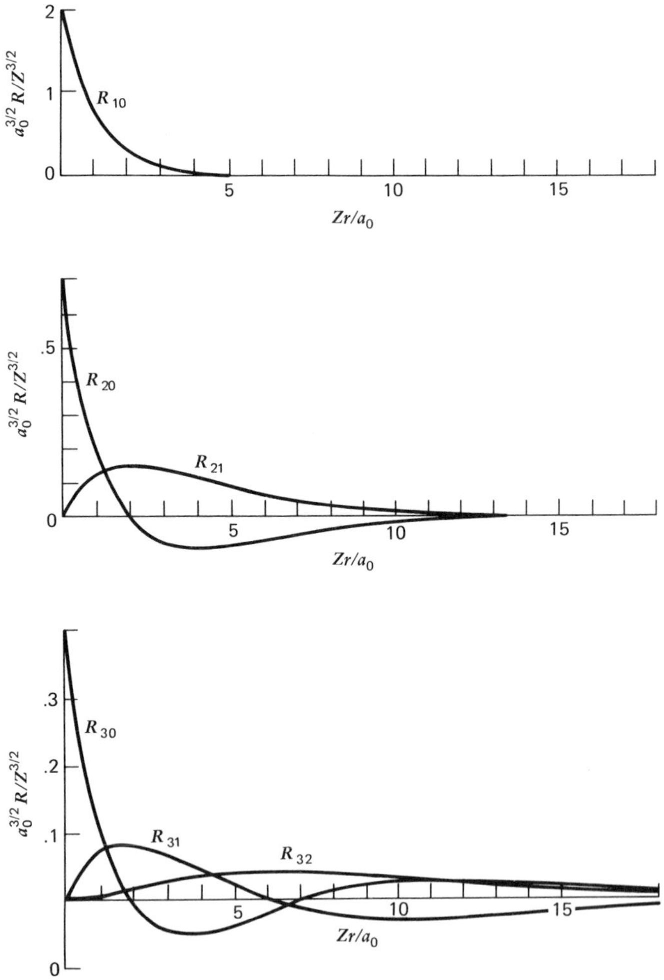

Figure 7.1 Graphs of the radial wavefunctions $R_{nl}(r)$ for $n=1$, 2, and 3, and $l=0,1,2$.

Table 7.1 A tabulation of some low-lying normalized hydrogenic eigenfunctions. The letters on the left correspond to the convention wherein s designates states with $l=0$, p $(l=1)$, d $(l=2)$, and so on.

s:	$u_{100}=\pi^{-1/2}\left(\frac{Z}{a_0'}\right)^{3/2}e^{-Zr/a_0'}$
s:	$u_{200}=(32\pi)^{-1/2}\left(\frac{Z}{a_0'}\right)^{3/2}\left(2-\frac{Zr}{a_0'}\right)e^{-Zr/2a_0'}$
p:	$u_{210}=(32\pi)^{-1/2}\left(\frac{Z}{a_0'}\right)^{3/2}\frac{Zr}{a_0'}e^{-Zr/2a_0'}\cos\theta$
p:	$u_{21\pm1}=(64\pi)^{-1/2}\left(\frac{Z}{a_0'}\right)^{3/2}\frac{Zr}{a_0'}e^{-Zr/2a_0'}\sin\theta\, e^{\pm i\varphi}$
s:	$u_{300}=\frac{(Z/a_0')^{3/2}}{81(3\pi)^{1/2}}\left(27-18\frac{Zr}{a_0'}+2\frac{Z^2r^2}{a_0'^2}\right)e^{-Zr/3a_0'}$
p:	$u_{310}=\frac{2^{1/2}(Z/a_0')^{3/2}}{81\pi^{1/2}}\left(6-\frac{Zr}{a_0'}\right)\frac{Zr}{a_0'}e^{-Zr/3a_0'}\cos\theta$
p:	$u_{31\pm1}=\frac{(Z/a_0')^{3/2}}{81\pi^{1/2}}\left(6-\frac{Zr}{a_0'}\right)\frac{Zr}{a_0'}e^{-Zr/3a_0'}\sin\theta\, e^{\pm i\varphi}$
d:	$u_{320}=\frac{(Z/a_0')^{3/2}}{81(6\pi)^{1/2}}\frac{Z^2r^2}{a_0'^2}e^{-Zr/3a_0'}(3\cos^2\theta-1)$
d:	$u_{32\pm1}=\frac{(Z/a_0')^{3/2}}{81\pi^{1/2}}\frac{Z^2r^2}{a_0'^2}e^{-Zr/3a_0'}\sin\theta\cos\theta\, e^{\pm i\varphi}$
d:	$u_{32\pm2}=\frac{(Z/a_0')^{3/2}}{162\pi^{1/2}}\frac{Z^2r^2}{a_0'^2}e^{-Zr/3a_0'}\sin^2\theta\, e^{\pm 2i\varphi}$

We conclude this section by listing in Table 7.1 some of the low-order normalized eigenfunctions of the hydrogenic atom. The parameter a_0' appearing in the table is related to the Bohr radius a_0 defined above by

$$a_0'=a_0(1+m_e/M_N)$$

where M_N is the mass of the nucleus. The correction factor $(1+m_e/M_N)$ accounts for the finite mass of the nucleus that, up to this point, was assumed to be infinite. This correction is considered in the next section.

7.3 NUCLEAR MASS CORRECTION OF THE HYDROGEN ATOM PROBLEM

In the treatment of the hydrogen atom of Section 7.2, the nuclear mass M was assumed to be infinite. This assumption was made tacitly by ignoring the kinetic energy term $\hat{\mathbf{p}}_N^2/2M$ in the Hamiltonian (7.13) ($\hat{\mathbf{p}}_N$ is the nuclear momentum operator).

In what follows we generalize the treatment of Section 7.2 to the case of a system of two charged particles with masses m_1 and m_2 separated by a distance $|\mathbf{r}_1 - \mathbf{r}_2|$. $\mathbf{r}_1$ and $\mathbf{r}_2$ are the locations of the masses m_1 and m_2, respectively, referred to some common origin. The complete Hamiltonian is

$$\hat{\mathcal{H}} = -\frac{\hbar^2}{2m_1}\nabla_1^2 - \frac{\hbar^2}{2m_2}\nabla_2^2 + V(|\mathbf{r}_1 - \mathbf{r}_2|) \tag{7.37}$$

In ∇_1^2 all the derivatives are to be taken with respect to the coordinates of particle 1. The interaction term V is assumed to depend only on the (scalar) distance $|\mathbf{r}_1 - \mathbf{r}_2|$ between the particles. We need to solve the eigenvalue problem

$$\hat{\mathcal{H}} u_E(\mathbf{r}_1, \mathbf{r}_2) = E u_E(\mathbf{r}_1, \mathbf{r}_2) \tag{7.38}$$

We transform the problem to a new set of variables $\mathbf{r}$ and $\mathbf{R}$ defined by

$$\begin{aligned} m_1 x_1 + m_2 x_2 &= MX, \text{ etc.} \\ x_1 - x_2 &= x, \text{ etc.} \\ M &= m_1 + m_2 \end{aligned} \tag{7.39}$$

so that $\mathbf{R}$ with components X, Y, Z is the address of the *center of mass*, while the vector $\mathbf{r}$ with components (x, y, z) gives the relative displacement of the two particles. It follows directly from (7.39) that

$$\begin{aligned} \nabla_1^2 &= \left(\frac{m_1}{M}\right)^2 \nabla_{\mathbf{R}}^2 + \nabla_{\mathbf{r}}^2 \\ \nabla_2^2 &= \left(\frac{m_2}{M}\right)^2 \nabla_{\mathbf{R}}^2 + \nabla_{\mathbf{r}}^2 \end{aligned} \tag{7.40}$$

Using (7.37) and (7.40) we can write the Schrödinger equation as

$$\left(-\frac{\hbar^2}{2M}\nabla_{\mathbf{R}}^2 - \frac{\hbar^2}{2\mu}\nabla_{\mathbf{r}}^2 + V(r)\right) U_E(\mathbf{r}, \mathbf{R}) = E U_E(\mathbf{r}, \mathbf{R}) \tag{7.41}$$

where μ, the *reduced mass*, is defined by

$$\frac{1}{\mu} = \frac{1}{m_1} + \frac{1}{m_2} \tag{7.42}$$

and where E is the energy eigenvalue and U_E is the corresponding eigenfunction. The eigenvalue problem (7.41), now expressed in the coordinate system $\mathbf{r}, \mathbf{R}$, can be treated by the method of separation of variables. We assume the solution $U_E(\mathbf{r}, \mathbf{R})$ to be given by a product of two functions

$$U_E(\mathbf{r}, \mathbf{R}) = u(\mathbf{r}) W(\mathbf{R}) \tag{7.43}$$

each a function of a single variable. Using (7.43) in (7.41) and dividing by uW gives

$$-\frac{\hbar^2}{2M}\frac{1}{W(\mathbf{R})}\nabla_{\mathbf{R}}^2 W(\mathbf{R}) - \frac{\hbar^2}{2\mu}\frac{1}{u(\mathbf{r})}\nabla_{\mathbf{r}}^2 u(\mathbf{r}) + V(r) = E \tag{7.44}$$

For (7.44) to hold for all $\mathbf{r}$ and $\mathbf{R}$, the $\mathbf{R}$ dependent part of the left side must be equal to some constant, say E', while the $\mathbf{r}$ dependent part is equal to E'':

$$-\frac{\hbar^2}{2M}\frac{1}{W(\mathbf{R})}\nabla_{\mathbf{R}}^2 W(R)=E' \tag{7.45}$$

$$-\frac{\hbar^2}{2\mu}\frac{1}{u(\mathbf{r})}\nabla_{\mathbf{r}}^2 u(\mathbf{r})+V(r)=E'' \tag{7.46}$$

and in order to satisfy (7.44),

$$E'+E''=E \tag{7.47}$$

The eigenvalue equation (7.41), which involves both $\mathbf{r}_1$ and $\mathbf{r}_2$, has thus been reduced to two simple equations—(7.45) and (7.46)—each involving a single variable. The solution of the first equation, (7.45), is

$$W(R)=e^{i\mathbf{K}\cdot\mathbf{R}} \tag{7.48}$$

where $\mathbf{K}$ is *any* real vector. The constant E' is

$$E'=\frac{\hbar^2K^2}{2M} \tag{7.49}$$

We identify $W(\mathbf{R})$, by analogy to (3.18), as the momentum eigenfunction corresponding to the center-of-mass momentum eigenvalue $\hbar\mathbf{K}$. The constant E' is thus the kinetic energy of the total mass M in the center-of-mass coordinate system ($\mathbf{R}$).

The remaining equation, (7.46), is identical in form to (7.1). It thus corresponds to the Schrödinger equation of a particle of mass μ moving in a potential field $V(r)$. In the special case of the hydrogenic atom with nuclear charge Ze, we take

$$V(r)=-Ze_M^2/r$$

The solution of (7.46) becomes identical to that carried out in Section 7.2, so that the eigenvalue E'' is taken directly from (7.28) after replacing m_e by μ:

$$E''=-\frac{\mu Z^2e_M^4}{2\hbar^2n^2}=-\frac{Z^2e_M^2}{2a_0'n^2} \tag{7.50}$$

$$a_0'=\frac{\hbar^2}{\mu e_M^2}=a_0\left(1+\frac{m_e}{M_N}\right) \tag{7.51}$$

We used (7.42) and associated m_1 with the electron mass m_e and m_2 with the mass of the nucleus M_N. In the case of hydrogen, the energy correction factor $(1+m_e/M_N)$ is $(1+1/1836)=1.000545$. The transition energies of the atom are thus reduced by a factor of $\sim 5\times 10^{-4}$, a change easily detected by spectroscopic techniques.

The total energy of the atom E consists of the sum (7.47) of the kinetic energy E' (in the center-of-mass system) and the internal excitation energy E'':

$$E = \frac{\hbar^2 K^2}{2M} - \frac{Z^2 e_M^2}{2a_0' n^2} \quad (n = 1, 2, \ldots) \tag{7.52}$$

Recoil Energies and Doppler Shifts

Excited hydrogen atoms (or any excited atomic system) can emit radiation in the process of undergoing a transition to a state of lower energy. It follows from (7.52) that the energy released in the process of making a transition from state n_1 to state n_2 depends on the translational energy $\hbar^2 K^2/2M$ of the center-of-mass system.

Consider an atom initially in a quantum state n_1 with a center-of-mass momentum $\hbar \mathbf{K}_1$, making a transition to the level n_2. In the process the center-of-mass momentum changes to $\hbar \mathbf{K}_2$, while a photon with energy $\hbar\omega$ and momentum $\hbar\mathbf{k}$ is emitted. The situation is depicted in Fig. 7.2. Since the total energy is conserved,

$$\frac{\hbar^2 K_1^2}{2M} + E_{n1} = \frac{\hbar^2 K_2^2}{2M} + E_{n2} + \hbar\omega \tag{7.53}$$

The conservation of momentum requires that

$$\mathbf{K}_1 = \mathbf{K}_2 + \mathbf{k} \tag{7.54}$$

so that

$$K_1^2 - K_2^2 = -k^2 + 2kK_1 \cos\alpha \tag{7.55}$$

where α is the angle between $\mathbf{k}$ and $\mathbf{K}_1$.

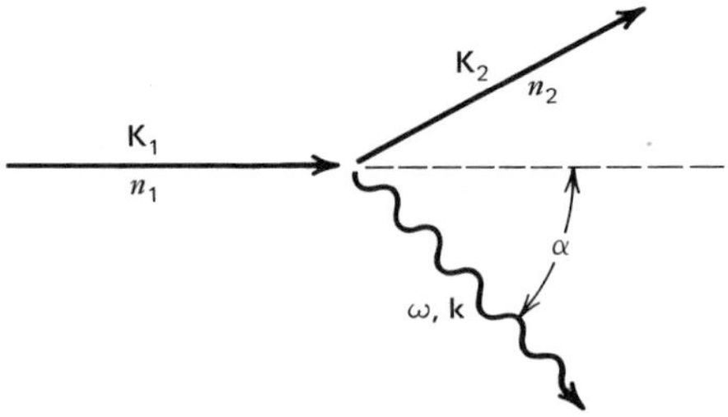

Figure 7.2 A hydrogen atom in a quantum state n_1 and with a center-of-mass momentum $\hbar\mathbf{K}_1$ emits a photon with momentum $\hbar\mathbf{k}$, thereby making a transition to the quantum state n_2. The final center-of-mass momentum of the atom is $\hbar\mathbf{K}_2$.

Solving (7.53) and (7.55) for $\hbar\omega$, the energy of the emitted photon, yields

$$\hbar\omega = \hbar(\omega_0 + \Delta\omega)$$

$$\omega_0 \equiv \frac{E_{n1} - E_{n2}}{\hbar}, \qquad \Delta\omega = \frac{\hbar}{2M}\left(2kK_1 \cos\alpha - k^2\right) \tag{7.56}$$

Using $\hbar K_1 = Mv$, $k = \omega/c$, where v is the initial velocity of the atom, gives

$$\Delta\omega = \omega\left(\frac{v}{c}\cos\alpha - \frac{\hbar\omega}{2Mc^2}\right) \tag{7.57}$$

so that for $\Delta\omega_0 \ll \omega_0$,

$$\omega \simeq \omega_0\left(1 + \frac{v}{c}\cos\alpha - \frac{\hbar\omega_0}{2Mc^2}\right) \tag{7.58}$$

The second term in parentheses is recognized as the *Doppler shift* of the emitted radiation due to the initial component $v\cos\alpha$ of the atom velocity along the direction of the emitted radiation. The last term $(-\hbar\omega_0^2/2Mc^2)$, often ignored, is the recoil correction to the emission frequency and accounts for the fact that even for $v=0$ (atom initially at rest) the finite momentum of the emitted photon causes the atom to recoil. The resulting kinetic energy must be subtracted from that of the emitted photon. In the Mösbauer effect[1] the recoil momentum due to the emission of γ rays in crystals is transmitted to the whole crystal. The recoil correction is thus $-\hbar\omega_0^2/2Mc^2$ with M the mass of the crystal and is exceedingly small.

Level Degeneracy

The energy of the hydrogen atom at rest $(K=0)$ depends on n alone[2]:

$$E_n = -\frac{Z^2 e_M^2}{2a_0' n^2}$$

Since with each n we may associate a set of quantum levels with l numbers $(l = 0, 1, 2, \ldots, n-1)$, the total number of levels all with energy E_n is

$$\sum_{l=0}^{n-1} (2l+1) = n^2 \tag{7.59}$$

Each l value possessing $2l+1(m = -l \cdots l)m$ levels. The set of levels associated with $n=2$, for example, is $(nlm) = 2s0, 2p1, 2p0, 2p-1$, where s denotes levels with $l=0$ and p with $l=1$.

---

[1]See, for example, F. C. Brown, *The Physics of Solids* (Benjamin, New York, 1967), p. 193.

[2]Later we shall see that this is not strictly true and that considerations of the intrinsic electron spin will lead to "fine structure" corrections.

Linear Combination of Eigenfunctions

Consider as an example the hydrogen atom $2p$ states $u_{21m}(r,\theta,\phi)$ given in Table 7.1 as

$$u_{211} = Cre^{-r/2a_0'}\sin\theta e^{i\phi} \tag{7.60}$$

$$u_{21-1} = Cre^{-r/2a_0'}\sin\theta e^{-i\phi} \tag{7.61}$$

$$u_{210} = \sqrt{2}\,Cre^{-r/2a_0'}\cos\theta \tag{7.62}$$

Since these three functions are degenerate—that is, they correspond to states of the same energy E—we can use them to form new orthonormal linear combinations that are also eigenfunctions of the original Hamiltonian with the same energy E.

One interesting choice leading to real eigenfunctions is

$$\begin{aligned}|2p_x\rangle &= \frac{1}{\sqrt{2}}(u_{211} + u_{21-1})\\ &= \frac{1}{\sqrt{2}}Cre^{-r/2a_0'}\sin\theta(e^{i\phi} + e^{-i\phi})\\ &= \sqrt{2}\,Ce^{-r/2a_0'}x\\ |2p_y\rangle &= \frac{1}{\sqrt{2}\,i}(u_{211} - u_{21-1})\\ &= \frac{1}{\sqrt{2}\,i}Cre^{-r/2a_0'}\sin\theta(e^{i\phi} - e^{-i\phi})\\ &= \sqrt{2}\,Ce^{-r/2a_0'}y\\ |2p_z\rangle &= u_{210} = \sqrt{2}\,Ce^{-r/2a_0'}z\end{aligned} \tag{7.63}$$

These are eigenfunctions of $\hat{\mathcal{H}}$ as well as of $\hat{\mathbf{l}}^2$ but no longer of $\hat{l}_z$. The new wavefunctions have electron probability distribution $|u(\mathbf{r})|^2$, which are very different from those of u_{211}, u_{21-1}, and u_{210}. Schematic polar plots of these functions are shown in Fig. 7.3. Note that in $|2p_x\rangle$, for example, the electron density is concentrated along the x axis.

Linear combination of degenerate eigenfunctions occurs when the atom is part of a crystal or a molecule. In such cases the wavefunctions need to reflect the local spatial symmetry and do so by "forming" appropriate linear combinations of hydrogenic wavefunctions. An example of such wavefunction hybridization in molecules will be considered next.

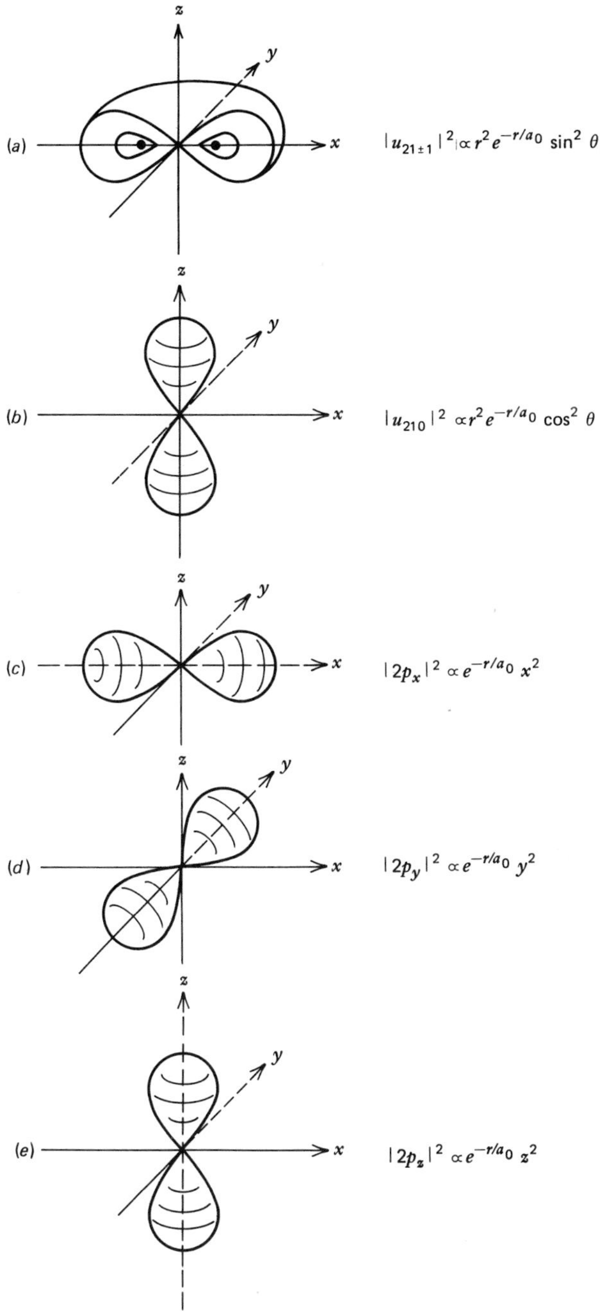

Figure 7.3 (a) and (b) are schematic plots corresponding to the probability distribution of the hydrogen wavefunctions $u_{21\pm1}$, u_{210}. (c), (d), and (e) correspond to linear combinations that result in distribution with probability concentrated along x, y, and z, respectively.

7.4 HYBRIDIZED WAVEFUNCTIONS AND MOLECULAR BONDING

Wavefunctions with various spatial properties can be synthesized by more complicated linear combinations involving both $s(l=0)$ and $p(l=1)$ wavefunctions. Take the case of methane, CH_4, as an example. A carbon atom is bonded to four hydrogen atoms by four tetrahedral bonds with a H—C—H bond angle of 109°28′, as shown in Fig. 7.4. This tetrahedral symmetry must be mirrored by the electron wavefunctions. Each of the four carbon bonds corresponds to a single electron wavefunction $\chi_1, \chi_2, \chi_3, \chi_4$, where

$$\chi_1 = \frac{1}{\sqrt{4}}(2s + 2p_x + 2p_y + 2p_z)$$

$$\chi_2 = \frac{1}{\sqrt{4}}(2s - 2p_x - 2p_y + 2p_z)$$

$$\chi_3 = \frac{1}{\sqrt{4}}(2s + 2p_x - 2p_y - 2p_z)$$

$$\chi_4 = \frac{1}{\sqrt{4}}(2s - 2p_x + 2p_y - 2p_z) \tag{7.64}$$

$2p_{x,y,z}$ represent the eigenfunctions $|2p_{x,y,z}\rangle$ as given by (7.63). The wavefunctions $\chi_1, \chi_2, \chi_3, \chi_4$ are called *hybridized orbitals*. The bonding of CH_4 is formed by "placing" each of the four outer electrons of the $2s^2 2p^2$ carbon configuration in one of the $\chi_1, \chi_2, \chi_3, \chi_4$ orbitals.

From equation (7.64) it is easy to verify that χ_1, χ_2, χ_3, and χ_4 are orthogonal to each other and have probability distributions with maxima

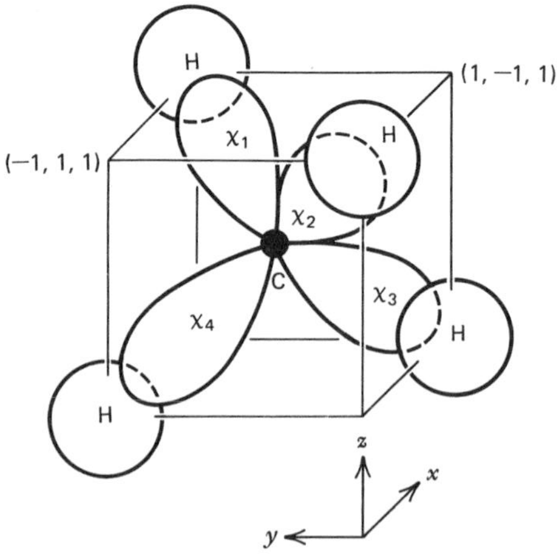

Figure 7.4 The structure and bonding of the methane (CH_4) molecule.

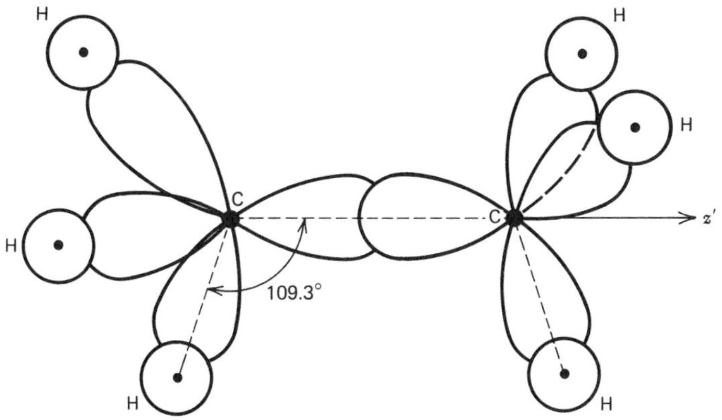

Figure 7.5 The tetrahedral hybridization of the carbon atom $2s^2 2p^2$ electrons is the basis for forming C—C and C—H bonds in ethane (C_2H_6).

along the (1,1,1), (−1,−1,+1), (1,−1,−1), and (−1,1,−1) directions, respectively. (The directions are defined with the position of the C atom in Fig. 7.4 as the origin.)

The spatial directionality of the hybridized wavefunctions (orbitals) is the basis for the understanding of molecular bonding. In ethane (C_2H_6), as an example, each of the two carbon atoms is in a configuration very similar to that of methane (Fig. 7.4) except that now two of the hybrid orbitals point along a common axis to form the C—C bond (z' axis) as shown in Fig. 7.5. The remaining three orbitals of each C atom form tetrahedral bonds with hydrogen atoms in their ground-state $1s$ orbital.

For a fuller discussion of molecular bonding and of the energy tradeoffs that stabilize these bonds, the reader should consult any basic text in physical chemistry.[3]

PROBLEMS

1. Show that the hybridized wavefunctions χ_1, χ_2, χ_3, and χ_4 of (7.64) have the directional properties shown in Fig. 7.4.
2. Check the orthogonality of χ_1, χ_2, χ_3 and χ_4.
3. Compare the $n=2 \rightarrow n=1$ transition energies in a hydrogen atom, a deuterium atom (one proton, one neutron, one electron), and a $(He^3)^+$ ion (2 protons, 1 neutron, 1 electron).
4. What is the recoil correction for the $n=2 \rightarrow n=1$ transition energy in the three cases of Problem 7.3?

[3]See, for example, M. Karplus and R. N. Porter, *Atoms and Molecules* (Benjamin, Menlo Park, Calif., 1970), Chapter 6.

5. Calculate the expectation value of $\langle 1/r \rangle_{n,l}$ in the state n, l of the hydrogen atom. Show that

$$\left\langle \frac{\hat{p}^2}{2m} \right\rangle_{n,l} = -\tfrac{1}{2}\langle V \rangle_{n,l}$$

where $\langle\ \rangle$ stands for the quantum mechanical expectation value of the observable inside the brackets, as discussed in Section 3.2.

CHAPTER EIGHT

Systems of Identical Particles

In classical physics it is permissible by measuring position and momentum to tag each member of an ensemble of identical particles at some time t and, in principle, to follow up exactly their subsequent evolution. We may thus predict their position and momentum at any later time $t' > t$.

The kind of measurement described above is not possible according to quantum mechanics. It is postulated that it is impossible to distinguish between configurations involving identical particles that differ from each other by a mere exchange of two (or more) particles. This restriction is shown in what follows to have a profound effect on the physical properties of systems of identical particles.

8.1 SYSTEMS OF TWO ELECTRONS

Consider a system of two noninteracting identical particles 1 and 2 that, taken individually, obey

$$\hat{\mathcal{H}}(1)\psi_\alpha(1) = E_1\psi_\alpha(1)$$
$$\hat{\mathcal{H}}(2)\psi_\beta(2) = E_2\psi_\beta(2) \qquad (8.1)$$

where ψ_α and ψ_β are any two eigenfunctions of the single particle Hamiltonian $\hat{\mathcal{H}}$. The notation $\psi_\alpha(1)$, as an example, describes the dependence of the wavefunction ψ_α of particle 1 on both the spatial and spin coordinates[1] of the particle. Since the particles are assumed to be noninteracting, the total system Hamiltonian is

$$\hat{\mathcal{H}}(1,2) = \hat{\mathcal{H}}(1) + \hat{\mathcal{H}}(2) \qquad (8.2)$$

[1]The student should look ahead at this point to Sections 9.8 and 18.2 and learn about the concept of spin.

It follows from (8.1) and (8.2) that $\psi(1,2)=\psi_\alpha(2)\psi_\beta(1)$ is a solution of the Schrödinger equation of the two-particle system

$$\hat{\mathcal{H}}(1,2)\psi(1,2)=(E_1+E_2)\psi(1,2)$$

with an eigenvalue (E_1+E_2). But this is also true for

$$\psi_\alpha(1)\psi_\beta(2) \quad \text{or} \quad \frac{1}{\sqrt{2}}\left[\psi_\alpha(1)\psi_\beta(2)\pm\psi_\alpha(2)\psi_\beta(1)\right]$$

The question arises then as to which of the four forms of $\psi(1,2)$ is the correct choice. To help settle this question consider the *permutation* operator $\hat{P}_{12}$ that effects an interchange of particles 1 and 2. It follows from (8.1) and (8.2) that

$$\hat{P}_{12}\hat{\mathcal{H}}(1,2)=\hat{\mathcal{H}}(1,2)\hat{P}_{12} \tag{8.3}$$

that is, $\hat{P}_{12}$ and $\hat{\mathcal{H}}(1,2)$ commute. It is therefore possible to choose common eigenfunctions for these two operators. Let $\psi(1,2)$ correspond to some eigenfunction of $\hat{P}_{12}$, that is,

$$\hat{P}_{12}\psi(1,2)=\lambda\psi(1,2) \tag{8.4a}$$

(λ is the eigenvalue of $\hat{P}_{12}$), and

$$\hat{P}_{12}^2\psi(1,2)=\lambda^2\psi(1,2)=\psi(1,2) \tag{8.4b}$$

since two permutations result in a restoration of the original state. From the second equality of (8.4*b*) it follows that

$$\lambda^2=1, \qquad \lambda=\pm 1$$

so that $\psi(1,2)$ must be either symmetric ($\lambda=1$) or antisymmetric ($\lambda=-1$) upon the interchange of the two particles. If we denote the symmetric wavefunction by ψ_s and the antisymmetric by ψ_a, we have

$$\begin{aligned}\hat{P}_{12}\psi_s(1,2)&=\psi_s(1,2)\\ \hat{P}_{12}\psi_a(1,2)&=-\psi_a(1,2)\end{aligned} \tag{8.5}$$

In the case of a system of two identical noninteracting particles, one in the state ψ_α and the other in ψ_β, the only possible symmetric and antisymmetric wavefunctions are, respectively,

$$\psi_s=\frac{1}{\sqrt{2}}\left[\psi_\alpha(1)\psi_\beta(2)+\psi_\alpha(2)\psi_\beta(1)\right] \tag{8.6a}$$

$$\psi_a=\frac{1}{\sqrt{2}}\left[\psi_\alpha(1)\psi_\beta(2)-\psi_\alpha(2)\psi_\beta(1)\right] \tag{8.6b}$$

Experiment shows that systems with half-odd-integral spin (i.e., $s_z=\frac{1}{2}\hbar,\frac{3}{2}\hbar,\frac{5}{2}\hbar,\ldots$), such as electrons, protons, and neutrons, are described by exchange antisymmetric wavefunctions. Such particles are called fermions.

Systems made up of photons, α particles, and other particles with integral spin ($s_z=0,\hbar,2\hbar,\ldots$) are described by symmetric wavefunctions. These particles are referred to as bosons.

The wavefunctions of a system of more than two fermions can be constructed in a manner similar to that of the two-electron ψ_a. It must change sign upon the interchange of any two particles and vanish if any two particles are associated with the same quantum state. The normalized wavefunction can be written compactly in the form of a Slater determinant:

$$\psi_a(1,2,\ldots,N)=\frac{1}{\sqrt{N!}}\begin{vmatrix} \psi_1(1) & \psi_1(2) & \ldots & \psi_1(N) \\ \psi_2(1) & \psi_2(2) & \ldots & \psi_2(N) \\ \vdots & & & \\ \psi_N(1) & \psi_N(2) & \ldots & \psi_N(N) \end{vmatrix} \tag{8.7}$$

where the numbers in parentheses denote the particle, while the subscripts $1,2,\ldots$ denote the eigenstate. The interchange of any two particles causes the sign of ψ_a to change, since it involves the interchange of the corresponding two columns. Another property of ψ_a is that if any two single-particle eigenfunctions are the same—that is, the eigenstates have the same set of eigen-numbers (including spin)—then $\psi_a=0$. This follows since the presence of two identical states will result in two identical rows in (8.7), which causes the determinant to vanish. The expansion of (8.7) has $N!$ terms, each consisting of a product of N single-particle wavefunctions. Each of the $N!$ product terms consists of a particular sequence in which each of the N particles is assigned to one of the N states. The $N!$ terms thus account for all the possible permutations of the N particles among the N states. We shall refer, loosely, to a term such as $\psi_1(3)\psi_5(4)\cdots\psi_n(m)$ as one where particle (3) is "placed" in state 1, particle (4) in state 5, and so on.

The antisymmetric character of the multi-fermion wavefunction results in a *vanishing of the eigenfunction* ψ_a when any two particles are assigned ("placed") *in the same eigenstate.* This fact is known as the Pauli exclusion principle.

We emphasize again that this principle results from the postulate of fermion antisymmetry and that its adoption as a basic tenet of quantum mechanics is in full agreement with theory.

The symmetric eigenfunctions that describe a system of bosons can be obtained by using a Slater determinant in which all the $(-)$ signs are replaced by $(+)$. Symmetric wavefunctions can therefore be formed in which two or all of the single-particle wavefunctions are the same. This can be verified in the simple case of a two-boson system described by (8.6*a*). Here the eigenfunction ψ_s does not vanish when we put $\alpha=\beta$.

The physical consequences of the wavefunction symmetrization are profound. Consider, as an example, the energy of a many-fermion system in the limit of zero temperature. Quantum statistics, a topic we take up in Chapter 15, tells us that as the temperature is lowered the particles will redistribute themselves among the available states in such a way as to *lower the total energy subject to the Pauli exclusion principle.* At $T=0$, the lowest energy of a system of fermions in a nondegenerate system—that is, one where only one eigenfunction is associated with each eigenvalue—can be obtained by placing one particle in the lowest state ψ_0 with energy E_0, and, since no two particles can be in the same quantum state, the second particle must be placed in the next highest state with energy E_1, and so on. Neglecting the interaction between particles, the total energy will be $E_{tot} = E_0 + E_1 + \cdots + E_N$.

In a system of N bosons the lowest energy state consists of one in which all the particles "occupy" the same ground state. (This is a loose way of stating that the one-particle wavefunctions making up ψ_s are all those of the ground state.) It follows directly that at $T=0$ the system of N bosons has a much lower energy than that of the fermions.

The above property of fermions is crucial in understanding the behavior of insulating, conducting, and semiconducting crystals, a topic we will take up in Chapter 17.

Another property of fermions apparent from (8.6*b*) is that when we put $(1)=(2)$, $\psi_a=0$. The probability of finding the two particles near each other is thus small. (This argument will be refined in the next section to show that the probability of finding two electrons at the same location with their spins parallel to each other is zero.) Thus fermions tend to "repel" each other while bosons prefer to bunch together.

The quantum statistical difference in the behavior of systems of fermions and bosons is considered in detail in Chapters 15 and 16.

8.2 THE HELIUM ATOM

In this section we apply the concepts of wavefunction symmetry just developed to the case of two electrons moving in the potential field of a nucleus with a charge Ze and mass M. (In the case of the helium nucleus, as an example, $Z=2$.) The total system Hamiltonian is given by

$$\mathcal{H} = -\frac{\hbar^2}{2\mu}\left(\nabla_1^2 + \nabla_2^2\right) - Ze_M^2\left(\frac{1}{r_1} + \frac{1}{r_2}\right) + \frac{e_M^2}{r_{12}} \tag{8.8}$$

The last term represents the Coulomb repulsive energy of the two electrons (hence the positive sign) and r_{12} is the distance between them. μ is the effective mass ($\mu^{-1} = m_e^{-1} + M^{-1}$), $e_M^2 \equiv e^2/4\pi\varepsilon_0$.

In the lowest order of approximation we neglect the interaction term e_M^2/r_{12}. The eigenvalue problem thus reduces to solving

$$\left[\frac{-\hbar^2}{2\mu}\left(\nabla_1^2 + \nabla_2^2\right) - Ze_M^2\left(\frac{1}{r_1} + \frac{1}{r_2}\right)\right]\psi(\mathbf{r}_1, \mathbf{r}_2, s_1, s_2) = E\psi(\mathbf{r}_1, \mathbf{r}_2, s_1, s_2) \tag{8.9}$$

where s_i denotes the spin state of the ith ($i=1,2$) electron. Since the Hamiltonian consists of the sum of two single-electron hydrogenlike Hamiltonians, one involving ∇_1 and $\mathbf{r}_1$, the other ∇_2 and $\mathbf{r}_2$, we can write the spatial part of the two-particle wavefunction as a product of two hydrogenic wavefunctions of the type derived in Chapter 7:

$$\psi(\mathbf{r}_1,\mathbf{r}_2,s_1,s_2)=u_{n_1l_1m_1}(\mathbf{r}_1)u_{n_2l_2m_2}(\mathbf{r}_2)\chi(s_1,s_2) \tag{8.10}$$

where

$$\left(\frac{-\hbar^2}{2\mu}\nabla_1^2-\frac{Ze_M^2}{r_1}\right)u_{n_1l_1m_1}(\mathbf{r}_1)=E_{n_1}u_{n_1l_1m_1}(\mathbf{r}_1) \tag{8.11}$$

$$E_{n_1}=-\frac{Z^2e_M^2}{2a_0'n_1^2} \tag{8.12}$$

and $\chi(s_1,s_2)$ is the spin-dependent wavefunction. The total energy E of $\psi(\mathbf{r}_1,\mathbf{r}_2)$ is

$$E=E_{n_1}+E_{n_2} \tag{8.13}$$

The ground state $\psi_0(1,2)$ of the heliumlike atom results when we take $n_1=1$, $n_2=1$; denoting u_{100} as u_{1s} we obtain from (8.10)

$$\psi_0(1,2)\equiv\psi_0(\mathbf{r}_1,\mathbf{r}_2,s_1,s_2)=u_{1s}(1)u_{1s}(2)\chi(s_1,s_2) \tag{8.14}$$

In order to make $\psi_0(1,2)$ exchange antisymmetric, we need to choose the spin-dependent wavefunction $\chi(s_1,s_2)$ to be antisymmetric with respect to the interchange of particles 1 and 2. This is accomplished by taking

$$\psi_0(1,2)=\frac{1}{\sqrt{2}}u_{1s}(1)u_{1s}(2)[\alpha(1)\beta(2)-\alpha(2)\beta(1)] \tag{8.15}$$

which has the form of (8.6*b*). Here α and β denote spin wavefunctions with $m_s=\frac{1}{2}$ and $m_s=-\frac{1}{2}$, respectively.

In the zero order approximation the energy of the ground state is obtained from (8.12) and (8.13) by taking $n_1=n_2=1$ as

$$E_0=-Z^2e_M^2/a_0' \tag{8.16}$$

The First-Order Correction to E_0

The first-order correction to the ground-state energy is obtained by using the zero-order wavefunction (8.15) to calculate the average (expectation) value of the interaction term e_M^2/r_{12}, heretofore neglected, and adding the result to (8.16). Using (3.10) we obtain

$$\Delta E_0=\left\langle\frac{e_M^2}{r_{12}}\right\rangle\equiv\left\langle\psi_0\left|\frac{e_M^2}{r_{12}}\right|\psi_0\right\rangle$$

$$=\iint u_{1s}^2(1)u_{1s}^2(2)\frac{e_M^2}{r_{12}}d^3\mathbf{r}_1d^3\mathbf{r}_2 \tag{8.17}$$

where $r_{12}=|\mathbf{r}_2-\mathbf{r}_1|$. The formal justification of the use of (8.17) to calculate first-order energy correction will be given in Chapter 11.

Using Table 7.1 we have

$$u_{1s}(1)=u_{100}(r_1)$$

$$=\frac{1}{\sqrt{\pi}}\left(\frac{Z}{a_0'}\right)^{3/2}e^{-Zr_1/a_0'}$$

so that (8.17) becomes

$$\Delta E_0=\frac{e_M^2}{\pi^2}\left(\frac{Z}{a_0'}\right)^6\iint e^{-2Z/a_0'(r_1+r_2)}\left(\frac{1}{r_{12}}\right)d^3\mathbf{r}_1\,d^3\mathbf{r}_2 \tag{8.18}$$

This integral can be evaluated by elementary methods (see Problem 8.3). The result is

$$\Delta E_0=\frac{5Ze_M^2}{8a_0'}$$

so that

$$E_0^{(1)}=-\frac{Ze_M^2}{a_0'}\left(Z-\tfrac{5}{8}\right) \tag{8.19}$$

The ionization energy of the atom—that is, the energy needed to remove one electron from the ground state to infinity—is the difference between $-(Z^2e_M^2)/2a_0'$, which is the energy of the one-electron system remaining after the ionization, and the initial energy (8.19):

$$E_{\text{ioniz}}=-\frac{Z^2e_M^2}{2a_0'}+\frac{Ze_M^2}{a_0'}\left(Z-\tfrac{5}{8}\right)$$

$$=\frac{Ze_M^2}{2a_0'}\left(Z-\tfrac{5}{4}\right) \tag{8.20}$$

A comparison of (8.20) with the observed ionization energies of a few heliumlike configurations is given in Table 8.1.

We note by comparing (8.20) with (7.28) that the effect of the interaction between the two electrons is to reduce the effective charge from Z to $\sqrt{Z(Z-\frac{5}{4})}$. This makes qualitative sense, since part of the charge distribution of electron 1 is always between that of charge 2 and the nucleus, and thus serves to partially shield (and reduce) the effective nuclear charge and vice versa.

Table 8.1

Configuration		*Experimental Value of* E_{ioniz} *(eV)*	*Eq. (8.20) (eV)*
He	$(Z=2)$	24.584	20.43
Li^{+}	$(Z=3)$	75.638	71.52
Be^{++}	$(Z=4)$	153.900	149.84
C^{++++}	$(Z=6)$	392.014	388.24

The Excited States

We can obtain the energy and wavefunction of the first excited state of the He atom by "putting" one electron in the ground state u_{1s} $(n=1)$ and the other in the first excited state u_{2s} $(n=2, l=0)$. We can form four antisymmetric wavefunctions by including both spatial and spin wave functions

$$\psi_s = \tfrac{1}{2}\left[u_{1s}(1)u_{2s}(2) + u_{1s}(2)u_{2s}(1)\right]\left[\alpha(1)\beta(2) - \alpha(2)\beta(1)\right] \quad (8.21)$$

$$\psi_{a1} = \tfrac{1}{2}\left[u_{1s}(1)u_{2s}(2) - u_{1s}(2)u_{2s}(1)\right]\left[\alpha(1)\beta(2) + \alpha(2)\beta(1)\right] \quad (8.22)$$

$$\psi_{a2} = \frac{1}{\sqrt{2}}\left[u_{1s}(1)u_{2s}(2) - u_{1s}(2)u_{2s}(1)\right]\alpha(1)\alpha(2) \quad (8.23)$$

$$\psi_{a3} = \frac{1}{\sqrt{2}}\left[u_{1s}(1)u_{2s}(2) - u_{1s}(2)u_{2s}(1)\right]\beta(1)\beta(2) \quad (8.24)$$

The subscript s designates the space-symmetric eigenfunction, and a the antisymmetric one. The total eigenfunctions (including both the space and spin coordinates) are, of course, antisymmetric upon the interchange of particles (1) and (2). The state ψ_s is called the *singlet* state and has a total spin angular momentum $S=0$. The two electrons thus have their spin antiparallel. (The name "para" is often used to designate this state.) The *triplet* of states ψ_{a1}, ψ_{a2}, and ψ_{a3} have each a total spin angular momentum quantum number $S=1$ ("ortho") and a projection $m_s=0$, 1 and -1, respectively. The two electrons have their spin parallel to each other. The three eigenfunctions differ only in the orientation of the total spin angular momentum vector. They are referred to, collectively, as the *ortho* states.

If we neglect the electron interaction term e_M^2/r_{12} in (8.8), then the states $\psi_s, \psi_{a1}, \psi_{a2}, \psi_{a3}$ are all degenerate. Their energy is

$$E^{(0)} = E_{1s} + E_{2s} \quad (8.25)$$

where

$$E_{ns} = -Z^2 e_M^2 / (2a_0' n^2)$$

The first-order corrected energy $E^{(1)}$ is obtained as in (8.17) if we add to (8.25) the correction due to the interaction term. This is equivalent to calculating the expectation value of the full Hamiltonian (8.8) using the zero-order wavefunctions [(8.21)–(8.24)]

$$E_s^{(1)} = \iint \psi_s^* \hat{\mathcal{H}} \psi_s \, d^3\mathbf{r}_1 \, d^3\mathbf{r}_2 \quad (8.26)$$

in the case of ψ_s, and

$$E_a^{(1)} = \iint \psi_a^* \hat{\mathcal{H}} \psi_a \, d^3\mathbf{r}_1 \, d^3\mathbf{r}_2 \quad (8.27)$$

for the triplet (ortho) states.

The result is

$$E_s^{(1)} = E_{1s} + E_{2s} + Q + A \quad (8.28)$$

$$E_a^{(2)} = E_{1s} + E_{2s} + Q - A \quad (8.29)$$

with

$$Q=\iint u_{1s}^2(1)\frac{e_M^2}{r_{12}}u_{2s}^2(2)\,d^3\mathbf{r}_1\,d^3\mathbf{r}_2 \tag{8.30}$$

$$A=\iint u_{1s}(1)u_{2s}(2)\frac{e_M^2}{r_{12}}u_{1s}(2)u_{2s}(1)\,d^3\mathbf{r}_1\,d^3\mathbf{r}_2 \tag{8.31}$$

The singlet state energy is thus higher by $2A$ compared with that of the triplet state. This reflects the fact that ψ_s is maximum when $\mathbf{r}_1=\mathbf{r}_2$, while $\psi_a(\mathbf{r}_1=\mathbf{r}_2)=0$. The crowding together of the two electrons in ψ_s thus increases the Coulomb repulsion contribution, $\langle e_M^2/r_{12}\rangle$, of the state.

PROBLEMS

1. Verify Eqs. (8.28) and (8.29).
2. Show that the total angular momentum quantum number of the three triplet states [(8.22)–(8.24)] have $m_s=0$, 1, and -1, respectively.
3. Perform the integration in (8.18).
4. Write out explicitly the determinantal form of the wavefunction of three electrons. Let the three single-particle eigenstates be $u_1|\frac{1}{2}\rangle$, $u_2|-\frac{1}{2}\rangle$, and $u_2|\frac{1}{2}\rangle$, where $\pm\frac{1}{2}$ refer to m_s, while u_i denotes the spatial part of the wavefunction.
5. Consider N electrons confined inside an infinite barrier box of volume V.
   (a) Obtain an expression for the energy of the highest occupied electron state—that is, the Fermi energy at $T=0$ K.
   (b) Obtain an expression for the total minimum energy of the system consistent with the Pauli exclusion principle.
   (c) Evaluate the energies of part (a) and (b) for the case $N=10^{22}$, $V=$ 1 cm^3

Hint: If the electrons are confined to a volume $V=a\times b\times c$, then the eigenfunction of an electron energy is of the form

$$u_{ltn}\propto\sin\left(l\frac{\pi}{a}x\right)\sin\left(t\frac{\pi}{b}y\right)\sin\left(n\frac{\pi}{c}z\right)$$

l, t, n is any triplet of integers. The eigen energy is

$$E_{nlm}=\frac{\hbar^2}{2m}\left[\left(\frac{l\pi}{a}\right)^2+\left(\frac{t\pi}{b}\right)^2+\left(\frac{n\pi}{c}\right)^2\right]$$

The above form of u_{ltn} satisfies the Schrödinger equation for a free electron and vanishes at the impenetrable boundaries. The student will benefit from reading Section 15.5.

CHAPTER NINE

Matrix Formulation of Quantum Mechanics

In Chapters 6 and 7 we solved the time-independent Schrödinger equation in order to obtain the eigenvalues (allowed energies) and eigenstates in a number of specific cases. In this chapter we show how to accomplish the same end using matrix methods.

The two approaches to quantum mechanics—the differential equation approach due to Schrödinger, and the matrix approach due to Heisenberg[1]—are formally equivalent, and in the actual practice of quantum mechanics are often used simultaneously.

Before undertaking the study of the matrix formulation of quantum mechanics, we review briefly some of the fundamental concepts and techniques of matrix theory.

9.1 SOME BASIC MATRIX PROPERTIES

It will be assumed in what follows that the student is familiar with the basic definitions and operations involving matrices. Consequently, the review of the necessary background of matrix algebra is brief.

The *kl* element of the product of the matrices A and B is

$$(AB)_{kl} = \sum_m A_{km} B_{ml}$$

[1]W. Heisenberg, *The Physical Principles of Quantum Theory* (transl. from German) (University of Chicago Press, 1930).

and by extension

$$(ABC)_{kl} = \sum_m \sum_n A_{km} B_{mn} C_{nl} \tag{9.1}$$

The Unit Matrix

The unit matrix I is defined as the matrix that, when multiplying a matrix B, leaves the latter unchanged:

$$BI = B, \qquad IB = B \tag{9.2}$$

It follows that $I_{kl} = \delta_{kl}$.

The Inverse Matrix

The inverse matrix B^{-1} of B is the matrix satisfying the conditions

$$B^{-1}B = BB^{-1} = I \tag{9.3}$$

The inverse of a product of matrices is equal to the product of the inverse matrices taken in a reverse order:

$$(AB)^{-1} = B^{-1}A^{-1} \tag{9.4}$$

The elements of A^{-1} are related to those of A by the relation

$$A_{kl}^{-1} = \frac{\text{cofactor of } A_{lk}}{\text{determinant of } A} \tag{9.5}$$

Notice the reverse order of k and l on both sides of the equality sign.

Hermitian Adjoint Matrices, Hermitian Matrices

The *Hermitian adjoint* of a matrix A, denoted as $A^\dagger$, is defined by

$$A_{kl}^\dagger = A_{lk}^* \tag{9.6}$$

A matrix is Hermitian when it is its own Hermitian adjoint—that is, when

$$A^\dagger = A \tag{9.7}$$

Equation (9.7) can be written, using (9.6), as

$$A_{kl}^\dagger = A_{kl} = A_{lk}^* \tag{9.8}$$

so that in a Hermitian matrix the interchange of rows and columns is equivalent to replacing each element by its complex conjugate. This applies, consequently, only to square matrices. It also follows that the matrix elements A_{kk} along the main diagonal are real.

Unitary Matrices

A *unitary* matrix is defined as the matrix whose Hermitian adjoint is equal to its inverse:

$$A^{\dagger} = A^{-1} \tag{9.9}$$

From (9.9) it follows that $(AA^{\dagger}) = I$ for a unitary matrix A, which when written in a component form becomes

$$\begin{aligned}(AA^{\dagger})_{kl} &= \sum_n A_{kn} A^{\dagger}_{nl} \\ &= \sum_n A_{kn} A^{*}_{ln} \\ &= \delta_{kl}\end{aligned} \tag{9.10}$$

where use was made of (9.6).

9.2 TRANSFORMATION OF A SQUARE MATRIX

A matrix A' derived from a square matrix A by the operation

$$A' = SAS^{-1} \tag{9.11}$$

is called the *transformation* of A by (the square matrix) S. Matrix equations are left invariant under a transformation of the individual matrices. A typical equation containing matrix products and sums such as

$$AB + CDE = F$$

can be rewritten as

$$SAS^{-1}SBS^{-1} + SCS^{-1}SDS^{-1}SES^{-1} = SFS^{-1}$$

since $S^{-1}S = I$. The last equality, according to (9.11), can be written as

$$A'B' + C'D'E' = F' \tag{9.12}$$

which is the same equation as that obeyed by the original (untransformed) matrices.

9.3 MATRIX DIAGONALIZATION

Of special interest in quantum mechanics is the transformation A (of a square matrix A'), which is diagonal. We illustrate in Section 9.6 how finding A_{kk} and the matrix elements of S takes the place, in the matrix formulation of quantum mechanics, of solving the wave equation. A is assumed to be diagonal with (unknown) elements $A_{kk} \equiv A_k$ so that

$$(SA'S^{-1})_{kl} = A_k \delta_{kl}$$

Postmultiplying the relation $SA'S^{-1} = A$ by S,

$$SA' = AS$$

and taking the kl element,

$$\sum_m S_{km} A'_{ml} = \sum_m A_{km} S_{ml} = A_k S_{kl}$$

or

$$\sum_m S_{km}(A'_{ml} - A_k \delta_{ml}) = 0 \qquad (9.13)$$

If S is a square N-dimensional matrix, we obtain N equations by fixing k and writing (9.13) for each value of l between 1 and N. These N simultaneous and homogeneous equations for the N unknowns $S_{k1},\ldots,S_{kN}$ have nontrivial solutions only when the determinant of the matrix made up of the coefficients vanishes:

$$\det[A'_{ml} - A_k \delta_{ml}] = 0 \qquad (9.14)$$

The N solutions $A_1,\ldots,A_N$ of A_k that result from solving (9.14) are the diagonal matrix elements sought. They are also referred to as the N *eigenvalues* of the matrix A'.

The remainder of the problem is to find the elements S_{km} of the transformation matrix. A given A_k is substituted in (9.13). The N homogeneous equations can be solved to yield $S_{k1}, S_{k2}, \cdots, S_{kN}$ to within an arbitrary multiplying constant. This procedure is repeated with each of the N A_k's, thus generating the matrix S_{km}. The additional condition necessary to determine S_{km} uniquely is, as shown in Section 9.4, that S be unitary so that, according to (9.10),

$$\sum_n S_{kn} S^*_{ln} = \delta_{kl}$$

9.4 REPRESENTATIONS OF OPERATORS AS MATRICES

In the matrix formulation of quantum mechanics, an arbitrary operator $\hat{A}$ is *represented* by a matrix A. The element km of the matrix is defined by

$$A_{km} = \int u^*_k(\mathbf{r}) \hat{A} u_m(\mathbf{r})\, d\mathbf{v} \qquad (9.15)$$

where $u_m(\mathbf{r})$ is any *arbitrary complete orthonormal set of functions*, and $d\mathbf{v}$ includes integration over spatial as well as spin coordinates. The representation A_{km} of an operator $\hat{A}$ is, consequently, not unique and depends on the (arbitrary) choice of the set $u_m(\mathbf{r})$. The operator $\hat{A}$ can thus have other representations as well. Let one such representation, obtained using the set $v_n(\mathbf{r})$, be A'. It follows that

$$A'_{km} = \int v^*_k \hat{A} v_m\, d\mathbf{v} \equiv \langle v_k | \hat{A} | v_m \rangle \qquad (9.16a)$$

It also follows that if we represent the operator $\hat{A}$ in terms of its own eigenfunctions ψ_n (i.e., $\hat{A}\psi_n = A_n \psi_n$), then the resulting matrix is diagonal and has the eigenvalues as its elements, since

$$\int \psi^*_n \hat{A} \psi_m\, d\mathbf{v} = A_n \delta_{nm} \qquad (9.16b)$$

A Unitary Transformation Matrix

We can expand an arbitrary member of the set v_n in terms of the set u_k:

$$v_n(\mathbf{r}) = \sum_k S_{kn} u_k(\mathbf{r})$$
$$= \sum_k |u_k\rangle\langle u_k|v_n\rangle \tag{9.17}$$

In the second equality we used relation (2.32*a*). The reverse expansion becomes

$$u_k = \sum_n S^*_{kn} v_n$$
$$= \sum_n |v_n\rangle\langle v_n|u_k\rangle \tag{9.18}$$

It follows that

$$S_{kn} = \int u_k^* v_n \, d\mathbf{v}$$
$$= \langle u_k|v_n\rangle \tag{9.19}$$

The set of numbers S_{kn} can be regarded as a matrix. We will refer to it as the *transformation* matrix from the function space v_n to u_k. The matrix S defined by (9.19) is a *unitary* matrix. As a proof we may show that $SS^\dagger = I$ (I is the unity matrix),

$$(SS^\dagger)_{kl} = \sum_n S_{kn} S^\dagger_{nl}$$
$$= \sum_n S_{kn} S^*_{ln}$$
$$= \sum_m \langle u_k|v_n\rangle\langle v_n|u_l\rangle$$
$$= \langle u_k|u_l\rangle$$
$$= \delta_{kl}$$

where we used relation (2.32*b*),

$$\sum_n |v_n\rangle\langle v_n| = \hat{I}$$

9.5 TRANSFORMATION OF OPERATOR REPRESENTATIONS

We considered in Section 9.4 two arbitrary representations of an operator $\hat{A}$, one in the function space v_n, the other in u_n:

$$A_{kl} = \int u_k^* \hat{A} u_l \, dv$$
$$A'_{kl} = \int v_k^* \hat{A} v_l \, dv$$
$$u_k = \sum_n S^*_{kn} v_n \tag{9.20}$$

We can derive the matrix A from A', and vice versa, by the transformation

$$A = SA'S^{-1} = SA'S^{\dagger}$$
$$A' = S^{\dagger}AS \tag{9.21}$$

where S is the unitary transformation matrix defined by (9.19).

The proof of Eq. (9.21) consists of replacing u_k^* and u_l in the first of Eqs. (9.20) by their expansion according to (9.18):

$$\begin{aligned} A_{kl} &= \int u_k^* \hat{A} u_l \, d\mathbf{v} \\ &= \int \Big(\sum_n S_{kn} v_n^* \Big) \hat{A} \sum_m S_{lm}^* v_m \, d\mathbf{v} \\ &= \sum_n \sum_m S_{kn} \int v_n^* \hat{A} v_m \, d\mathbf{v} \, S_{lm}^* \\ &= \sum_n \sum_m S_{kn} A'_{nm} S_{ml}^{\dagger} \\ &= (SA'S^{\dagger})_{kl} \end{aligned}$$

If the matrix A' is Hermitian, so is the matrix A. In other words, *the Hermiticity of a matrix is preserved under unitary transformations.* The proof consists of showing that $A_{kl} = A_{lk}^*$, where A is given by (9.21):

$$\begin{aligned} A_{kl} &= \sum_{mn} S_{km} A'_{mn} S_{nl}^{\dagger} \\ &= \sum_{mn} S_{km} (A'_{nm})^* S_{ln}^* \\ A_{lk} &= \sum_{mn} S_{lm} A'_{mn} S_{kn}^* \end{aligned}$$

where the definition of a Hermitian matrix $A'_{mn} = (A'_{mn})^*$ was used. Taking the complex conjugate of the last equation and interchanging m and n leads to

$$\begin{aligned} (A_{lk})^* &= \sum_{mn} S_{ln}^* (A'_{nm})^* S_{km} \\ &= A_{kl} \end{aligned}$$

so that, according to (9.8), A is Hermitian.

A corollary of this result is the following: *The eigenvalues of a Hermitian matrix are real.* This result follows from the fact that any unitary transformation of a Hermitian matrix is Hermitian, and thus obeying (9.8) must possess real diagonal elements. This applies also to the diagonal transformation whose elements are the eigenvalues.

Another important result, and one used in this chapter, is the following: *Two Hermitian matrices that are related by a unitary transformation have the same eigenvalues.*

9.6 DERIVING THE EIGENFUNCTIONS AND EIGENVALUES OF AN OPERATOR BY THE MATRIX METHOD

Finding the eigenfunctions and eigenvalues of an arbitrary Hermitian operator $\hat{A}$ consists of solving the differential equation

$$\hat{A}u_n = A_n u_n \tag{9.22}$$

The set of functions u_n is referred to as the eigenfunctions of $\hat{A}$, and the (real) numbers A_n are its eigenvalues. As an example of an eigenvalue problem we have already solved, we may take Eq. (5.8). The eigenfunctions u_n are given by (5.29), while the eigenvalues are, according to (5.16), $E_n = \hbar\omega(n+\frac{1}{2})$. A second example is provided by the operator $\hat{\mathbf{L}}^2$. The eigenfunctions and eigenvalues are given by Eq. (6.37) as

$$\hat{\mathbf{L}}^2 Y_l^m(\theta,\phi) = \hbar^2 l(l+1) Y_l^m(\theta,\phi) \tag{9.23}$$

An alternate approach for obtaining the eigenfunctions and eigenvalues of a Hermitian operator $\hat{A}$ is to start with the matrix representation A' of $\hat{A}$ in some complete, orthonormal, but otherwise arbitrary, set of functions $v_k(\mathbf{r})$. The matrix A' as defined by

$$A'_{kl} = \int v_k^* \hat{A} v_l \, d\mathbf{v}$$

is then diagonalized using the methods of Section 9.3. The result is the diagonal matrix

$$A = SA'S^{-1}$$

whose elements are, according to (9.16*b*) and the last statement in Section 9.5, the sought eigenvalues. The solution also yields the matrix S, which transforms A' to A. This matrix can be used according to (9.18) to generate the eigenfunctions u_k of $\hat{A}$:

$$u_k = \sum_n S_{kn}^* v_n$$

The functions u_k are indeed the eigenfunctions of $\hat{A}$, since the matrix representation $A_{kl} = \int u_k^* \hat{A} u_l \, d\mathbf{v}$ is diagonal. This procedure is commonly used in atomic physics to obtain the eigenfunctions and eigenvalues, since it lends itself readily to numerical methods.

9.7 MATRIX ELEMENTS OF THE ANGULAR MOMENTUM OPERATORS

As an illustration of the ideas developed in this chapter, we derive below some matrix elements involving the angular momentum operator. These play a central role in later developments concerned with the addition of angular momenta and with the orbital and spin magnetic moments.

Let $\hat{\mathbf{L}}$ be the angular momentum operator of our physical system. The raising and lowering operators $\hat{\mathbf{L}}^+$ and $\hat{\mathbf{L}}^-$ respectively, are defined by[2]

$$\hat{L}^+ \equiv \hat{L}_x + i\hat{L}_y, \qquad \hat{L}_x \equiv \tfrac{1}{2}(\hat{L}^+ + \hat{L}^-)$$

$$\hat{L}^- \equiv \hat{L}_x - i\hat{L}_y, \qquad \hat{L}_y \equiv \frac{1}{2i}(\hat{L}^+ - \hat{L}^-) \tag{9.24}$$

so that $\hat{L}^+ = (\hat{L}^-)^\dagger$. These operators obey the commutation relations

$$[\hat{L}^\pm, \hat{L}_z] = \mp \hbar \hat{L}^\pm$$

$$[\hat{\mathbf{L}}^2, \hat{L}^\pm] = 0$$

$$[\hat{L}^+, \hat{L}^-] = 2\hbar L_z \tag{9.25}$$

These relations can be proved with the aid of Eqs. (6.46–6.48). As an illustration, consider the first of Eqs. (9.25):

$$\begin{aligned}[\hat{L}^\pm, \hat{L}_z] &= \left[(\hat{L}_x \pm i\hat{L}_y), \hat{L}_z\right] \\ &= [\hat{L}_x, \hat{L}_z] \pm i(\hat{L}_y, \hat{L}_z] \\ &= -i\hbar \hat{L}_y \mp \hbar \hat{L}_x \\ &= \mp \hbar \hat{L}^\pm\end{aligned}$$

The result of operating with $\hat{L}^\pm$ on the eigenfunctions $Y_j^m(\theta, \phi)$ of $\hat{\mathbf{L}}^2$ can be studied by the following development:

$$\begin{aligned}\hat{L}_z(\hat{L}^\pm Y_j^m) &= (\hat{L}^\pm \hat{L}_z \pm \hbar \hat{L}^\pm) Y_j^m \\ &= (m\hbar \hat{L}^\pm \pm \hbar \hat{L}^\pm) Y_j^m \\ &= (m \pm 1)\hbar (\hat{L}^\pm Y_j^m)\end{aligned}$$

where we used the relation $\hat{L}_z Y_j^m = m\hbar Y_j^m$ and the first of Eqs. (9.25). The last equality states that $\hat{L}^\pm Y_j^m$ is an eigenfunction of $\hat{L}_z$ with eigenvalues $(m \pm 1)\hbar$. These have been shown to be the functions $Y_j^{m\pm1}$. We can thus write directly

$$\hat{L}^\pm Y_j^m(\theta, \phi) = \hbar C_m^\pm Y_j^{m\pm1}(\theta, \phi) \tag{9.26}$$

where the constants $C_m^\pm$ remain to be evaluated.

---

[2]In this section we denote angular momentum operators by capital letters rather than by lower case letters as in Chapter 6. This is done to emphasize the fact that they may apply to a system of particles and not necessarily to a single particle.

Consider next the matrix elements of $\hat{\mathbf{L}}^2, \hat{L}_z$, and $\hat{L}^{\pm}$ in the function space $Y_j^m(\theta, \phi)$. Using Eqs. (6.37) results in

$$(\hat{\mathbf{L}}^2)_{j,m,j',m'} = \int_0^{\pi}\int_0^{2\pi}\left[Y_j^m(\theta,\phi)\right]^* \hat{\mathbf{L}}^2 Y_{j'}^{m'}(\theta,\phi)\sin\theta\, d\theta\, d\phi$$

$$= j(j+1)\hbar^2\delta_{j,j'}\delta_{m,m'}$$

$$(\hat{L}_z)_{j,m,j',m'} = m\hbar\delta_{j,j'}\delta_{m,m'}$$

$$(\hat{L}^+)_{j,m+1;j,m} = C_m^+\hbar \equiv C_m\hbar$$

$$(\hat{L}^-)_{j,m;j,m+1} = C_{m+1}^-\hbar = (C_m^+)^*\hbar = C_m^*\hbar \tag{9.27}$$

The last two equations involving $(\hat{L}^{\pm})$ were derived using Eq. (9.26). The proof of the relation $C_{m+1}^- = (C_m^+)^*$ is left as an exercise (Problem 8 at the end of this chapter).

In order to derive the constant C_m, consider the mth diagonal element of the third of Eqs. (9.25):

$$(L^+L^- - L^-L^+)_{m,m} = 2\hbar(L_z)_{m,m,} = 2m\hbar^2$$

where the subscript j is omitted since matrix elements involving states with a different j have been shown in (9.27) to be zero. All the relations developed below will thus assume a constant j. Expanding the matrix elements gives

$$\sum_{m'}\hat{L}^+_{m,m'}\hat{L}^-_{m',m} - \hat{L}^-_{m,m'}\hat{L}^+_{m',m} = \hat{L}^+_{m,m-1}\hat{L}^-_{m-1,m} - \hat{L}^-_{m,m+1}\hat{L}^+_{m+1,m} = 2m\hbar^2 \tag{9.28}$$

or

$$C_{m-1}C^*_{m-1} - C^*_mC_m = 2m$$

where use has been made of the last two equations of (9.27). The last relation is rewritten as

$$|C_{m-1}|^2 - |C_m|^2 = 2m \tag{9.29}$$

To evaluate $|C_m|$ we use the relation

$$\hat{\mathbf{L}}^2 = \hat{L}_z^2 + \tfrac{1}{2}(\hat{L}^+\hat{L}^- + \hat{L}^-\hat{L}^+) \tag{9.30}$$

and take the m, m diagonal matrix elements of both sides.

The matrix elements of the first two terms are taken directly from (9.27), the result being

$$j(j+1)\hbar^2 = m^2\hbar^2 + \tfrac{1}{2}\sum_{m'}\hat{L}^+_{m,m'}\hat{L}^-_{m',m} + \hat{L}^-_{m,m'}\hat{L}^+_{m',m}$$

$$= m^2\hbar^2 + \tfrac{1}{2}\hbar^2(C_{m-1}C^*_{m-1} + C^*_mC_m) \tag{9.31}$$

Combining this result with (9.29) gives

$$C_m = \sqrt{j(j+1) - m(m+1)}$$

where the arbitrary phase of the wavefunction Y_j^m is assumed to be such that

C_m as defined by (9.27) is real and positive. Substituting C_m in Eq. (9.26) gives

$$\begin{aligned} \hat{L}^+ Y_j^m &= \hbar\sqrt{j(j+1)-m(m+1)}\, Y_j^{m+1} \\ \hat{L}^- Y_j^m &= \hbar\sqrt{j(j+1)-m(m-1)}\, Y_j^{m-1} \end{aligned} \tag{9.32}$$

or

$$\begin{aligned} (\hat{L}^-)_{j,m;j,m+1} &= (\hat{L}^+)_{j,m+1;j,m} \\ &= \hbar\sqrt{j(j+1)-m(m+1)} \end{aligned} \tag{9.33}$$

These are the desired results.

Using Eqs. (9.27) and (9.33) we can construct the matrices corresponding to the various angular momentum operators in the space $Y_j^m(\theta,\phi)$. Since the matrix elements involving states with different j values are zero, we may limit ourselves to the submatrix within a constant j manifold. (The term manifold will be applied often in this context to describe the subspace made up of the $2j+1$ eigenfunctions $Y_j^m(\theta,\phi)$ with $-j \le m \le j$.) Choosing $j=1$, as an example, results in the following set of 3×3 matrices:

$$\begin{array}{cc} m= & 1 \quad 0 \quad -1 \\ L_z = & \begin{array}{|ccc|} \hline 1 & 0 & 0 \\ 0 & 0 & 0 \\ 0 & 0 & -1 \\ \hline \end{array} \;\hbar \end{array} \tag{9.34}$$

$$\begin{array}{cc} & 1 \quad 0 \quad -1 \\ L^2 = & \begin{array}{|ccc|} \hline 1 & 0 & 0 \\ 0 & 1 & 0 \\ 0 & 0 & 1 \\ \hline \end{array} \;2\hbar^2 \end{array} \tag{9.35}$$

$$\begin{array}{cc} & 1 \quad 0 \quad -1 \\ L^+ = & \begin{array}{|ccc|} \hline 0 & 1 & 0 \\ 0 & 0 & 1 \\ 0 & 0 & 0 \\ \hline \end{array} \;\sqrt{2}\,\hbar \end{array} \tag{9.36}$$

$$\begin{array}{cc} & 1 \quad 0 \quad -1 \\ L^- = & \begin{array}{|ccc|} \hline 0 & 0 & 0 \\ 1 & 0 & 0 \\ 0 & 1 & 0 \\ \hline \end{array} \;\sqrt{2}\,\hbar \end{array} \tag{9.37}$$

$$\begin{array}{cc} & 1 \quad 0 \quad -1 \\ L_x = & \begin{array}{|ccc|} \hline 0 & 1 & 0 \\ 1 & 0 & 1 \\ 0 & 1 & 0 \\ \hline \end{array} \;\dfrac{\hbar}{\sqrt{2}} \end{array} \tag{9.38}$$

$$m = \quad 1 \quad 0 \quad -1$$

$$L_y = \begin{vmatrix} 0 & -i & 0 \\ i & 0 & -i \\ 0 & i & 0 \end{vmatrix} \frac{\hbar}{\sqrt{2}} \tag{9.39}$$

The three eigenfunctions Y_1^1, Y_1^0, Y_1^{-1} can be represented in the form of column matrices:

$$Y_1^1 = \begin{vmatrix} 1 \\ 0 \\ 0 \end{vmatrix}, \qquad Y_1^0 = \begin{vmatrix} 0 \\ 1 \\ 0 \end{vmatrix}, \qquad Y_1^{-1} = \begin{vmatrix} 0 \\ 0 \\ 1 \end{vmatrix} \tag{9.40}$$

The eigenvalue relation $\hat{L}_z Y_1^{-1} = -\hbar Y_1^{-1}$, as an example, becomes

$$\hbar \begin{vmatrix} 1 & 0 & 0 \\ 0 & 0 & 0 \\ 0 & 0 & -1 \end{vmatrix} \begin{vmatrix} 0 \\ 0 \\ 1 \end{vmatrix} = -\hbar \begin{vmatrix} 0 \\ 0 \\ 1 \end{vmatrix} \tag{9.41}$$

so that the eigenvalue equations (6.37) are satisfied if we replace the operators by their matrix representations (9.34–38) and the eigenfunctions by column matrices as in (9.40).

9.8 SPIN ANGULAR MOMENTUM

Up to this point the treatment of the angular momentum operators and their eigenfunctions was based on solving the eigenvalue problem (6.37)

$$\hat{\mathbf{L}}^2(\theta,\phi) Y_l^m(\theta,\phi) = \hbar^2 (l+1) l Y_l^m(\theta,\phi)$$

Since the operators were functions of the spatial variables (θ, ϕ), the solutions $Y_l^m(\theta,\phi)$ had to be single valued in real space. This, as shown in Section 6.2, forced m and l to assume integral values.

If the starting point for evolving the theory is taken as the commutation relationship $\mathbf{L}\times\mathbf{L} = i\hbar\mathbf{L}$ (6.49), the restriction on l and m mentioned above does not hold. This would be true if particles had, in addition to orbital angular momentum $\hat{\mathbf{L}} = \hat{\mathbf{r}} \times \hat{\mathbf{p}}$, some intrinsic angular momentum that does not depend on the spatial coordinates. Such angular momentum operators would commute with any Hamiltonian that depends only on $\mathbf{r}$ and $\mathbf{p}$, and would, consequently, be constants of the motion. This intrinsic angular momentum $\mathbf{S}$ is called spin, as distinguished from the orbital, angular momentum $\mathbf{L}$. The electron and proton both are found experimentally to possess spin angular momentum $s = \frac{1}{2}$ so that $m_s = \pm 1/2$. In treating spin angular momentum operators we stipulate that they obey Eq. (6.49) so that the matrix representations and the operator manipulations are identical to those of the orbital angular momentum operators except that S can assume half-odd integral values as well as integral values. The total eigenfunction specifying the state of a free electron is thus a function of $\mathbf{r}$ and an additional coordinate specifying the projection of the spin angular momentum along any arbitrary (z) direction:

$$\psi = \psi(r, \theta, \phi, S_z)$$

while the total angular momentum operator is given by

$$\hat{\mathbf{J}} = \hat{\mathbf{L}} + \hat{\mathbf{S}} \tag{9.42}$$

We can also use the formalism of Section 9.7 to express the (intrinsic) spin angular momentum operators and eigenvectors in a matrix representation. The eigenvectors become

$$|m_s = \tfrac{1}{2}\rangle = \begin{vmatrix} 1 \\ 0 \end{vmatrix}, \qquad |m_s = -\tfrac{1}{2}\rangle = \begin{vmatrix} 0 \\ 1 \end{vmatrix} \tag{9.43}$$

while the operators are

$$\begin{matrix} & m_s = \tfrac{1}{2} & -\tfrac{1}{2} \\ L_x = & \begin{bmatrix} 0 & 1 \\ 1 & 0 \end{bmatrix} & \dfrac{\hbar}{2} \end{matrix} \qquad \begin{matrix} & m_s = \tfrac{1}{2} & -\tfrac{1}{2} \\ L_y = & \begin{bmatrix} 0 & -i \\ i & 0 \end{bmatrix} & \dfrac{\hbar}{2} \end{matrix}$$

$$\begin{matrix} & m_s = \tfrac{1}{2} & -\tfrac{1}{2} \\ L_z = & \begin{bmatrix} 1 & 0 \\ 0 & -1 \end{bmatrix} & \dfrac{\hbar}{2} \end{matrix} \tag{9.44}$$

The three matrices of (9.44) (without the $\hbar/2$ factor) are known as the Pauli spin matrices σ_x, σ_y, and σ_z, respectively, so that the spin angular momentum matrices are given by $L_x = \frac{1}{2}\hbar\sigma_x$, $L_y = \frac{1}{2}\hbar\sigma_y$, and $L_z = \frac{1}{2}\hbar\sigma_z$. The operators $\hat{L}^{\pm} = \hat{L}_x \pm i\hat{L}_y$ are represented by

$$\begin{matrix} & m_s = \tfrac{1}{2} & -\tfrac{1}{2} \\ L^+ = & \begin{bmatrix} 0 & 1 \\ 0 & 0 \end{bmatrix} & \hbar \end{matrix} \qquad \begin{matrix} & m_s = \tfrac{1}{2} & -\tfrac{1}{2} \\ L^- = & \begin{bmatrix} 0 & 0 \\ 1 & 0 \end{bmatrix} & \hbar \end{matrix} \tag{9.45}$$

Note that unlike L_x, L_y, and L_z, the operators L^+ and L^- are not Hermitian [i.e., $L^{\pm}_{12} \neq (L^{\pm}_{21})^*$].

9.9 ADDITION OF ANGULAR MOMENTA

Consider two "particles" with total angular momentum quantum numbers j_1 and j_2. The sets of operators describing these particles commute with each other. Let the eigenfunctions of $\hat{\mathbf{L}}_1^2$ and $\hat{\mathbf{L}}_2^2$ be $\alpha_{j_1}^{m_1}$ and $\beta_{j_2}^{m_2}$. We can form a representation by taking all the possible product functions $\alpha_{j_1}^{m_1}\beta_{j_1}^{m_1}$. There are $(2j_1+1)(2j_2+1)$ such functions. These functions are simultaneous eigenfunctions of $\hat{\mathbf{L}}_1^2, \hat{\mathbf{L}}_2^2, \hat{L}_{1z}, \hat{L}_{2z}$.

We can form another representation in which $\hat{\mathbf{L}}_1^2$, $\hat{\mathbf{L}}_2^2$, $\hat{\mathbf{L}}^2$, and $\hat{\mathbf{L}}_z$ are diagonal, where $\hat{\mathbf{L}}$ is the sum angular momentum operator

$$\hat{\mathbf{L}} = \hat{\mathbf{L}}_1 + \hat{\mathbf{L}}_2 \tag{9.46}$$

The eigenvalues of $\hat{L}_z$ and $\hat{L}^2$ will correspond, according to the basic postulates of quantum mechanics, to the possible values of the z component and the squared magnitude, respectively, of the angular momentum operator $\hat{\mathbf{L}} = \hat{\mathbf{L}}_1 + \hat{\mathbf{L}}_2$. This procedure amounts to the *addition* of angular momenta. Since both sets of eigenfunctions span the same space, they must be connected by a unitary transformation. The eigenfunctions ψ_j^m of $\hat{\mathbf{L}}^2$ and $\hat{L}_z$ can consequently be expanded as a linear superposition of the $\alpha_{j_1}^{m_1}\beta_{j_2}^{m_2}$ functions:

$$\psi_j^m = \sum_{m_1}\sum_{m_2} (j_1 j_2 m_1 m_2 | j_1 j_2 jm)\alpha_{j_1}^{m_1}\beta_{j_2}^{m_2} \tag{9.47}$$

where the expansion coefficient $(j_1 j_2 m_1 m_2 | j_1 j_2 jm)$ is, according to (9.20), the element of the transformation matrix relating the set of ψ_j^m to the set $\alpha_{j_1}^{m_1}\alpha_{j_2}^{m_2}$. It is clear that only the product wavefunctions $\alpha_{j_1}^{m_1}\beta_{j_2}^{m_2}$, where $m_1 + m_2 = m$, are to be included in the summation on the right side of (9.47). This can be verified by operating on both sides with $\hat{L}_z = \hat{L}_{1z} + \hat{L}_{2z}$. We thus have

$$\psi_j^m = \sum_{m} (j_1 j_2 m_1 (m - m_1) | j_1 j_2 jm)\alpha_{j_1}^{m_1}\beta_{j_2}^{m-m_1} \tag{9.48}$$

Another property of the (|) coefficients is

$$(j_1 j_2 j_1 j_2 | j_1 j_2 (j_1 + j_2)(j_1 + j_2)) = 1 \tag{9.49}$$

so that

$$\psi_{j_1+j_2}^{j_1+j_2} = \alpha_{j_1}^{j_1}\beta_{j_2}^{j_2} \tag{9.50}$$

This results from the fact that, according to (9.48), $m = m_1 + m_2 = j_1 + j_2$, and there is only one product function, given by (9.50), where $m_1 + m_2 = j_1 + j_2$. Since $j \geq m$, it follows that $j = j_1 + j_2$.

It follows from the above discussion that the maximum value of j is equal to $j_1 + j_2$. With each value of j there are $2j+1$ wavefunctions with $-j \leq m \leq j$. Since the total number of wavefunctions ψ_j^m must be equal to that of the function space $\alpha_{j_1}^{m_1}\beta_{j_2}^{m_2}$—namely, $(2j_1+1)(2j_2+1)$—it follows that the smallest value of j is $|j_1 - j_2|$. This can be verified from the summation

$$\sum_{j=|j_1-j_2|}^{j_1+j_2} (2j+1) = (2j_1+1)(2j_2+1)$$

The coefficients (|) are known as the Clebsch–Gordan coefficients. Condon and Shortley,[3] as an example, tabulate these coefficients for a number of j_1's and j_2's.

---

[3] E. V. Condon and G. H. Shortley, *The Theory of Atomic Spectra* (Cambridge University Press, New York, 1959), p. 73.

For simple cases it is quite easy to derive the eigenfunctions of $\hat{L}^2$ and $\hat{L}_z$ (where $\hat{\mathbf{L}} = \hat{\mathbf{L}}_1 + \hat{\mathbf{L}}_2$) starting with (9.50). Consider, for example, "adding" the angular momenta of two particles with $j_1 = j_2 = 1$. We start with

$$\psi_2^2 = \alpha_1^1 \beta_1^1 \tag{9.51}$$

Applying $L^- = L_1^- + L_2^-$ to both sides and using (9.32) results in

$$L^- \psi_2^2 = 2\psi_2^1 = \sqrt{2}\left(\alpha_1^0 \beta_1^1 + \alpha_1^1 \beta_1^0\right) \tag{9.52}$$

$$\psi_2^1 = \frac{1}{\sqrt{2}}\left(\alpha_1^0 \beta_1^1 + \alpha_1^1 \beta_1^0\right) \tag{9.53}$$

The next function ψ_1^1 can be written according to (9.48) as

$$\psi_1^1 = a\alpha_1^0 \beta_1^1 + b\alpha_1^1 \beta_1^0 \tag{9.54}$$

Requiring that $\langle \psi_1^1 | \psi_1^1 \rangle$ be unity gives (for real a and b)

$$a^2 + b^2 = 1 \tag{9.55}$$

While the condition $\langle \psi_1^1 | \psi_2^1 \rangle = 0$ gives

$$a + b = 0 \tag{9.56}$$

The last two equations are satisfied by

$$a = \pm \frac{1}{\sqrt{2}}$$

$$b = \mp \frac{1}{\sqrt{2}} \tag{9.57}$$

which, choosing arbitrarily the upper sign gives

$$\psi_1^1 = \frac{1}{\sqrt{2}}\left(\alpha_1^0 \beta_1^1 - \alpha_1^1 \beta_1^0\right) \tag{9.58}$$

This procedure can be used to generate the remaining eigenfunctions.

PROBLEMS

1. A function f may be expanded in terms of two arbitrary complete orthonormal sets v_n and u_n in the form

$$f = \Sigma f_n u_n = \Sigma f_n' v_n$$

The set v_n can be expanded as

$$v_n(r) = \sum_k S_{kn} u_k$$

Show that the unitarity of the matrix S can be derived by requiring that $\int f^* f \, d\mathbf{v}$ be independent of the set (v_n or u_n) in which it is expanded.

2. Show that the trace of a square matrix A,

$$\operatorname{tr} A = \sum_k A_{kk}$$

is invariant under unitary transformations—that is, that

$$\operatorname{tr} A = \operatorname{tr}(SAS^{-1})$$

where S is any unitary matrix.

3. Prove that $(AB)^{\dagger} = B^{\dagger}A^{\dagger}$.

4. Show that, if $A_{kl} = \int v_k^* \hat{A} v_l \, d\mathbf{v} = A_{lk}^*$ where v_n is any orthonormal function set, the operator $\hat{A}$ is Hermitian.

5. Show that the matrix representation of a product of operators is equal to the product of the individual matrices.

6. Check the result of Problem 5 against the solution obtained in the problem section of Chapter 5 for the matrix elements of x^2:

$$x_{mn}^2 = \int u_m x^2 u_n \, d\mathbf{v}$$

7. Show that the necessary and sufficient condition that two matrices commute is that the same transformation diagonalizes each one of them.

8. Prove that $c_{m+1}^- = (c_m^+)^*$, where c_{m+1}^- and c_m^+ are defined by Eq. (9.27).

9. Prove that $(1/\sqrt{2})(\alpha_1^0\beta_1^1 - \alpha_1^1\beta_1^0)$ is an eigenfunction of $\hat{\mathbf{L}}^2$, where $\hat{\mathbf{L}} = \hat{\mathbf{L}}_1 + \hat{\mathbf{L}}_2$. What is the eigenvalue?

10. Generate the set of eigenfunctions ψ_j^m resulting from the addition of two angular momenta $j_1 = \frac{1}{2}$ and $j_2 = 1$.

11. Show that for spin $\frac{1}{2}$ $(s = \frac{1}{2})$,

$$S_x S_y + S_y S_x = 0$$

12. Prove that the eigenvectors (9.40) are orthonormal. Show, using matrix representation, that they are eigenvectors of $\hat{\mathbf{L}}^2$.

13. (a) Show that the state ψ_s of Eq. (8.21) has spin angular momentum quantum number $S = 0$.
    (b) Show that the states ψ_{a1}, ψ_{a2}, and ψ_{a3} of Eqs. (8.22)–(8.24) have spin quantum numbers as stated. [The spin quantum numbers are defined by $\hat{\mathbf{S}}_z\psi = m_s\hbar\psi$ and $\hat{\mathbf{S}}^2\psi = s(s+1)\hbar^2\psi$, where $\hat{\mathbf{S}} = \hat{\mathbf{s}}_1 + \hat{\mathbf{s}}_2$.]

14. Prove Eq. (9.21) using the Dirac notation. (Recall that $\Sigma_n |v_n\rangle\langle v_n| = \hat{I}$.)

CHAPTER TEN

The Time Dependent Schrödinger Equation

Up to this point we considered problems in which the essential characteristics of the particle, such as its energy and probability function $\psi(x)$, were not functions of time. The eigenfunction of a particle with energy E_n was obtained by solving the time-independent Schrödinger equation

$$\left(-\frac{\hbar^2}{2m}\nabla^2+V(\mathbf{r})\right)u_n(\mathbf{r})=E_n u_n(\mathbf{r}) \tag{10.1}$$

In the case of a free particle [$V(\mathbf{r})=0$], we found in Section 2.2 that the solution of (10.1) is [see (2.6)] in the form of a plane wave

$$u(\mathbf{r},\mathbf{p})\propto e^{i\mathbf{p}\cdot\mathbf{r}/\hbar} \tag{10.2}$$

We have, however, learned that certain experimental situations such as electron diffraction in crystals, for example, can only be explained by postulating a wavelike wavefunction that, by its nature, must be time-dependent [see (1.20)]:

$$\psi(\mathbf{r},t)\propto e^{i(\mathbf{p}\cdot\mathbf{r}/\hbar-Et/\hbar)} \tag{10.3}$$

We conclude that the time-independent Schrödinger equation does not tell the whole story. What is needed is thus an extension of the Schrödinger equation that accounts explicitly for the effects of time. A clue to the form of the sought equation is obtained by realizing that (10.3) satisfies

$$\left(-\frac{\hbar^2}{2m}\nabla^2+V(\mathbf{r},t)\right)\psi(\mathbf{r},t)=i\hbar\frac{\partial}{\partial t}\psi(\mathbf{r},t) \tag{10.4}$$

or, in general,

$$\hat{\mathcal{H}}(\mathbf{r},-i\hbar\nabla,t)\psi(\mathbf{r},t)=i\hbar\frac{\partial}{\partial t}\psi(\mathbf{r},t) \tag{10.5}$$

Equation (10.5) is the time-dependent Schrödinger equation. The justification for its use is: (a) When the Hamiltonian does not depend on time the predictions of the theory are identical with that of the time-independent Schrödinger equation; (b) the prediction of the theory agrees with experiment; and (c) in the limit $\hbar\to 0$ with classical physics.

We find by substitution that in the special case when $\hat{\mathcal{H}}$ does not depend explicitly on time, the general solution of (10.5) can be taken as

$$\psi(\mathbf{r},t)=\sum_n a_n u_n(\mathbf{r})e^{-iE_nt/\hbar} \tag{10.6}$$

where u_n is the eigenfunction of $\hat{\mathcal{H}}$ with energy E_n:

$$\hat{\mathcal{H}}u_n(\mathbf{r})=E_nu_n(\mathbf{r}) \tag{10.7}$$

and a_n does not depend on time.

The probability of finding the system in the state u_n with energy E_n is then

$$\begin{aligned}P_n&=|\langle\psi(t)|u_n\rangle|^2\\&=|a_n|^2\end{aligned}$$

10.1 THE STATISTICAL INTERPRETATION OF $\psi(\mathbf{r},t)$

The interpretation of $\psi(\mathbf{r},t)$ is similar to that postulated in Section 3.1 for $\psi(\mathbf{r})$. The probability $P(\mathbf{r},t)$ of finding a particle inside a differential volume $d^3\mathbf{r}$ at time t is

$$P(\mathbf{r},t)=|\psi(\mathbf{r},t)|^2\,d^3\mathbf{r} \tag{10.8}$$

Since the probability of finding the particle somewhere in space is unity, we must have

$$\int|\psi(\mathbf{r},t)|^2\,d^3\mathbf{r}=1$$

so that

$$\frac{d}{dt}\int|\psi(\mathbf{r},t)|^2\,d^3\mathbf{r}=0 \tag{10.9}$$

This last statement of the conservation of probability is proved as follows:

$$\frac{d}{dt}\int \psi^* \psi \, d^3\mathbf{r} = \int \frac{\partial}{\partial t}(\psi^* \psi)\, d^3\mathbf{r}$$

$$= \int \left(\psi \frac{\partial \psi^*}{\partial t} + \frac{\partial \psi}{\partial t} \psi^* \right) d^3\mathbf{r}$$

Substituting for $\partial\psi/\partial t$ and $\partial\psi^*/\partial t$ from (10.4) leads to

$$\frac{d}{dt}\int_V |\psi|^2 \, d^3\mathbf{r} = \frac{i\hbar}{2m}\int_V (\psi^* \nabla^2 \psi - \psi \nabla^2 \psi^*)\, d^3\mathbf{r} \tag{10.10}$$

We use Green's theorem[1] to rewrite the last integral as

$$\frac{d}{dt}\int_V |\psi|^2 \, d^3\mathbf{r} = \frac{i\hbar}{2m}\int_V (\psi^* \nabla^2 \psi - \psi \nabla^2 \psi^*)\, d^3\mathbf{r}$$

$$= \frac{i\hbar}{2m}\int_A (\psi^* \nabla \psi - \psi \nabla \psi^*)\cdot \hat{n}\, da \tag{10.11}$$

Since ψ vanishes at infinity, the last integral is zero when V extends over all space. This completes the proof.

10.2 EXPECTATION VALUES OF OPERATORS

The expectation value of some physical quantity whose quantum mechanical operator is $\hat{A}$ is obtained according to (3.12) by

$$\langle A \rangle = \int \psi^*(\mathbf{r}, t) \hat{A} \psi(\mathbf{r}, t)\, d^3\mathbf{r} \tag{10.12}$$

and is, in general, time dependent. The time evolution of $\langle A \rangle$ can also be expressed in a convenient differential form. To do so we start with the Schrödinger equation (10.5) using the Dirac notation:

$$\frac{1}{i\hbar}\hat{\mathcal{H}}|\psi\rangle = |\dot{\psi}\rangle \quad \left(|\dot{\psi}\rangle \equiv \frac{\partial}{\partial t}|\psi\rangle \right) \tag{10.13}$$

so that

$$-\frac{1}{i\hbar}\langle \psi|\hat{\mathcal{H}} = \langle \dot{\psi}| \tag{10.14}$$

and (10.12) is written as

$$\langle A \rangle = \langle \psi | \hat{A} | \psi \rangle \tag{10.15}$$

---

[1] Green's theorem states that for any two scalar functions $f(r)$ and $g(r)$,

$$\int_V (f \nabla^2 g - g \nabla^2 f)\, d^3\mathbf{r} = \int_A (f \nabla g - g \nabla f)\cdot \mathbf{n}\, da$$

A is the surface bounding the arbitrary volume V. $\mathbf{n}$ is the unit outward normal vector, $d^3\mathbf{r}$ and da are, respectively, differential volume and area elements.

Taking the time derivative of (10.15),

$$\frac{d}{dt}\langle A\rangle=\langle\dot{\psi}|\hat{A}|\psi\rangle+\langle\psi|\hat{A}|\dot{\psi}\rangle+\langle\psi|\dot{\hat{A}}|\psi\rangle$$

$$=-\frac{1}{i\hbar}\langle\psi|\hat{\mathcal{H}}\hat{A}|\psi\rangle+\frac{1}{i\hbar}\langle\psi|\hat{A}\hat{\mathcal{H}}|\psi\rangle+\langle\psi|\dot{\hat{A}}|\psi\rangle \qquad (10.16)$$

The last result can be rewritten as

$$\frac{d}{dt}\langle\hat{A}\rangle=\frac{i}{\hbar}\langle[\hat{\mathcal{H}},\hat{A}]\rangle+\left\langle\frac{\partial\hat{A}}{\partial t}\right\rangle \qquad (10.17)$$

which is the sought result. We next make use of (10.17) to prove some important results.

Ehrenfest's Theorem

According to Ehrenfest's theorem the classical equations

$$m\frac{d\mathbf{r}}{dt}=\mathbf{p} \quad \text{and} \quad \frac{d\mathbf{p}}{dt}=-\nabla V \qquad (10.18)$$

[$V(\mathbf{r})$ is the potential energy function] are also valid in quantum mechanics, provided we replace all the classical quantities by the expectation values of their corresponding quantum mechanical operators.

Proof: We apply (10.17) to the operator x.

$$\frac{d\langle x\rangle}{dt}=\frac{i}{\hbar}\langle[\hat{\mathcal{H}},x]\rangle$$

$$=\frac{i}{\hbar}\left\langle\left[\left(-\frac{\hbar^2}{2m}\nabla^2+V(\mathbf{r})\right),x\right]\right\rangle$$

$$=\frac{i}{\hbar}\left\langle\left[-\frac{\hbar^2}{2m}\nabla^2,x\right]\right\rangle$$

$$=-\frac{i\hbar}{2m}\int\psi^*(\nabla^2x-x\nabla^2)\psi\,d^3\mathbf{r}$$

but $\nabla^2x=x\nabla^2-2(\partial/\partial x)$; therefore,

$$\frac{d\langle x\rangle}{dt}=-\frac{i\hbar}{m}\int\psi^*\frac{\partial}{\partial x}\psi\,d^3\mathbf{r}$$

$$=\frac{1}{m}\langle p_x\rangle$$

where we used $\hat{p}_x=-i\hbar(\partial/\partial x)$ and $[V(\mathbf{r}),x]=0$. This completes the proof of the first of relations (10.18).

The second proof starts, again, with (10.17) applied to $\hat{p}_x$:

$$\begin{aligned}
\frac{d\langle \hat{p}_x\rangle}{dt} &= \frac{i}{\hbar}\langle[\hat{\mathcal{H}}, \hat{p}_x]\rangle \\
&= \frac{i}{\hbar}\left\langle\left[\left(-\frac{\hbar^2}{2m}\nabla^2+V\right), -i\hbar\frac{\partial}{\partial x}\right]\right\rangle \\
&= \left\langle\left[V, \frac{\partial}{\partial x}\right]\right\rangle \\
&= \int\left[\psi^*\left(V\frac{\partial}{\partial x}\right)\psi - \psi^*\frac{\partial}{\partial x}(V\psi)\right]d^3\mathbf{r} \\
&= -\int\psi^*\frac{\partial V}{\partial x}\psi\, d^3\mathbf{r} \\
&= -\left\langle\frac{\partial V}{\partial x}\right\rangle
\end{aligned}$$

where we used $[\nabla^2, \partial/\partial x]=0$.

This completes the proof.

PROBLEMS

1. If the dominant term in $\psi(\mathbf{r}, t)$ $(r\rightarrow\infty)$ varies as r^{-n}, what values can n possess in order that the integral in (10.11) taken over the surface at infinity is to vanish?
2. Show that if $\psi(\mathbf{r}, t)$ defined by

$$\psi(\mathbf{r}, t)=\left(\frac{1}{2\pi\hbar}\right)^{3/2}\int_{-\infty}^{+\infty} e^{i\mathbf{p}\cdot\mathbf{r}/\hbar}\Phi(\mathbf{p}, t)\, d^3\mathbf{p}.$$

is to satisfy Schrödinger's equation, $\Phi(\mathbf{p}, t)$ satisfies the equation

$$\left(\frac{p^2}{2m}+V\left(\mathbf{r}\rightarrow i\hbar\nabla_p, \mathbf{p}, t\right)\right)\Phi(\mathbf{p}, t)=i\hbar\frac{\partial\Phi(\mathbf{p}, t)}{\partial t}$$

where $\mathbf{r}\rightarrow i\hbar\nabla_p$ means that x_i is to be replaced by $i\hbar(\partial/\partial p_i)$.

Hint: Show that

$$\int_{-\infty}^{+\infty}\frac{\partial\Phi}{\partial p}e^{ipx/\hbar}\, dp = -\frac{ix}{\hbar}\int_{-\infty}^{+\infty}\Phi e^{ipx/\hbar}\, dp$$

for $\Phi(-\infty)=\Phi(+\infty)=0$.

CHAPTER ELEVEN

Perturbation Theory

In the first part of this book we dealt with a variety of problems in which the eigenfunctions and eigenvalues of some operators were obtained. Special emphasis was placed on solutions of the energy eigenvalue problem (the time-independent Schrödinger equation):

$$\hat{\mathcal{H}}_0 u_m = E_m u_m$$

In this chapter we consider the effect on the energies E_m and on the eigenfunction u_m of small perturbations of the Hamiltonian $\hat{\mathcal{H}}_0$. Such perturbations arise in practice from the presence of electric and magnetic fields or from the interactions with other particles when these effects are not included in the unperturbed Hamiltonian $\hat{\mathcal{H}}_0$. Since exact solutions of the full Schrödinger equation are seldom possible, the perturbation methods discussed below are some of the main practical tools in quantum mechanics.

11.1 TIME-INDEPENDENT PERTURBATION THEORY

The problem we pose is the following: Given a Hamiltonian $\hat{\mathcal{H}}_0$, its eigenfunctions u_m, and the eigenvalues E_m so that

$$\hat{\mathcal{H}}_0 u_m = E_m u_m \tag{11.1}$$

What are the new eigenfunctions and eigenvalues when the Hamiltonian is perturbed from $\hat{\mathcal{H}}_0$ to $\hat{\mathcal{H}}_0 + \hat{\mathcal{H}}'$? One method of solution would be to diagonalize the matrix of $\hat{\mathcal{H}}_0 + \hat{\mathcal{H}}'$ in some arbitrary representation as discussed in Section 9.3. This method is often used in practice. If $\hat{\mathcal{H}}_0 \gg \hat{\mathcal{H}}'$, we can employ perturbation techniques and obtain expressions for the perturbation of u_m and E_m to any desired order. This is the concern of this section.

First-Order Perturbation

The Hamiltonian operator is taken as $\hat{\mathcal{H}}_0 + \lambda\hat{\mathcal{H}}'$, where $0<\lambda<1$ is a parameter that "turns the perturbation on" ($\lambda=1$) or "off" ($\lambda=0$). We are looking for the energies W and functions ψ that satisfy

$$(\hat{\mathcal{H}}_0 + \lambda\hat{\mathcal{H}}')\psi = W\psi \tag{11.2}$$

Expanding ψ and W in a power series in λ,

$$\psi = \psi_0 + \lambda\psi_1 + \lambda^2\psi_2 + \cdots$$
$$W = W_0 + \lambda W_1 + \lambda^2 W_2 + \cdots \tag{11.3}$$

and substituting in (11.2), gives

$$(\hat{\mathcal{H}}_0 + \lambda\hat{\mathcal{H}}')(\psi_0 + \lambda\psi_1 + \lambda^2\psi_2 + \cdots) = (W_0 + \lambda W_1 + \lambda^2 W_2 + \cdots) \times (\psi_0 + \lambda\psi_1 + \lambda^2\psi_2 + \cdots)$$

Equating the coefficients for λ^0, λ^1, and λ^2 on both sides of the last equation gives

$$\hat{\mathcal{H}}_0\psi_0 = W_0\psi_0$$
$$\hat{\mathcal{H}}_0\psi_1 + \hat{\mathcal{H}}'\psi_0 = W_0\psi_1 + W_1\psi_0 \tag{11.4a}$$
$$\hat{\mathcal{H}}_0\psi_2 + \hat{\mathcal{H}}'\psi_1 = W_0\psi_2 + W_1\psi_1 + W_2\psi_0$$

respectively. Comparing the first of Eqs. (11.4*a*) with (11.1) identifies the zero-order solutions as

$$\psi_0 = u_m$$
$$W_0 = E_m \tag{11.4b}$$

where u_m and E_m are the eigenfunctions and eigenvalues at the absence of perturbation. Next we expand ψ_1 in terms of u_n as

$$\psi_1 = \sum_n a_n^{(1)} u_n \tag{11.5}$$

and substitute it in the second of Eqs. (11.4*a*). The result is

$$\sum_n a_n^{(1)} E_n u_n + \hat{\mathcal{H}}' u_m = E_m \sum_n a_n^{(1)} u_n + W_1 u_m$$

premultiplying by u_k^*, and integrating and recalling that $\langle u_n | u_k \rangle = \delta_{kn}$, gives

$$E_k a_k^{(1)} + \mathcal{H}'_{km} = E_m a_k^{(1)} + W_1 \delta_{km} \tag{11.6}$$

which for $k \neq m$ yields

$$a_k^{(1)} = \frac{\mathcal{H}'_{km}}{E_m - E_k} \qquad (k \neq m) \tag{11.7}$$

Putting $k = m$ in (11.6) gives

$$W_1 = \mathcal{H}'_{mm} \tag{11.8}$$

According to (11.3) and (11.4*b*), W_1 is the first-order correction to energy E_m. We still need to evaluate $a_m^{(1)}$. This is done by requiring that the first-order corrected wavefunction $\psi = u_m + \psi_1$ be normalized to unity:

$$\int \left(u_m + \lambda \sum_n a_n^{(1)} u_n \right)^* \left(u_m + \lambda \sum_s a_s^{(1)} u_s \right) dv = 1 + \lambda a_m^{(1)} + \lambda a_m^{*(1)} + \lambda^2 \sum_n a_n^{(1)} a_n^{*(1)}$$

$$= 1 \qquad (11.9)$$

which, neglecting the second-order term, gives $a_m^{(1)} = 0$ as a possible solution. The eigenfunction and eigenvalue to first-order perturbation are thus given as

$$\psi = u_m + \sum_{k \neq m} \frac{\mathcal{H}'_{km}}{E_m - E_k} u_k \qquad (11.10a)$$

$$W = E_m + \mathcal{H}'_{mm} \qquad (11.10b)$$

Second-Order Perturbation

Our aim here is to obtain expressions for W_2 and ψ_2. The second order correction to the eigenfunction, ψ_2, may be expanded as

$$\psi_2 = \sum_n a_n^{(2)} u_n$$

This expansion is next used in the third of Eqs. (11.4*a*)

$$\sum_n a_n^{(2)} E_n u_n + \hat{\mathcal{H}}' \sum_n a_n^{(1)} u_n = \sum_n a_n^{(2)} E_m u_n + W_1 \psi_1 + W_2 u_m$$

Substituting for ψ_1 its expansion according to (11.5), then multiplying by u_k^* and integrating, results in

$$a_k^{(2)} E_k + \sum_n a_n^{(1)} \mathcal{H}'_{kn} = a_k^{(2)} E_m + W_1 a_k^{(1)} + W_2 \delta_{mk} \qquad (11.11)$$

Setting $k = m$ gives

$$W_2 = \sum_n a_n^{(1)} \mathcal{H}'_{mn} - W_1 a_m^{(1)}$$

$$= \sum_{n \neq m} a_n^{(1)} \mathcal{H}'_{mn} + a_m^{(1)} \mathcal{H}'_{mm} - W_1 a_m^{(1)}$$

Using (11.7) for $a_n^{(1)}$ and (11.8) for W_1, the last two terms cancel each other with the result

$$W_2 = \sum_{n \neq m} \frac{|\mathcal{H}'_{mn}|^2}{E_m - E_n} \qquad (11.12)$$

Going back to (11.11) for the case $k \neq m$, using (11.7), (11.8), and the result $a_m^{(1)} = 0$, gives

$$\underset{k \neq m}{a_k^{(2)}} = \sum_{n \neq m} \frac{\mathcal{H}'_{kn} \mathcal{H}'_{nm}}{(E_m - E_n)(E_m - E_k)} - \frac{\mathcal{H}'_{mm} \mathcal{H}'_{km}}{(E_m - E_k)^2}$$

To find $a_m^{(2)}$ we go back to the normalization integral (11.9). Adding the second-order correction to ψ gives

$$\int\left(u_m+\lambda\sum_n a_n^{(1)}u_n+\lambda^2\sum_n a_n^{(2)}u_n\right)^*\left(u_m+\lambda\sum_s a_s^{(1)}u_s+\lambda^2\sum_s a_s^{(2)}u_s\right)dv=1$$

Using the result $a_m^{(1)}=0$, the last equation yields

$$a_m^{(2)}=-\tfrac{1}{2}\sum_n|a_n^{(1)}|^2$$

$$=-\tfrac{1}{2}\sum_{n\neq m}\frac{|\mathcal{H}'_{mn}|^2}{(E_m-E_n)^2} \tag{11.13}$$

Finally, we let $\lambda\to 1$ and write the eigenfunction and the energy, to second order, as

$$\psi=u_m+\sum_{k\neq m}\frac{\mathcal{H}'_{km}}{E_m-E_k}u_k$$

$$+\sum_{k\neq m}\left[\left(\sum_{n\neq m}\frac{\mathcal{H}'_{kn}\mathcal{H}'_{nm}}{(E_m-E_n)(E_m-E_k)}-\frac{\mathcal{H}'_{mm}\mathcal{H}'_{km}}{(E_m-E_k)^2}\right)u_k-\frac{|\mathcal{H}'_{km}|^2}{2(E_m-E_k)^2}u_m\right] \tag{11.14}$$

$$W=E_m+\mathcal{H}'_{mm}+\sum_{n\neq m}\frac{|\mathcal{H}'_{mn}|^2}{E_m-E_n} \tag{11.15}$$

Notice that the second-order correction tends, according to (11.12), to increase the energy separation $|E_m-E_n|$. This fact is often expressed in the physics jargon as "energy levels repel each other."

11.2 TIME-DEPENDENT PERTURBATION THEORY

Time-dependent perturbation theory is the main analytical tool for treating the transitions of quantum mechanical systems from one energy state to another.

We have shown [see (10.6)] that if the Hamiltonian of a system does not depend on time, the general solution of the Schrödinger equation

$$\hat{\mathcal{H}}_0\psi(\mathbf{r},t)=i\hbar\frac{\partial\psi(\mathbf{r},t)}{\partial t} \tag{11.16}$$

is in the form of

$$\psi(\mathbf{r},t)=\sum_n a_n u_n(\mathbf{r})e^{-iE_nt/\hbar} \tag{11.17}$$

where the coefficients a_n are constant and $\hat{\mathcal{H}}_0 u_n=E_n u_n$. If the system is found to possess at some time, say $t=0$, an energy E_m, then we have

$$\left.\begin{aligned} a_m &= 1 \\ a_n &= 0 \end{aligned}\right\} \quad (n \neq m) \tag{11.18}$$

for all subsequent times.

Let us assume next that the system is perturbed in such a way that the Hamiltonian is modified from $\hat{\mathcal{H}}_0$ to

$$\hat{\mathcal{H}}(t) = \hat{\mathcal{H}}_0 + \hat{\mathcal{H}}'(t) \tag{11.19}$$

The wavefunction $\psi(t)$ is now a solution of the Schrödinger equation

$$\left[\hat{\mathcal{H}}_0 + \hat{\mathcal{H}}'(t)\right]\psi = i\hbar\frac{\partial\psi}{\partial t} \tag{11.20}$$

At some particular time t, we may, using the completeness property, expand $\psi(\mathbf{r}, t)$ in terms of u_n:

$$\psi(\mathbf{r}, t) = \sum_n a_n(t)u_n(\mathbf{r})e^{-iE_nt/\hbar} \tag{11.21}$$

Since the Hamiltonian is time dependent, the coefficients a_n, unlike (11.17), are now functions of time. The significance of this time dependence is of fundamental importance. Let us assume that a measurement of the unperturbed energy at some time, say $t=0$, yields E_m. We thus have

$$\left.\begin{aligned} a_m(0) &= 1 \\ a_n(0) &= 0 \end{aligned}\right\} \quad (n \neq m) \tag{11.22}$$

Since the coefficients a_n evolve with time, a subsequent measurement of the energy, say at time t, may yield the value E_k. The probability of such an event is $|a_k(t)|^2$, which is thus the probability of finding the system in the state k at time t given that at $t=0$ it occupied the state m. The solution of the time-dependent Schrödinger equation thus provides a description of the manner in which the probability of finding the system in the various eigenstates u_n of $\hat{\mathcal{H}}_0$ evolves with time under the influence of the perturbation $\hat{\mathcal{H}}'(t)$.

To describe the evolution of the system we thus need to solve for the coefficients $a_n(t)$. We substitute (11.21) in (11.20), obtaining

$$\sum_n u_n\left[a_n\left(-\frac{iE_n}{\hbar}\right)e^{-iE_nt/\hbar} + \dot{a}_n e^{-iE_nt/\hbar}\right] = -\frac{i}{\hbar}\sum_n a_n\left(\hat{\mathcal{H}}_0 + \hat{\mathcal{H}}'\right)u_n e^{-iE_nt/\hbar}$$

which after multiplying by u_k^* and integrating becomes

$$\dot{a}_k = -\frac{i}{\hbar}\sum_n a_n \mathcal{H}'_{kn}(t)e^{i\omega_{kn}t} \tag{11.23}$$

where ω_{kn} is defined by

$$\omega_{kn} \equiv \frac{E_k - E_n}{\hbar}$$

Up to this point the analysis is exact, and solving Eqs. (11.23) is fully equivalent to a solution of the Schrödinger equation. In a manner similar to that used in Section 11.1, we introduce the "turning on" parameter λ by taking the perturbation as $\lambda\hat{\mathcal{H}}'$ so that the Hamiltonian becomes

$$\hat{\mathcal{H}}_0 + \lambda\hat{\mathcal{H}}'(t)$$

The power-series expansion for a_n is written as

$$a_n = a_n^{(0)} + \lambda a_n^{(1)} + \lambda^2 a_n^{(2)} + \cdots$$

which, when substituted in (11.23), becomes

$$\dot{a}_k^{(0)} + \lambda\dot{a}_k^{(1)} + \lambda^2\dot{a}_k^{(2)} + \cdots = -\frac{i}{\hbar}\sum_n \left(a_n^{(0)} + \lambda a_n^{(1)} + \lambda^2 a_n^{(2)} + \cdots\right)\lambda\mathcal{H}'_{kn}e^{i\omega_{kn}t}$$

Equating the same powers of λ results in the set of relations

$$\begin{aligned}
\dot{a}_k^{(0)} &= 0\\
\dot{a}_k^{(1)} &= -\frac{i}{\hbar}\sum_n a_n^{(0)}\mathcal{H}'_{kn}(t)e^{i\omega_{kn}t}\\
\dot{a}_k^{(2)} &= -\frac{i}{\hbar}\sum_n a_n^{(1)}\mathcal{H}'_{kn}(t)e^{i\omega_{kn}t}\\
&\vdots\\
\dot{a}_k^{(s)} &= -\frac{i}{\hbar}\sum_n a_n^{(s-1)}\mathcal{H}'_{kn}(t)e^{i\omega_{kn}t}
\end{aligned}\tag{11.24}$$

The solution of the zero-order equation is $a_k^{(0)} = \text{constant}$. The $a_k^{(0)}$ are thus the initial values for the problem. These are chosen as

$$\begin{aligned}
a_m^{(0)} &= 1\\
a_n^{(0)} &= 0 \quad (n \neq m)
\end{aligned}$$

so that at $t=0$ the system is known with certainty to occupy a state with energy E_m. The second of Eqs. (11.24) now reduces to

$$\dot{a}_k^{(1)} = -\frac{i}{\hbar}\mathcal{H}'_{km}e^{i\omega_{km}t} \tag{11.25}$$

$|a_k^{(1)}(t)|^2$ is the probability to first order of finding the system at time t in the state k given that at $t=0$ it is in the state m.

Harmonic Perturbation

As a special case we consider a perturbation that varies sinusoidally with time:

$$\hat{\mathcal{H}}'(t) = \hat{H}'e^{-i\omega t} + (\hat{H}')^\dagger e^{i\omega t} \tag{11.26a}$$

The breakdown of $\hat{\mathcal{H}}'(t)$ into the two parts is done so as to ensure its Hermiticity. The result of substituting $\hat{\mathcal{H}}'(t)$ into (11.25) and performing the integration is

$$a_k^{(1)}(t) = \int_0^t \left(-\frac{i}{\hbar}\right) H'_{km}(t') e^{i\omega_{km}t'}\, dt'$$

$$= -\hbar^{-1}\left(H'_{km}\frac{e^{i(\omega_{km}-\omega)t}-1}{\omega_{km}-\omega} + H'^{*}_{mk}\frac{e^{i(\omega_{km}+\omega)t}-1}{\omega_{km}+\omega}\right) \qquad (11.26b)$$

where the lower limit of the integration is zero, since $a_{k\neq m}^{(1)}(0)=0$. We limit ourselves next to a case in which ω is nearly equal to $|\omega_{km}|$—that is, $\hbar\omega \simeq |E_k - E_m|$. The transition probability from the state m to k is then

$$|a_k^{(1)}|^2 \simeq \frac{4|H'_{km}|^2}{\hbar^2}\,\frac{\sin^2[\frac{1}{2}(\omega_{km}\pm\omega)t]}{(\omega_{km}\pm\omega)^2} \qquad (11.27)$$

where the $(-)$ sign is to be used when $\omega_{km}\simeq\omega$ while the $(+)$ sign applies when $\omega_{mk}\simeq\omega$. The cross terms with a denominator involving the product $(\omega_{km}+\omega)(\omega_{km}-\omega)$ have been left out since, for the conditions of interest, $|\omega_{km}|\simeq\omega$, their contribution can be neglected. The first expression on the right side of (11.26*b*) dominates when $E_k > E_m$ and $E_k - E_m \sim \hbar\omega$, while the second expression dominates when $E_k < E_m$ and $E_m - E_k \sim \hbar\omega$. The harmonic perturbation can thus cause both upward and downward transitions from state m to states k, separated in energy by $\sim\hbar\omega$.

To be specific, let us calculate the transition probability from m to a *group* of states clustered about state k, where $E_k > E_m$. Let the density of these final states per unit of ω_{km} be $\rho(\omega_{km})$. Since $\omega_{km}\simeq\omega$, we use the $(-)$ sign in (11.27) and obtain

$$|a_k^{(1)}|^2 = \frac{1}{\hbar^2}\int_{-\infty}^{+\infty} |H'_{km}|^2 \frac{\sin^2[\frac{1}{2}(\omega_{km}-\omega)t]}{[\frac{1}{2}(\omega_{km}-\omega)]^2}\rho(\omega_{km})\, d\omega_{km} \qquad (11.28)$$

If $|H'_{km}|^2$ is not a strong function of the final state k, we can take it outside the integral sign. The remaining integrand is then a product of two functions:

$$(1)\quad g(\omega_{km},t) = \frac{\sin^2[\frac{1}{2}(\omega_{km}-\omega)t]}{[\frac{1}{2}(\omega_{km}-\omega)]^2} \quad \text{and} \quad (2)\quad \rho(\omega_{km})$$

These functions are plotted in Fig. 11.1, the independent variable being ω_{km}. The interval in ω_{km}, where the function

$$g(\omega_{km},t) = \frac{\sin^2[\frac{1}{2}(\omega_{km}-\omega)t]}{[\frac{1}{2}(\omega_{km}-\omega)]^2} \qquad (11.29)$$

is appreciable, is $\sim 2\pi/t$ and can be made arbitrarily small by increasing the time of observation t. The area under this function is

$$\int_{-\infty}^{\infty} \frac{\sin^2[\frac{1}{2}(\omega_{km}-\omega)t]}{[\frac{1}{2}(\omega_{km}-\omega)]^2}\, d\omega_{km} = 2\pi t \qquad (11.30)$$

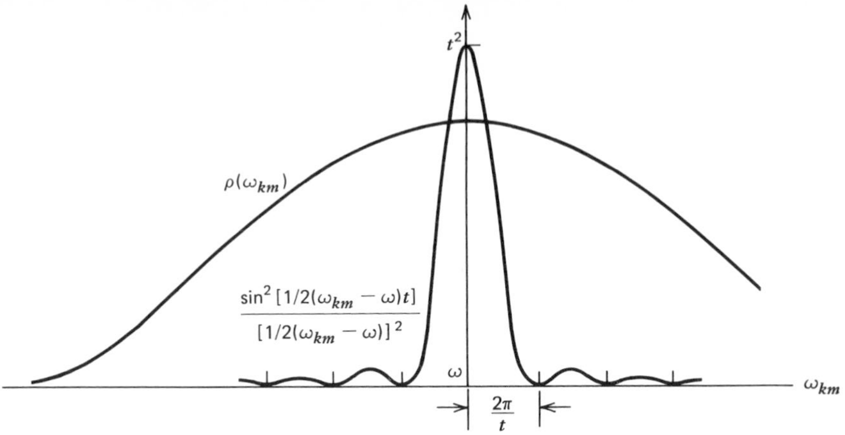

Figure 11.1 The two components: the density of final states function $p(\omega_{km})$ and the transition probability function $g(\omega_{km}, t) \equiv \sin^2[\frac{1}{2}(\omega_{km}-\omega)t]/[\frac{1}{2}(\omega_{km}-\omega)]^2$ involved in the integral of Eq. (11.28).

Let us consider the case when t is big enough so that $2\pi/t$, the width of $g(\omega_{km}, t)$, is very much smaller than $\Delta\omega$, the width of $\rho(\omega_{km})$. The integral (11.28) becomes

$$|a_k^{(1)}(t)|^2 = \frac{1}{\hbar^2}|H'_{km}|^2 \rho(\omega_{km}=\omega)\int_{-\infty}^{\infty} \frac{\sin^2\left[\frac{1}{2}(\omega_{km}-\omega)t\right]}{\left[\frac{1}{2}(\omega_{km}-\omega)\right]^2} d\omega_{km}$$
$$= \frac{2\pi}{\hbar^2}|H'_{km}|^2 \rho(\omega_{km}=\omega)t \qquad (11.31)$$

and the transition rate per unit time is

$$W_{m\to k} = \frac{d}{dt}|a_k^{(1)}(t)|^2$$
$$= \frac{2\pi}{\hbar}|H'_{km}|^2 \rho(E=E_m \pm \hbar\omega) \qquad (11.32)$$

where $\rho(E)$ is the density of final states expressed as a function of energy $[\rho(\omega)=\hbar\rho(E)]$. The minus sign is to be used when $E_k < E_m$ and is due to the second term on the right side of (11.26).

The result (11.32) is consistent with writing the transition rate from $|m\rangle$ to $|k\rangle$, where $|k\rangle$ is a single state *within* a *continuum*

$$W_{m\to k} = \frac{d}{dt}|a_k^{(1)}(t)|^2$$
$$= \frac{2\pi}{\hbar}|H'_{km}|^2 \delta(E_k - E_m - \hbar\omega) \qquad (11.33)$$

If $E_k - E_m \simeq -\hbar\omega$ we replace the argument of the δ function by $(E_k - E_m + \hbar\omega)$. In writing (11.33), we used the fact that for sufficiently long t, $\sin^2(xt/2)/(x/2)^2 \to 2\pi t\delta(x)$.

Equation (11.32) is known as Fermi's Golden Rule. We must remember when applying (11.32) that it applies to transitions from a *single* state m to a *continuum* of states k. If the final state k is single and not part of a continuum, we need to go back to (11.27).

Step Function Perturbation

A second case of interest is one in which the perturbation has the form of a step function applied at $t=0$, that is,

$$\hat{\mathcal{H}}'(t)=0 \quad (t\leq 0)$$
$$\hat{\mathcal{H}}'(t)=H' \quad (t\geq 0) \tag{11.34}$$

This situation may be regarded as a limiting case of the harmonic perturbation discussed above with $\omega \to 0$.

Using the second of Eqs. (11.24) with $a_n^{(1)}(0)=\delta_{nm}$ (i.e., the system is initially in the state m) and repeating the steps leading to (11.32), yields

$$\begin{aligned} W_{m\to k} &= \frac{2\pi}{\hbar}|H'_{km}|^2\delta(E_m - E_k) \\ &= \frac{1}{\hbar^2}|H'_{km}|^2\delta(\nu_m - \nu_k) \end{aligned} \tag{11.35}$$

The form of $W_{m\to k}$ is similar to that of (11.33). The important difference is in the argument of the delta function, which involves an initial state (m) and a final state of the same energy. It must be emphasized that (11.35) applies, as does (11.33), to a case where the single state k is part of a continuum. The total transition rate out of $|m\rangle$ is obtained by summing $W_{m\to k}$ over all final states.

Limits of Validity of the Golden Rule

Two conditions were used in deriving Eqs. (11.31) and (11.33). The first was that $2\pi/t$ be small compared with the width $2\pi\Delta\nu$ of $\rho(\omega_{km})$. The second condition results from our use of first-order perturbation theory and requires that $|a_k^{(1)}(t)|^2 \ll 1$; otherwise higher-order terms must be considered. This second condition can be stated using (11.27) as

$$\frac{H'_{km}}{\hbar} \ll \frac{1}{t} \tag{11.36}$$

Its physical significance is that the results of first-order perturbation theory are only valid for times short enough so that the probability for transitions out of the initial state m is very small compared with unity. Combining these two conditions leads to

$$\frac{|H'_{km}|}{\hbar} \ll \frac{1}{t} \ll \Delta\nu$$

as the validity limits for Eqs. (11.33) and (11.35). Cases in which the last condition is not fulfilled have to be treated separately. This is done in Section 13.1.

11.3 THE DENSITY MATRIX FORMALISM

The density matrix formalism to be introduced in this section is one of the most powerful and widely used methods for describing the time evolution of large systems of indistinguishable atomic particles, because it is constructed in such a way that it is especially convenient for ensemble averaging. We will put it to work in Chapter 13 in describing absorption and dispersion of electromagnetic radiation in atomic media.

Consider the wavefunction $\psi(\mathbf{r}, t)$ of a single isolated atomic system. This function satisfies the time-dependent Schrödinger equation

$$\hat{\mathcal{H}}(t)\psi(\mathbf{r}, t) = i\hbar \frac{\partial \psi(\mathbf{r}, t)}{\partial t} \tag{11.37}$$

$\psi(\mathbf{r}, t)$ can be expanded in some *arbitrary* but complete orthonormal set $u_n(\mathbf{r})$ according to

$$\begin{aligned} \psi(\mathbf{r}, t) &= \sum_n C_n(t) u_n(\mathbf{r}) \\ &= \sum_n C_n |n\rangle \end{aligned} \tag{11.38}$$

Using

$$\langle n | m \rangle = \delta_{nm}$$

we obtain from (11.38)

$$C_n = \int u_n^* \psi(\mathbf{r}, t)\, d^3\mathbf{r} \equiv \langle n | \psi \rangle \tag{11.39}$$

The expectation value of some observable A is given by (10.13) as

$$\langle A \rangle = \int \psi^* \hat{A} \psi \, d^3\mathbf{r} = \langle \psi | \hat{A} | \psi \rangle \tag{11.40}$$

which, using (2.32*b*), can be written as

$$\begin{aligned} \langle A \rangle &= \sum_n \sum_m \langle \psi | m \rangle \langle m | \hat{A} | n \rangle \langle n | \psi \rangle \\ &= \sum_n \sum_m C_m^* C_n A_{mn} \end{aligned} \tag{11.41}$$

Equation (11.41) applies to a single isolated atomic entity (to be called, in what follows, the atom). In most real systems the observations involve a very large number of identical atoms. In such cases the measured quantity is not $\langle A \rangle$ but involves an averaging of $\langle A \rangle$ over the ensemble of similar particles. We denote this average by a bar on top of the affected variables:

$$\overline{\langle A\rangle} = \sum_n \sum_m \overline{C_m^* C_n}\, A_{mn} \tag{11.42}$$

We find it convenient to define

$$\rho_{nm} \equiv \overline{C_n C_m^*} \tag{11.43}$$

$$= \overline{\langle n|\psi\rangle\langle\psi|m\rangle} \tag{11.44}$$

so that

$$\overline{\langle A\rangle} = \sum_n \sum_m \rho_{nm} A_{mn} = \sum_n (\rho A)_{mn}$$

$$= \mathrm{tr}(\rho A) \tag{11.45}$$

It follows directly from (11.44) and the definition of matrix products that ρ_{nm} may be viewed formally as the *nm* matrix element of the operator

$$\hat{\rho} = \overline{|\psi\rangle\langle\psi|} \tag{11.46}$$

which is referred to as the *density* operator.

We note that according to (11.43) $\rho_{mn} = \rho_{nm}^*$ so that the operator $\hat{\rho}$ is Hermitian.

It is often advantageous to use a differential equation for obtaining $\hat{\rho}(t)$. We start with (11.46):

$$\frac{d\hat{\rho}}{dt} = \overline{|\dot{\psi}\rangle\langle\psi|} + \overline{|\psi\rangle\langle\dot{\psi}|} \tag{11.47a}$$

We can rewrite (11.37) as

$$|\dot{\psi}\rangle = \frac{1}{i\hbar}\hat{\mathcal{H}}|\psi\rangle \tag{11.47b}$$

Using the Hermiticity of $\hat{\mathcal{H}}$, we write[1]

$$\langle\dot{\psi}| = -\frac{1}{i\hbar}\langle\psi|\hat{\mathcal{H}} \tag{11.47c}$$

which substituted in (11.47) gives

$$\frac{d\hat{\rho}}{dt} = \frac{1}{i\hbar}\hat{\mathcal{H}}\overline{|\psi\rangle\langle\psi|} - \frac{1}{i\hbar}\overline{|\psi\rangle\langle\psi|}\hat{\mathcal{H}}$$

$$= \frac{1}{i\hbar}\left[\hat{\mathcal{H}}, \overline{|\psi\rangle\langle\psi|}\right]$$

Using (11.46),

$$\frac{d\hat{\rho}}{dt} = \frac{1}{i\hbar}\left[\hat{\mathcal{H}}, \hat{\rho}\right] \tag{11.48}$$

---

[1]Equation (11.47*c*) is a formal and concise way of stating that since $\hat{\mathcal{H}}$ is Hermitian,

$$\langle\dot{\psi}|f\rangle = \left\langle \frac{1}{i\hbar}\hat{\mathcal{H}}\psi \,\middle|\, f\right\rangle = -\frac{1}{i\hbar}\langle\psi|\hat{\mathcal{H}}|f\rangle$$

where f and ψ are arbitrary state functions.

In practice one solves the series of equations

$$\frac{d\rho_{nm}}{dt}=\frac{1}{i\hbar}\left[\hat{\mathcal{H}},\hat{\rho}\right]_{nm}+\text{relaxation terms} \tag{11.49}$$

The relaxation terms are added phenomenologically to account for the ensemble aspects of the problem. A representative example of such an application will be found in Chapter 13, where (11.49) will be used to describe the absorption and dispersion of electromagnetic waves in atomic media.

PROBLEMS

1. According to Eq. (11.27) or Fig. 11.1, a transition can take place due to an electrical field oscillating at a radian frequency ω between two states k and m where $E_k - E_m = \omega + \delta$. The energy discrepancy δ can be as large as $\sim 2\pi/t$ where t is the observation time.

   Is this result a violation of the law of conservation of energy? Is it consistent with the uncertainty principle relating the measurement of time and energy?

2. Consider a circularly polarized electric field

   $$E_x = E_0 \cos\omega t$$
   $$E_y = E_0 \sin\omega t$$

   interacting with hydrogenic atoms initially in the state $|n, l, m=0\rangle$ and causing an induced transition to the state $|n', l', m'\rangle$.

   What are the necessary relations between n', l', m' and n, l, m for a transition to take place when:
   (a) $E_{n'} > E_n$
   (b) $E_{n'} < E_n$

3. Same as Problem 2 except that the sense of circular polarization of the applied is reversed; that is,

   $$E_x = E_0 \cos\omega t$$
   $$E_y = -E_0 \sin\omega t$$

4. Same as Problem 2 except that now the field is linearly polarized, in the z direction; that is,

   $$\mathbf{E} = \hat{z}E_0 \cos\omega t$$

5. (a) Show that one can resolve a linearly polarized electric field, say

   $$\mathbf{E} = \hat{x}E_0 \cos\omega t$$

   into two oppositely circularly polarized fields in the $x-y$ plane.
   (b) What are the selection rules, that is, the necessary relations between n, l, m and n', l', m' for transitions to take place between states $|n, l, m$ and $|n', l', m'\rangle$ due to this field?

CHAPTER TWELVE

The Interaction of Electromagnetic Radiation with Atomic Systems

In this chapter we consider the interaction of atomic systems with electromagnetic fields. We will apply these concepts to describe the processes of spontaneous and induced transitions and the phenomena of absorption and amplification of radiation. The material of this chapter will serve as background for the treatment of laser oscillators.

12.1 SOME BASIC ELECTROMAGNETIC BACKGROUND

The Maxwell equations describing the propagation of electromagnetic fields are

$$\nabla \times \mathbf{E} = -\frac{\partial \mathbf{B}}{\partial t}, \qquad \nabla \cdot \mathbf{D} = \rho$$

$$\nabla \times \mathbf{H} = \mathbf{J} + \frac{\partial \mathbf{D}}{\partial t}, \qquad \nabla \cdot \mathbf{B} = 0 \tag{12.1}$$

where $\mathbf{J}$ is the current density; the other symbols have their conventional definitions. In a homogeneous isotropic medium $\mathbf{B}$ and $\mathbf{D}$ are related to $\mathbf{H}$ and $\mathbf{E}$ by

$$\mathbf{B} = \mu \mathbf{H}$$

$$\mathbf{D} = \varepsilon \mathbf{E} \tag{12.2}$$

where μ and ε are, respectively, the magnetic permeability and the dielectric constant of the medium.

Taking the curl of the first of Eqs. (12.1) and then using the second equation as well as (12.2) results in the case of $\mathbf{J}=0$ in

$$\nabla \times \nabla \times \mathbf{E} = -\mu\varepsilon \frac{\partial^2 \mathbf{E}}{\partial t^2}$$

Using the vector identity $\nabla \times (\nabla \times \mathbf{A}) = -\nabla^2 \mathbf{A} + \nabla \nabla \cdot \mathbf{A}$ in Cartesian coordinates and the fact that in a charge-free medium $\nabla \cdot \mathbf{E} = 0$, the last equation becomes

$$\nabla^2 \mathbf{E} - \mu\varepsilon \frac{\partial^2 \mathbf{E}}{\partial t^2} = 0 \tag{12.3}$$

Equation (12.3) admits solutions of the type

$$\mathbf{E} = \mathbf{E}_0 e^{-i(\omega t - \mathbf{k}\cdot\mathbf{r})} \quad \left(k = \omega\sqrt{\mu\varepsilon}\right) \tag{12.4}$$

corresponding to a wave propagating along the arbitrary $\mathbf{k}$ direction with a wavelength

$$\lambda = 2\pi/k \tag{12.5}$$

with an oscillation frequency

$$\nu = \omega/2\pi \tag{12.6}$$

and a phase velocity

$$v_p = \frac{\omega}{k} = \frac{1}{\sqrt{\mu\varepsilon}} = \frac{c}{\sqrt{(\mu/\mu_0)(\varepsilon/\varepsilon_0)}} = \frac{c}{n} \tag{12.7}$$

where $n = \sqrt{(\mu/\mu_0)(\varepsilon/\varepsilon_0)}$ is the index of refraction of the medium, and $c = (\mu_0\varepsilon_0)^{-1/2}$ is the velocity of light in a vacuum.

An examination of (12.1) shows that since $\nabla \cdot \mathbf{E} = 0$ these equations can be satisfied only if we take $\mathbf{E}_0$ to be normal to the direction of propagation $\mathbf{k}$. The magnetic field $\mathbf{H}$ of the wave (12.4) is obtained from the second of equations (12.1):

$$\mathbf{H} = \sqrt{\varepsilon/\mu}\,(\mathbf{a}_k \times \mathbf{E}_0) e^{-i(\omega t - \mathbf{k}\cdot\mathbf{r})} \tag{12.8}$$

where $\mathbf{a}_k$ is a unit vector in the $\mathbf{k}$ direction. $\mathbf{H}$ is thus perpendicular to $\mathbf{E}$ as well as to $\mathbf{k}$. [The student may derive (12.8) by merely replacing ∇ by $i\mathbf{k}$ in Maxwell's equations, a substitution that is formally correct for waves of the form (12.4). Alternatively, he may verify (12.8) for some simple case by taking $\mathbf{E}_0$ to lie, as an example, in the x direction, and then using the relation

$$\nabla \times \mathbf{E} = -\mu \frac{\partial \mathbf{H}}{\partial t}$$

to obtain $\mathbf{H}$.]

The relative directions of $\mathbf{E}$, $\mathbf{H}$, and $\mathbf{k}$ are illustrated in Fig. 12.1.

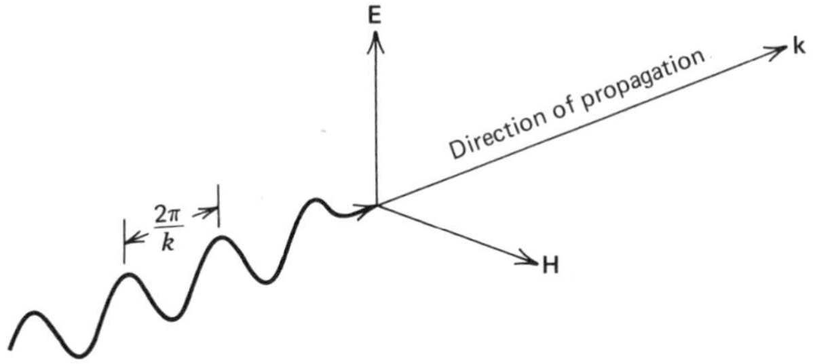

Figure 12.1 The relative direction of **E**, **H** and **k** of a monochromatic plane wave with wavelength $\lambda = 2\pi/k$.

The Energy of Electromagnetic Fields

If we take the scalar product of the first of Eqs. (12.1) with **H**, that of the second equation with **E**, and subtract the second equation from the first, we obtain

$$\mathbf{E}\cdot\nabla\times\mathbf{H}-\mathbf{H}\cdot\nabla\times\mathbf{E}=\left(\mathbf{E}\cdot\frac{\partial\mathbf{D}}{\partial t}+\mathbf{H}\cdot\frac{\partial\mathbf{B}}{\partial t}\right)$$

$$=\frac{1}{2}\frac{\partial}{\partial t}(\varepsilon\mathbf{E}\cdot\mathbf{E}+\mu\mathbf{H}\cdot\mathbf{H})$$

Next we use the vector identity

$$\nabla\cdot(\mathbf{A}\times\mathbf{B})=\mathbf{B}\cdot\nabla\times\mathbf{A}-\mathbf{A}\cdot\nabla\times\mathbf{B}$$

$$-\nabla\cdot(\mathbf{E}\times\mathbf{H})=\frac{1}{2}\frac{\partial}{\partial t}(\varepsilon\mathbf{E}\cdot\mathbf{E}+\mu\mathbf{H}\cdot\mathbf{H}) \tag{12.9}$$

and applying the Stokes theorem,

$$\int_V(\nabla\cdot\mathbf{A})\,d^3\mathbf{r}=\int\mathbf{A}\cdot\mathbf{n}\,da \tag{12.10}$$

to (12.9)

$$-\int_S(\mathbf{E}\times\mathbf{H})\cdot\mathbf{n}\,da=\int_V\frac{\partial}{\partial t}\left(\frac{\varepsilon}{2}\mathbf{E}\cdot\mathbf{E}+\frac{\mu}{2}\mathbf{H}\cdot\mathbf{H}\right)d^3\mathbf{r} \tag{12.11}$$

where V is any *arbitrary* volume, S is the surface bounding V, **n** is the outward normal unit vector on S, and da is the differential surface area element. The magnitude of $\mathbf{E}\times\mathbf{H}$—the Poynting vector—corresponds to the power flow per unit area. The direction of $\mathbf{E}\times\mathbf{H}$ is that of the power flow. The integral over S in (12.11) is thus equal to the total power flow into the volume V. Conservation of energy decrees that, if the medium is lossless, this power be equal to the rate of increase of the stored energy. It thus follows that the total energy stored in V is

$$\mathcal{E}=\tfrac{1}{2}\int_V(\varepsilon\mathbf{E}\cdot\mathbf{E}+\mu\mathbf{H}\cdot\mathbf{H})\,d^3\mathbf{r} \tag{12.12}$$

12.2 QUANTIZATION OF ELECTROMAGNETIC MODES

We consider next the modes of a plane wave resonator. To understand how a resonator works, consider the configuration shown in Fig. 12.2, which consists of two plane parallel conducting walls and an electromagnetic field in the intervening space. Let the electric field of the mode be parallel to the y axis. Since E_y needs to be zero at the perfectly conducting walls at $z=0$ and $z=L$, we take the mode field as

$$\mathbf{E}_l = \mathbf{j}\sqrt{(2/V\varepsilon)}\, p_l(t) \sin k_l z \qquad (12.13a)$$

where V is the total volume occupied by the mode, $\mathbf{j}$ is a unit vector along the y axis. A mode in the form of (12.13a) can be synthesized by a superposition of two traveling plane waves of the form (12.4). To insure that $\mathbf{E}$ vanishes at $z=L$, we must restrict k_l to a member of the set $l\pi/L$, where l is some integer. The integer l is thus equal to the number of half wavelengths contained in L. The magnetic field H must be maximum at the walls and perpendicular to $\mathbf{E}$. This can be satisfied by

$$\mathbf{H}_l = \mathbf{i}\omega_l\sqrt{(2/V\mu)}\, q_l(t) \cos k_l z \qquad (12.13b)$$

where $\mathbf{i}$ is a unit vector along the x direction and $\omega_l = k_l(\mu\varepsilon)^{-1/2}$. If we substitute (12.13) in the first of Maxwell's equations (12.1), we obtain

$$p_l = \frac{dq_l}{dt} \qquad (12.14)$$

Using the second of Eqs. (12.1) gives

$$\omega_l^2 q_l = -\frac{dp_l}{dt} \qquad (12.15)$$

From (12.14) and (12.15) one obtains

$$\omega_l^2 q_l = -\ddot{q}_l, \qquad \omega_l^2 p_l = -\ddot{p}_l$$

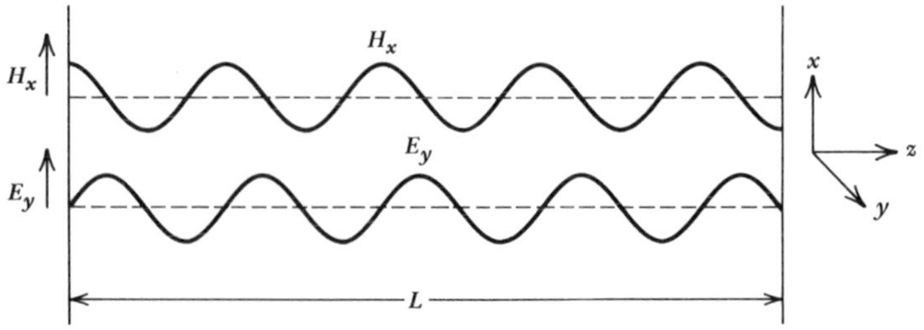

Figure 12.2 A mode of the electromagnetic field established between two perfectly reflecting walls spaced by L. In order for the electric field E_y to vanish at $z=0$ and $z=L$, the distance L must correspond to an integral number of half wavelengths—that is, $L=l(\lambda/2)$ or, equivalently, $k_l L = m\pi$.

It follows that, classically, the time dependence of $q_l(t)$ is of the form

$$q_l(t)=q_l(0)\cos\omega_l t+\frac{\dot{q}_l(0)}{\omega_l}\sin\omega_l t \tag{12.16}$$

Using (12.13), the mode energy (12.12) becomes

$$\begin{aligned}\mathcal{E}_l&=\tfrac{1}{2}\int_V(\mu\mathbf{H}_l\cdot\mathbf{H}_l+\varepsilon\mathbf{E}_l\cdot\mathbf{E}_l)\,dv\\&=\tfrac{1}{2}\left[p_l^2(t)+\omega_l^2q_l^2(t)\right]\end{aligned} \tag{12.17}$$

Compare (12.17) to the expression (5.8) for the Hamiltonian of a harmonic oscillator:

$$\hat{\mathcal{H}}_{\text{ho}}=\frac{\hat{p}^2}{2m}+\tfrac{1}{2}\omega^2mx^2 \tag{12.18}$$

where $K=\omega^2m$. We note that the field Hamiltonian $\mathcal{E}_l$ in (12.17) is similar to (12.18) provided we associate $p_l\to\hat{p}$, $q_l\to x$, and take the mass $m=1$. The quantization of the electromagnetic mode is thus accomplished by treating it formally, as a harmonic oscillator, and considering $\hat{p}_l$ and $\hat{q}_l$ as a conjugate momentum pair[1] that obeys, as do $\hat{p}$ and x in the case of a particle, the commutation relationship

$$[\hat{p}_l,\hat{p}_m]=[\hat{q}_l,\hat{q}_m]=0 \tag{12.19}$$

$$[\hat{q}_l,\hat{p}_m]=i\hbar\delta_{l,m} \tag{12.20}$$

The formal association of $\hat{p}_l$ and $\hat{q}_l$ with p and x is based on more than just the similarity of the Hamiltonian in (12.17) and (12.18). The equations of motion obeyed by both sets of conjugate momenta are fundamentally similar. Those describing p and x are obtained applying the Hamiltonian equations to (12.18):

$$\begin{aligned}\frac{d\hat{p}}{dt}&=-\frac{\partial\hat{\mathcal{H}}_{\text{ho}}}{\partial x}=-\omega^2mx\\\frac{dx}{dt}&=\frac{\partial\hat{\mathcal{H}}_{\text{ho}}}{\partial p}=\frac{\hat{p}}{m}\end{aligned} \tag{12.21}$$

Repeating the procedure with the electromagnetic mode, the Hamiltonian in (12.17) yields

$$\begin{aligned}\frac{d\hat{p}_l}{dt}&=-\frac{\partial\hat{\mathcal{H}}_l}{\partial q_l}=-\omega_l^2\hat{q}_l\\\frac{d\hat{q}_l}{dt}&=\frac{\partial\hat{\mathcal{H}}}{\partial p_l}=\hat{p}_l\end{aligned} \tag{12.22}$$

which are the same as (12.14) and (12.15) derived from Maxwell's equations. We thus establish the formal correspondence between $\hat{p}$ and x on the one hand and $\hat{p}_l$ and $\hat{q}_l$ on the other.

---

[1]From this point on we consider p_l and q_l as operators, and hence designate them by $\hat{p}_l$ and $\hat{q}_l$.

Electromagnetic Creation and Annihilation Operators

As in the case of the harmonic oscillator (see Section 5.3), we find it useful to introduce $\hat{a}_l$—the mode annihilation operator—and $\hat{a}_l^\dagger$—the creation operator—which are defined by

$$\hat{a}_l = \left(\frac{1}{2\hbar\omega_l}\right)^{1/2}(\omega_l\hat{q}_l + i\hat{p}_l)$$

$$\hat{a}_l^\dagger = \left(\frac{1}{2\hbar\omega_l}\right)^{1/2}(\omega_l\hat{q}_l - i\hat{p}_l) \tag{12.23}$$

or

$$\hat{p}_l = i\left(\frac{\hbar\omega_l}{2}\right)^{1/2}(\hat{a}_l^\dagger - \hat{a}_l)$$

$$\hat{q}_l = \left(\frac{\hbar}{2\omega_l}\right)^{1/2}(\hat{a}_l^\dagger + \hat{a}_l) \tag{12.24a}$$

Unlike $\hat{p}_l$ and $\hat{q}_l$, the operators $\hat{a}_l$ and $\hat{a}_l^\dagger$ are not Hermitian. They are, however, the Hermitian adjoint of each other.

If we express the mode fields $\mathbf{E}_l$ and $\mathbf{H}_l$ [Eqs. (12.13*a*) and (12.13*b*)] in terms of the operators $\hat{a}_l^\dagger$ and $\hat{a}_l$, we obtain

$$\hat{E}_{l_y} = i\sqrt{\frac{\hbar\omega_l}{V\varepsilon}}\,(\hat{a}_l^\dagger - \hat{a}_l)\sin k_l z$$

$$\hat{\mathcal{H}}_{l_x} = \sqrt{\frac{\hbar\omega_l}{V\mu}}\,(\hat{a}_l^\dagger + \hat{a}_l)\cos k_l z \tag{12.24b}$$

so that the electric and magnetic fields themselves correspond to quantum mechanical operators.

Using the commutation relations (12.19), (12.20), and (12.23), we obtain

$$[\hat{a}_l, \hat{a}_m] = [\hat{a}_l^\dagger, \hat{a}_m^\dagger] = 0$$

$$[\hat{a}_l, \hat{a}_m^\dagger] = \delta_{l,m} \tag{12.25}$$

The mode Hamiltonian in (12.17) may be expressed in terms of the operators $\hat{a}_l^\dagger$ and $\hat{a}_l$:

$$\begin{aligned}\hat{\mathcal{H}}_l &= \tfrac{1}{2}(\hat{p}_l^2 + \omega_l^2\hat{q}_l^2)\\ &= \left(-\frac{\hbar\omega_l}{2}(\hat{a}_l^\dagger - \hat{a}_l)(\hat{a}_l^\dagger - \hat{a}_l) + \frac{\hbar\omega_l}{2}(\hat{a}_l^\dagger + \hat{a}_l)(\hat{a}_l^\dagger + \hat{a}_l)\right)\\ &= \frac{\hbar\omega_l}{2}(\hat{a}_l^\dagger\hat{a}_l + \hat{a}_l\hat{a}_l^\dagger)\end{aligned}$$

Using the last of (12.25) to replace $\hat{a}_l\hat{a}_l^\dagger$ by $1 + \hat{a}_l^\dagger\hat{a}_l$ results in

$$\hat{\mathcal{H}}_l = \hbar\omega_l(\hat{a}_l^\dagger\hat{a}_l + \tfrac{1}{2}) \tag{12.26}$$

The formal analogy between the radiation mode and the harmonic oscillator enables us to associate with the oscillator an eigenfunction $|n_l\rangle$ that, like its counterpart u_n in Section 5.3, obeys the following relations:

$$\hat{a}_l^\dagger|n_l\rangle=(n_l+1)^{1/2}|n_l+1\rangle$$
$$\hat{a}_l|n\rangle=(n_l)^{1/2}|n_l-1\rangle \tag{12.27}$$

The energy eigenvalue equation for the radiation mode becomes

$$\begin{aligned}\hat{\mathcal{H}}_l|n_l\rangle&=\hbar\omega_l\left(\hat{a}_l^\dagger\hat{a}_l+\tfrac{1}{2}\right)|n_l\rangle\\&=\hbar\omega_l\left(n_l+\tfrac{1}{2}\right)|n_l\rangle\end{aligned} \tag{12.28}$$

The total mode energy is thus restricted to the set

$$E_{n_l}=\hbar\omega_l\left(n_l+\tfrac{1}{2}\right) \tag{12.29}$$

and is quantized. This quantization of the energy of a radiation mode that was postulated by Planck in 1900 (see the discussion in Section 1.2) is thus found here to result directly from the formal extension of the quantum mechanical postulates to electromagnetic radiation.

Traveling Wave Quantization

Instead of the standing wave quantization leading to Eq. (12.24*b*), we can use a plane wave quantization and take the electric and magnetic field operators of a *single* mode as

$$\mathbf{E}_{\mathbf{k}\lambda}(\mathbf{r},t)=i\mathbf{e}_{\mathbf{k}\lambda}\sqrt{\frac{\hbar\omega_k}{2V\varepsilon}}\left[a_{\mathbf{k}\lambda}^\dagger(t)e^{-i\mathbf{k}\cdot\mathbf{r}}-a_{\mathbf{k}\lambda}(t)e^{i\mathbf{k}\cdot\mathbf{r}}\right]$$

$$\mathbf{H}_{\mathbf{k}\lambda}(\mathbf{r},t)=\left(\mathbf{e}_{\mathbf{k}\lambda}\times\frac{\mathbf{k}}{k}\right)\sqrt{\frac{\hbar\omega_k}{2V\mu}}\left[a_{\mathbf{k}\lambda}^\dagger(t)e^{-i\mathbf{k}\cdot\mathbf{r}}+a_{\mathbf{k}\lambda}e^{i\mathbf{k}\cdot\mathbf{r}}\right] \tag{12.30}$$

The direction of propagation of the wave is $\mathbf{k}$. The direction of the electric field is that of the unit vector $\mathbf{e}_{\mathbf{k}\lambda}$, which is normal to $\mathbf{k}$. The two allowed directions of polarization for a given $\mathbf{k}$ are denoted by λ ($\lambda=1,2$) and the direction of $\mathbf{H}_{\mathbf{k}\lambda}$ is parallel to $\mathbf{e}_{\mathbf{k}\lambda}\times(\mathbf{k}/k)$ and is thus normal to both $\mathbf{k}$ and $\mathbf{E}_{\mathbf{k}\lambda}$.

12.3 BLACK BODY RADIATION

In Section 1.2 we described how one of the early realizations that classical physics was inadequate resulted from its failure to provide a satisfactory accounting of thermal radiation density (or radiation power). Let us use our newly gathered knowledge and apply it to the same problem. We consider the thermal radiation field inside a large cubic box of sides L that is at thermal equilibrium at temperature T.

Using a three-dimensional extension of the arguments used in connection with (12.13*a*), we write the field corresponding to a single electromagnetic mode of the enclosure in the form of

$$E_{lmn} \propto \cos(k_x x)\sin(k_y y)\sin(k_z z)\cos \omega t$$

The field vanishes at the walls provided

$$k_x = p\frac{\pi}{L}, \qquad k_y = l\frac{\pi}{L}, \qquad k_z = m\frac{\pi}{L} \tag{12.31}$$

where p, l, m, are any integers. If we substitute the assumed form of the mode in the wave equation (12.3), we find

$$\begin{aligned} k^2 &= \omega^2 \mu \varepsilon \\ &= \frac{\pi^2}{L^2}(p^2 + l^2 + m^2) \end{aligned} \tag{12.32}$$

Each triplet of integers p, l, m thus defines a mode whose frequency ω is given by (12.32).

Next we need to determine the density of modes (number of modes per unit frequency). We start by dividing the space whose coordinate axes are labeled by k_x, k_y, k_z into elemental volumes. According to (12.31) we can associate a unique mode with each triplet of positive integers p, l, m (changing the sign of one of these integers does not generate a mode that is linearly independent, since it only differs in sign from that with a positive integer), so that with each mode we may associate a volume $dk_x\,dk_y\,dk_z = (\pi/L)^3$ in k_x, k_y, k_z space. This volume is situated at the tip of the vector extending from

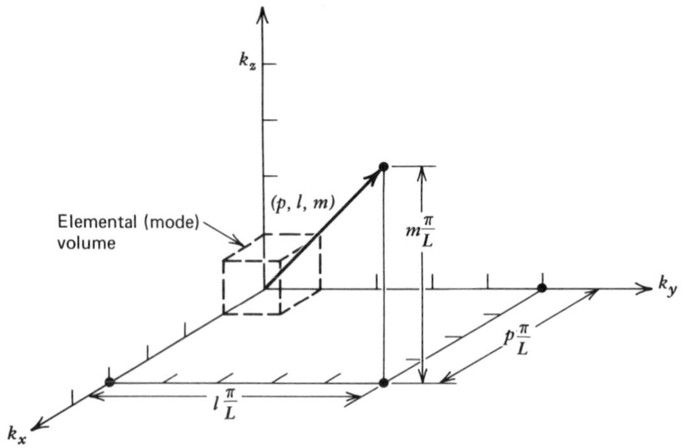

Figure 12.3 Each triplet p, l, m defines an electromagnetic mode with a propagation constant $\mathbf{k} = [p(\pi/L), l(\pi/L), m(\pi/L)]$. With each mode we may associate a unique volume $(\pi/L)^3$. One such volume is shown next to the origin.

the origin to the point $[p(\pi/L), l(\pi/L), m(\pi/L)]$ in **k** space, as shown in Fig. 12.3.

To find the number N_k of electromagnetic modes whose **k** vectors have magnitudes between 0 and k, we divide the total volume in **k** space occupied by these modes $[\frac{1}{8}(\frac{4}{3}\pi k^3)]$ by the volume per mode $[(\pi/L)^3]$. (The factor $\frac{1}{8}$ is due to the fact that k_x, k_y, k_z are positive and so are restricted to one octant of **k** space. We then multiply the result by a factor of 2 to account for the two directions of independent field polarization associated with each direction **k**.) The result is

$$N_k = \frac{k^3 V}{3\pi^2} \tag{12.33}$$

where $V = L^3$ is the volume of the enclosure. Using (12.32) and $\mu\varepsilon \equiv (n^2/c^2)$ (n is the index of refraction), we have

$$\begin{aligned} k^2 &= \omega^2 \mu\varepsilon \\ &= \frac{4\pi^2\nu^2}{c^2} n^2 \end{aligned} \tag{12.34}$$

Using (12.34) we obtain for N_ν the number of modes whose frequencies lie between 0 and ν:

$$N_\nu = \frac{8\pi\nu^3 n^3}{3c^3} V$$

The mode density—the number of modes per unit volume per unit frequency interval—is obtained finally as

$$\begin{aligned} p(\nu) &= \frac{1}{V}\frac{dN_\nu}{d\nu} \\ &= \frac{8\pi n^3 \nu^2}{c^3} \end{aligned} \tag{12.35}$$

The average thermal energy per mode is (see the derivation below)

$$E_{\text{av}} = \frac{h\nu}{e^{h\nu/kT} - 1} \tag{12.36}$$

To find the black body energy density due to modes whose frequencies lie between ν and $\nu + d\nu$, we multiply the density $p(\nu)$ of these modes by E_{av} obtaining

$$\rho(\nu) = \frac{8\pi n^3 h\nu^3}{c^3(e^{h\nu/kT} - 1)} \tag{12.37}$$

where $\rho(\nu)$ is the black body energy density per unit frequency. If we want to obtain the expression for the black body spectral intensity $I(\nu)$ (watts per m^2 per unit frequency), we need to multiply (12.37) by the velocity of light[2]

[2]For a single direction of propagation, the relation

$$\text{Intensity} = (\text{velocity of energy propagation}) \times (\text{energy density})$$

is always valid.

(c/n) and perform an averaging over the continuum of directions represented by the modes. The result is

$$I(\nu)=\frac{2\pi n^2 h\nu^3}{c^2(e^{h\nu/kT}-1)} \tag{12.38}$$

Derivation of the Average Energy per Mode

Here we derive the expression for E_{av}, the average thermal equilibrium energy per radiation mode.

The energy of a mode in the quantum state n is

$$E_n=(n+\tfrac{1}{2})\hbar\omega \tag{12.39}$$

The probability that the mode will be found in the state n is given by the Boltzmann factor [see (16.2)]:

$$p(n)=\frac{e^{-E_n/kT}}{\sum_{s=0}^{\infty} e^{-E_s/kT}} \tag{12.40}$$

The average thermal excitation energy per mode is thus

$$E_{av}=\sum_{n=0}^{\infty} E_n p(n)=\frac{\sum_n (n+\frac{1}{2})\hbar\omega \exp(-(n+\frac{1}{2})\beta\hbar\omega)}{\sum_s \exp(-(s+\frac{1}{2})\beta\hbar\omega)} \tag{12.41}$$

where $\beta\equiv(kT)^{-1}$. The denominator of (12.41) contains as a factor a geometric series whose sum is

$$\sum_{s=0}^{\infty} e^{-s\hbar\omega\beta}=\frac{1}{1-e^{-\hbar\omega\beta}}$$

Taking the derivative with respect to β gives

$$\sum_s s\hbar\omega e^{-s\hbar\omega\beta}=\frac{\hbar\omega e^{-\hbar\omega\beta}}{(1-e^{-\hbar\omega\beta})^2}$$

Substituting the last result in (12.41) yields

$$E_{av}=\frac{\hbar\omega}{2}+\frac{\hbar\omega}{e^{\hbar\omega/kT}-1} \tag{12.42}$$

The term $\hbar\omega/2$ represents the zero point energy of the mode—that is, the energy of the lowest $(n=0)$ quantum state. This energy cannot be extracted and is thus not available. It is consequently omitted from thermal considerations so that one writes

$$E_{av}=\frac{\hbar\omega}{e^{\hbar\omega/kT}-1} \tag{12.43}$$

The result was quoted in (12.36). We note that the classical result

$$E_{av}=kT \tag{12.43}$$

obtains in the high-temperature limit $kT\gg\hbar\omega$.

12.4 INDUCED TRANSITIONS IN COLLISION DOMINATED ATOMIC SYSTEMS

When an atomic system in an excited energy state is subject to electromagnetic radiation, it may undergo transitions to lower or higher lying states at a rate proportional to that of the incident radiation density. We next derive the rate for such a process. We will use a semiclassical approach in which the atoms are treated quantum mechanically and the radiation field classically.

Let the system occupy at $t=0$, the start of our observation period, the state 2, which is separated from some lower lying state 1 by $E_2 - E_1 \equiv \hbar\omega_{21}$. The situation is depicted in Fig. 12.4. Let the electric field due to the incident radiation at the position $\mathbf{r}$ of the atom be

$$\mathbf{E}(\mathbf{r}, t) = \mathbf{E}_0(\mathbf{r}) \cos \omega t \tag{12.44}$$

Since the potential energy of an electron in a field $\mathbf{E}$ is

$$V = -\mathbf{eE} \cdot \mathbf{r} \tag{12.45}$$

we take the perturbation Hamiltonian due to the field as

$$\mathcal{H}'(t) = -\frac{eE_0 y}{2}\left(e^{i\omega t} + e^{-i\omega t}\right) \tag{12.46}$$

where we assumed the field was polarized along the y direction. The perturbation Hamiltonian (12.46) is cast in the form of (11.26); we thus use directly the result (11.32) to obtain the downward $(2 \to 1)$ or upward $(1 \to 2)$ transition rate[3]

$$\begin{aligned} W_{1\to 2} = W_{2\to 1} \equiv W_i &= \frac{2\pi}{\hbar} |H'_{12}|^2 \delta(E_2 - E_1 - \hbar\omega) \\ &= \frac{\pi e^2 E_0^2}{2\hbar} |y_{12}|^2 \delta(E_2 - E_1 - \hbar\omega) \end{aligned} \tag{12.47}$$

where

$$e y_{12} \equiv e \int u_1^* y u_2 \, dv \tag{12.48}$$

If the exact value of the energy separation $E_2 - E_1$ of the transition is not certain (the reasons for this uncertainty are discussed at the end of this section), we employ a "lineshape" function $g(E)$ so defined that

$$g(E)\, dE = \text{probability of finding } E_2 - E_1 \text{ between } E \text{ and } E + dE, \tag{12.49}$$

It follows that $\int_{-\infty}^{\infty} g(E)\, dE = 1$. The induced transition rate W_i is obtained by

[3]The use of (11.32) limits us (see the discussion at the end of Section 11.2) to situations where the atoms undergo many collisions before making a radiative transition. This is true in most experimental situations that involve low intensity fields. When the induced transition rate, which is proportional to the field amplitude, is fast compared to the collision rate, the use of (11.32) is not justified. This situation is treated in Section 13.1

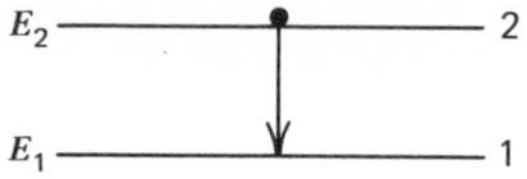

Figure 12.4 Two levels of an atomic system used in the derivation of the induced transition rate. (Other levels not shown.) The system occupies the state 2 at $t=0$.

summing (12.47) over all possible values of $E_2 - E_1$ and weighing each contribution by the probability $g(E_2 - E_1)$:

$$W_i = \frac{\pi e^2 E_0^2 |y_{12}^2|}{2\hbar} \int_{-\infty}^{\infty} \delta(E_2 - E_1 - \hbar\omega) g(E_2 - E_1) d(E_2 - E_1)$$

$$= \frac{\pi e^2 E_0^2 |y_{12}|^2}{2\hbar} g(\hbar\omega)$$

We may convert to a lineshape function $g(\nu)$ in the frequency domain by using $g(\nu)d\nu = g(E)d(h\nu)$ (conservation of probability) and write

$$W_i = \frac{E_0^2 e^2 |y_{12}|^2}{4\hbar^2} g(\nu) \tag{12.50}$$

In practice one deals most often with the field intensity I_ν (watts-m^2) rather than E_0^2. We use the relation

$$I_\nu = \left(\frac{\text{energy}}{\text{volume}}\right) \times (\text{energy velocity})$$

$$= \left(\frac{\varepsilon_0 n^2 E_0^2}{2}\right)\left(\frac{c}{n}\right)$$

$$= \frac{cn\varepsilon_0 E_0^2}{2} \tag{12.51}$$

and rewrite (12.50) as

$$W_i = \frac{e^2 |y_{12}|^2}{2\hbar^2 cn\varepsilon_0} g(\nu) I_\nu \tag{12.52a}$$

In the next section we show that the matrix element y_{12} is related to the spontaneous lifetime t_{spont} of the $2 \rightarrow 1$ transition by

$$\frac{1}{t_{\text{spont}}} = \frac{16\pi^3 n^3 e^2 |y_{12}|^2 \nu_0^3}{\varepsilon h c^3}$$

so that W_i may also be written as

$$W_i = \frac{\lambda^3 g(\nu)}{8\pi h c n^2 t_{\text{spont}}} I_\nu \tag{12.52b}$$

Equation (12.52a, b) plays an important role in treating phenomena of light absorption or emission. Let us consider its consequences in some detail.

The transition rate from level 2 to 1 is proportional to the incident intensity. It is also proportional to the (square of) matrix element $e^2 y_{12}^2$ as given by (12.48). It follows immediately that if u_1 and u_2 have the same parity so that the product $u_1^* u_2$ is an even function, then $y_{12}=0$ and the rate W_i is zero. The transition is said to be "dipole-forbidden." An example of a forbidden electric dipole transition is that connecting the $n=3, l=2$ hydrogenic state with the ground state $n=0, l=0$ since, according to Section 6.3, the parity of the state is even(odd) when l is even(odd). The dependence of $W_{2\leftrightarrow 1}$ on $g(\nu)$ requires a broader discussion. Up to this point we considered the state energies $E_2 - E_1$ as perfectly well-defined numbers, since the energies E_i are the eigenvalues of a differential equation—the Schrödinger equation. In real atomic systems there are *always* mechanisms that cause the value of the eigen energies, and hence the value of $E_2 - E_1$, to be uncertain—that is, make it impossible to specify $E_2 - E_1$ exactly. One of the most common reasons is the interruption of the lifetime of the atom in level 1 or 2 by collisions or by spontaneous transitions. This results, according to the uncertainty principle (3.35), in a smearing of $E_2 - E_1$ by

$$\Delta E \sim \hbar / \tau \tag{12.53}$$

where τ is the lifetime. In the case of isolated atoms the spectral width ΔE of the lineshape function $g(E)$ is indeed given by (12.53).

12.5 SPONTANEOUS TRANSITIONS

We have shown in the last section that an atom subject to an electromagnetic field will be *induced* to undergo transitions to other levels at a rate that, according to (12.52), is *proportional* to the intensity.

In this section we show that, in addition to the induced process, an atom can undergo transitions to *lower states even when no external radiation is present.* These transitions are called "spontaneous" since their rate is independent of the field intensity I_ν.

The presence of such transitions was inferred by Einstein[4] using essentially classical arguments. His reasoning ran as follows: Imagine a large system of identical atoms that are in *thermal equilibrium* with black body radiation at some temperature T. The average numbers of atoms occupying some two levels, say, 2 and 1 (two of the eigenstates of the atom) obey the Boltzmann relation (16.3), which, assuming equal degeneracy, $g_2 = g_1$, is given by

$$\frac{N_2}{N_1} = e^{-(E_2 - E_1)/kT} \tag{12.54}$$

If $E_2 > E_1$, then $N_2 < N_1$. The total number of atoms undergoing transition from level 2 to 1 in any time interval must be equal in equilibrium to that making transitions from 1 to 2, since N_2 and N_1 are constant with time. This requirement can be reconciled with the result $N_2 < N_1$ only by assuming that

---

[4] A. Einstein, Die Quantentheorie der Strahlung, *Phys. Lett.* **18**, 121 (1917).

the downward (2→1) transition rate per atom is greater than that for the 1→2 (upward) transition. Einstein satisfied this requirement by postulating

$$W_{2\to1} = B\rho(\nu) + A$$
$$W_{1\to2} = B\rho(\nu) \tag{12.55}$$

In words: The downward (2→1) transition rate (per atom) is the sum of an induced term $B\rho(\nu)$ proportional to the radiation density $\rho(\nu)$ and a spontaneous rate A. The upward rate is just $B\rho(\nu)$. In order that the number of upward and downward transitions during any time period be equal, we need to satisfy

$$N_2[B\rho(\nu) + A] = N_1 B\rho(\nu) \tag{12.56}$$

so that

$$\frac{N_2}{N_1} = \frac{B\rho(\nu)}{B\rho(\nu) + A}$$
$$= e^{-(E_2 - E_1)/kT}$$
$$= e^{-h\nu/kT} \tag{12.57}$$

where $h\nu = E_2 - E_1$ and the last equality reflects thermal equilibrium for the energy density of thermal radiation. If we use (12.37)

$$\rho(\nu) = \frac{8\pi n^3 h\nu^3}{c^3(e^{h\nu/kT} - 1)}$$

in conjunction with (12.57), we obtain

$$\frac{A}{B} = \frac{8\pi n^3 h\nu^3}{c^3} \tag{12.58}$$

The spontaneous transition rate per atom A is called the "Einstein A coefficient."

Let us assume that by means of some agency we succeeded in elevating a large number of atoms to some excited state 2 that normally would have a negligibly small population. If no appreciable radiation is present $[\rho(\nu) = 0]$, then, according to the first of (12.55), the average number of atoms making 2→1 transitions per unit time is $N_2 A$ so that the average population N_2 will decrease according to

$$\frac{dN_2}{dt} = -N_2 A$$

and

$$N_2(t) = N_2(0)e^{-At}$$
$$= N_2(0)e^{-t/t_{sp}}$$

The Einstein coefficient A is thus the inverse of the spontaneous (decay) lifetime t_{spont}. In Section 12.6 we show that each spontaneous decay is

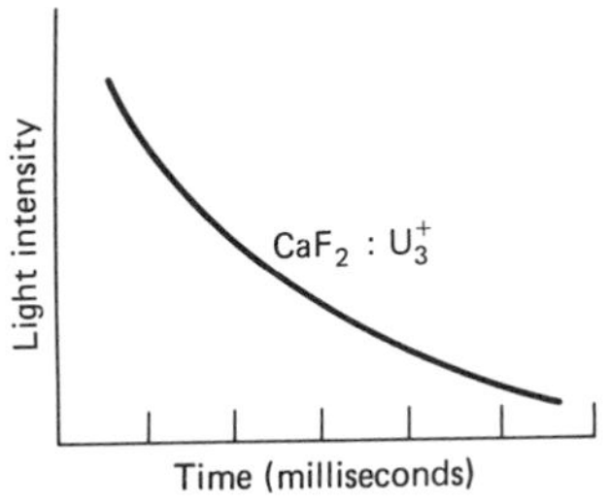

Figure 12.5 Spontaneous decay of an excited population monitored via the emitted radiation.

accompanied by the emission of one photon at a frequency $\nu \sim (E_2 - E_1)/h$. We may thus monitor the decay of N_2 and so measure t_{spont} by recording the temporal decay of the spontaneously emitted emission following the excitation of a large number of atoms to an excited state. An example of such a decay is shown in Fig. 12.5.

12.6 QUANTUM MECHANICAL DERIVATION OF THE SPONTANEOUS TRANSITION RATE *A*

To derive the expression for the spontaneous transition rate of an atom out of some excited state $|2\rangle$ due to transition to a lower state $|1\rangle$, we will assume first that the atom interacts with a single radiation mode. The resulting expression for the transition rate per mode will then be summed up over all

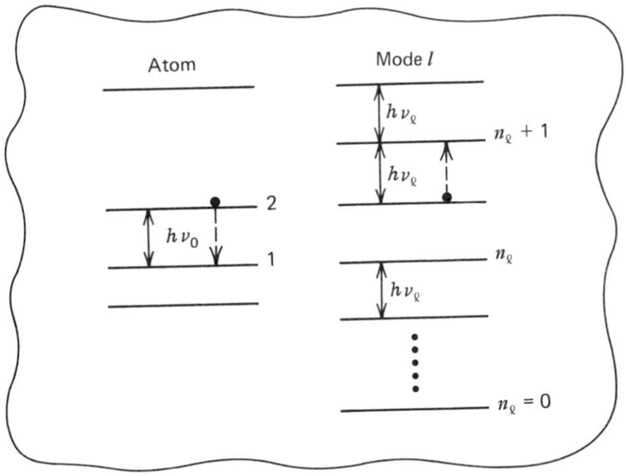

Figure 12.6 The basic quantum mechanical system used in the derivation of the expression for the spontaneous transition rate. One atom initially in the state $|2\rangle$ and a radiation mode l initially in $|n_l\rangle$. The transition causes the atom to undergo a downward transition to $|1\rangle$ while the radiation mode is elevated to $|n_l + 1\rangle$.

the radiation modes in order to obtain the total spontaneous rate for transition from $|2\rangle$ to $|1\rangle$. The quantum mechanical system consists of one atom and one radiation mode l as shown in Fig. 12.6. The atom is initially in an excited state $|2\rangle$, while the radiation mode l is in the state $|n_l\rangle$. The initial state is designated as

$$|2, n_l\rangle = |2\rangle|n_l\rangle \tag{12.59}$$

After the transition, the atom is in the lower state $|1\rangle$, while the mode l has gained a quantum of radiation and is now in the state $|n_l+1\rangle$. The final state is thus

$$|1, n_l+1\rangle = |1\rangle|n_l+1\rangle \tag{12.60}$$

The initial and final energies are

$$E_i = E_2 + \hbar\omega_l\left(n_l + \tfrac{1}{2}\right)$$
$$E_f = E_1 + \hbar\omega_l\left(n_l + \tfrac{3}{2}\right)$$

so that

$$E_i - E_f = E_2 - E_1 - \hbar\omega_l \tag{12.61}$$

The interaction Hamiltonian is given by (12.45)

$$\mathcal{H}' = -eE_{l,y}(z,t)\,y \tag{12.62}$$

where we take the mode field as polarized along y and propagating along z. Next, we use the operator expansion (12.30) of the field to rewrite (12.62) as

$$\hat{\mathcal{H}}' = -ie\sqrt{\frac{\hbar\omega_l}{V\varepsilon}}\left(\hat{a}_l^\dagger e^{-i\mathbf{k}\cdot\mathbf{r}} - \hat{a}_l e^{i\mathbf{k}\cdot\mathbf{r}}\right)y \tag{12.63}$$

Since the quantum mechanical system includes the electromagnetic field, the time dependence of $\hat{\mathcal{H}}'$ is implicit. The appropriate transition rate is consequently given by (11.36):

$$W^{(l)} = \left(\frac{2\pi}{\hbar}\right)\left(\frac{e^2\hbar\omega_l}{2V\varepsilon}\right)$$

$$\left|\langle 1, n_l+1|\left(\left(a_l^\dagger e^{-i\mathbf{k}\cdot\mathbf{r}}\right) - \left(a_l e^{i\mathbf{k}\cdot\mathbf{r}}\right)\right)y|2, n_l\rangle\right|^2 \delta(E_2 - E_1 - \hbar\omega_l) \tag{12.64}$$

Using the relations

$$\begin{aligned}
\langle n_l+1|a_l^\dagger|n_l\rangle &= \sqrt{n_l+1}\\
\langle n_l+1|a_l|n_l\rangle &= 0\\
\langle 1|\,y|2\rangle &\equiv y_{12}
\end{aligned} \tag{12.65}$$

in (12.64) leads to

$$W^{(l)} = \frac{2\pi e^2\omega_l}{V\varepsilon}|\,y_{12}|^2(n_l+1)\delta(E_2 - E_1 - \hbar\omega_l) \tag{12.66}$$

where the superscript (l) indicates that the transition is that due to the interaction with mode l. It is customary to separate $W^{(l)}$ into two components:

$$W^{(l)} = W_{\text{ind}}^{(l)} + W_{\text{spont}}^{(l)}$$

where

$$W_{\text{ind}}^{(l)} = \frac{2\pi e^2 \omega_l}{V\varepsilon} | y_{12}|^2 n_l \delta(E_2 - E_1 - \hbar\omega_l) \tag{12.67}$$

is proportional to the mode radiation energy density $(n_l \hbar\omega_l / V)$ and is referred to as the induced transition rate.

$$W_{\text{spont}}^{(l)} = \frac{2\pi e^2 \omega_l}{V\varepsilon} | y_{12}|^2 \delta(E_2 - E_1 - \hbar\omega_l) \tag{12.68}$$

is independent of the mode energy and is called the spontaneous transition rate. We note that

$$\frac{W_{\text{ind}}^{(l)}}{W_{\text{spont}}^{(l)}} = n_l \tag{12.69}$$

that is, the ratio of the induced transition rate to the spontaneous rate, both involving mode l, is equal to n_l, the number of radiation quanta (photons) in the mode. If $n_l = 0$—that is, the mode is in its ground quantum state—then the induced rate $W_{\text{ind}}^{(l)}$ is zero and the total transition rate is equal to $W_{\text{spont}}^{(l)}$.

To obtain the total spontaneous transition rate, we sum (12.68) over all modes l. The number of such modes per unit frequency interval in a volume V was obtained in (12.35) as

$$p(\nu_l) = \frac{8\pi \nu_l^2 n^3 V}{c^3}$$

The total spontaneous transition rate is thus

$$W_{\text{spont}} = \int_0^\infty W_{\text{spont}}^{(l)} p(\nu_l)\, d\nu_l$$

$$= \frac{16 n^3 \pi^3 e^2 | y_{12}|^2 \nu_0^3}{\varepsilon h c^3} \tag{12.70}$$

In deriving (12.70) we took $(E_2 - E_1) \equiv h\nu_0$. A slightly more accurate derivation that accounts for the different electric field polarization of the different modes requires that we replace $| y_{12}|^2$ by

$$| y_{12}|^2 \to \tfrac{1}{3}\left(|x_{12}|^2 + | y_{12}|^2 + |z_{12}|^2\right)$$

so that the final expression for the spontaneous transition rate is

$$W_{\text{spont}} = \frac{1}{t_{\text{spont}}}$$

$$= \frac{16\pi^3 n^3 e^2 \nu_0^3}{3\varepsilon h c^3}\left(|x_{12}|^2 + | y_{12}|^2 + |z_{12}|^2\right) \tag{12.71}$$

We recall that W_{spont} is also designated by the symbol A as in (12.58).

EXAMPLE: The spontaneous lifetime of the $|2p\rangle \rightarrow |1s\rangle$ transition in atomic hydrogen

As an example, we calculate in this section the spontaneous lifetime for a transition from the hydrogenic state $n=2, l=1$ ($|2p\rangle$) to the state $n=1, l=0$ ($|1s\rangle$).

The upper state $n=2, l=1$ is threefold degenerate with states characterized by $m=1,0,-1$. Alternatively, the three upper states can be represented by [see (7.63)]

$$
\begin{aligned}
|x\rangle &= \frac{1}{\sqrt{2}}(u_{211}+u_{21-1}) = \frac{1}{\sqrt{32\pi}}\left(\frac{Z}{a_0}\right)^{5/2} e^{-zr/2a_0}x \\
|y\rangle &= \frac{1}{\sqrt{2}}(u_{211}-u_{21-1}) = \frac{1}{\sqrt{32\pi}}\left(\frac{Z}{a_0}\right)^{5/2} e^{-zr/2a_0}y \qquad (12.72)\\
|z\rangle &= u_{210} = \frac{1}{\sqrt{32\pi}}\left(\frac{Z}{a_0}\right)^{5/2} e^{-zr/2a_0}z
\end{aligned}
$$

The ground state is

$$
|1s\rangle \equiv |1\rangle = \frac{1}{\sqrt{\pi}}\left(\frac{Z}{a_0}\right)^{3/2} e^{-zr/a_0} \qquad (12.73)
$$

The spontaneous transition rate given by (12.71) as

$$
W_{\text{spont}} = \frac{16\pi^3 n^3 e^2 \nu_0^3}{3\varepsilon h c^3}\left(|x_{12}|^2+|y_{12}|^2+|z_{12}|^2\right) \qquad (12.74)
$$

depends in general on which of the three degenerate upper ($n=2$) states is occupied initially by the atom. To be specific, we assume that the atom is in the excited state $|x\rangle$. The transition matrix elements are then obtained from (12.72) and (12.73):

$$
\begin{aligned}
y_{12} &= \langle 1|y|x\rangle \propto \int_V xy\, e^{-3zr/2a_0}\, d^3\mathbf{r} = 0 \\
z_{12} &= \langle 1|z|x\rangle \propto \int_V zx\, e^{-3zr/2a_0}\, d^3\mathbf{r} = 0 \qquad (12.75)\\
x_{12} &= \langle 1|x|x\rangle \\
&= \frac{1}{\sqrt{32}}\left(\frac{1}{\pi}\right)\left(\frac{1}{a_0}\right)^4 \int_0^\infty r^4 e^{-3r/2a_0}\, dr \int_0^\pi \sin^3\theta\, d\theta \int_0^{2\pi} \cos^2\phi\, d\phi \\
&= \frac{128\sqrt{2}}{243}\frac{a_0}{Z} = 0.7450\frac{a_0}{Z} \qquad (12.76)
\end{aligned}
$$

We substitute, next, the values of x_{12}, y_{12}, z_{12} determined above in (12.74) using in the process the fact that in a hydrogenic transition (see Eq. 7.50)

$$h\nu_0(n=2\rightarrow n=1)=\frac{3}{4}\frac{\mu^2 Z^2 e^4}{32\pi^2\varepsilon_0 h^2}$$

so that $\nu_0=3.288\times 10^{15}Z^2$. The result is

$$W_{\substack{\text{spont}\\2\rightarrow 1}}=6.27\times 10^8 Z^4$$

in the case of hydrogen $Z=1$, so that

$$t_{\text{spont}}=W_{\text{spont}}^{-1}=1.60\times 10^{-9}s \tag{12.77}$$

Had we taken the initial excited state as $|y\rangle$ or $|z\rangle$, the result would have been identical to (12.77), since the relevant matrix elements are equal in all three cases ($\langle z|z|1\rangle=\langle y|y|1\rangle=\langle x|x|1\rangle$). We may thus associate (12.77) with the spontaneous lifetime of the $n=2\,l=1$ state.

PROBLEMS

1. Show that the spontaneous lifetime of the states $|y\rangle$ and $|z\rangle$ [see (12.12)] is the same as that of $|x\rangle$.
2. Calculate the spontaneous lifetime for $n=3\,l=2\rightarrow n=2\,l=1$ transition in atomic hydrogen.
3. Show that the expression for the induced transition rate (12.67) is consistent with (12.50).
4. (a) Calculate the spontaneous lifetime due to a $n\rightarrow n-1$ and $n\rightarrow n-2$ transitions of an electron moving in a quadratic potential well $V=\frac{1}{2}m\omega_0^2x^2$. Take $\hbar\omega_0=1$ eV. Assume the electron is in the state n.
   (b) What is the lifetime in the state n due to $n\rightarrow n-2$ transition when the potential is given by $V(x)=\frac{1}{2}m\omega_0^2x^2+bx^3$, where $b=2\times 10^{29}$ eV-m^{-3}?
5. Derive the expression (12.38) for the black body radiation intensity starting with the expression for the energy density.
6. Quantize the simple electrical circuit consisting of an inductance L in parallel with a capacitance C. Specifically:
   (a) What are the canonically conjugate momenta in this case?
   (b) What is the form of the total circuit Hamiltonian when expressed in terms of the conjugate momenta and in terms of the creation and annihilation operators $\hat{a}^\dagger$ and $\hat{a}$?

Clue: Somewhere in the course of the solution one should obtain

$$\hat{a}^\dagger=\sqrt{\frac{C}{2\hbar\omega}}\left(V(t)-i\sqrt{\frac{L}{C}}\,I(t)\right)$$

$$\hat{a}=\sqrt{\frac{C}{2\hbar\omega}}\left(V(t)+i\sqrt{\frac{L}{C}}\,I(t)\right)$$

where $I(t)$ and $V(t)$ are, respectively, the instantaneous current and voltage in the circuit, and $\omega^2 \equiv (LC)^{-1}$.

7. (a) Using the results of Problem 6, derive expressions for the root-mean-square voltage and current due to zero-point vibrations of the LC circuit. Calculate them for the case of $L = 2.53 \times 10^{-6}$ henry and $C = 10^{-12}$ farad.
   (b) Assuming that the results of Problems 6 and 7 are still valid when the circuit contains a resistance R in parallel with L (and C), show that the average power dissipated in the resistance is

$$P \sim \hbar \omega B$$

   where $B = (RC)^{-1}$ is the bandwidth (or equivalently the inverse of the decay time constant) of the circuit.

CHAPTER THIRTEEN

Absorption and Dispersion of Radiation in Atomic Media

Electromagnetic radiation propagating in a material medium interacts with the atoms (or molecules) and modifies their quantum states. This interaction is manifested most often as absorption of radiation and by dispersion—a dependence of the phase velocity of the wave on its frequency. Under nonthermal equilibrium conditions of inverted atomic populations, the incident radiation may even gain in intensity (i.e., negative absorption) as is the case in laser amplifiers and oscillators.

In Section 11.2 we have shown how the interaction of an atom with a time-harmonic perturbation, such as that of electromagnetic radiation fields, can cause the atom to undergo transitions between two quantum states, say n and m, provided the photon energy $\hbar\omega$ is nearly equal to the energy difference $|E_n - E_m|$ of the states.

In the analytical treatment of this subject, one often distinguishes between two main cases:

1. The first case is the collisionless regime in which the atom undergoes a transition in a time that is very much shorter than the collision time of the atom. (The word collision is used to describe any perturbative interaction of the atom with the external world.) Since the time for a transition will be shown below to be $\sim \pi\hbar/\mu E_0$ (here μ is the dipole matrix element for the transition, and E_0 is the field amplitude), the interaction is in the collisionless regime when $\pi\hbar/\mu E_0 \ll \tau_{\text{collision}}$.

The advent of lasers with their large electric fields made it possible to satisfy this condition under laboratory conditions. The theory relevant to this case is developed in Section 13.1.

2. The second case—to which we will refer as the collision dominated regime—obtains when

$$\pi\hbar/\mu E_0 \gg \tau_{\text{collision}}$$

This is the case in the great majority of the experimental situations we encounter, and in most natural phenomena such as absorption and dispersion. It also applies to the interaction between atoms and the field inside laser resonators. This case will be considered in detail in Sections 13.2, 13.3, and 13.4.

13.1 THE TIME EVOLUTION OF A COLLISIONLESS TWO-LEVEL ATOM

Most of the practical problems involving the interaction of radiation fields with atomic systems are treated using perturbation methods. There is one model problem, however, that can be solved by essentially exact methods. This is the problem of an isolated, hence collisionless, atomic entity (atom, molecule, or single electron) with two quantum states, designated as 1 and 2, interacting with a classical monochromatic radiation field. Real atoms that possess an infinity of energy levels can closely approximate the postulated model when only one atomic transition is (nearly) resonant[1] with the applied frequency.

The eigenfunctions of the two levels are taken as u_1 and u_2, the unperturbed Hamiltonian as $\hat{\mathcal{H}}_0$, and the corresponding unperturbed energies at E_1 and E_2, so that

$$\hat{\mathcal{H}}_0 u_1(\mathbf{r}) = E_1 u_1(\mathbf{r})$$
$$\hat{\mathcal{H}}_0 u_2(\mathbf{r}) = E_2 u_2(\mathbf{r}) \tag{13.1}$$

where $\mathbf{r}$ represents both spatial and spin coordinates of the electron. In the presence of an electric field

$$\mathcal{E}(t) = E_0 \cos \omega t$$

the perturbation can be represented, as in (12.46), by a Hamiltonian

$$\hat{\mathcal{H}}' = -\hat{\mu} E_0 \cos \omega t \tag{13.2}$$

where $\hat{\mu}$ is the projection of the electric dipole operator $e\mathbf{r}$ along the field direction, so that the total Hamiltonian is $\hat{\mathcal{H}}_0 + \hat{\mathcal{H}}'$. (In the case of electric dipole transitions due to, say, a y-polarized electromagnetic field, we have

---

[1]"Resonance" here means that the photon energy $\hbar\omega$ is nearly equal to the transition energy $|E_2 - E_1|$.

$\hat{\mu} = ey$). In the presence of the perturbation $\hat{\mathcal{H}}'$ we take the wavefunction as a linear superposition of u_1 and u_2:

$$\psi(t) = a_1(t) u_1 e^{-i\omega_1 t} + a_2(t) u_2 e^{-i\omega_2 t} \tag{13.3}$$

where $\omega_{1,2} = E_{1,2}/\hbar$, and $a_1(t)$ and $a_2(t)$ are to be determined. $\psi(t)$ satisfies the Schrödinger time-dependent equation $\hat{\mathcal{H}}\psi = i\hbar\dot{\psi}$, which, in this case, becomes

$$\left(\hat{\mathcal{H}}_0 - \frac{\hat{\mu} E_0}{2}\left(e^{i\omega t} + e^{-i\omega t}\right)\right)\psi = i\hbar \frac{\partial \psi}{\partial t} \tag{13.4}$$

Note that if $E_0 = 0$, then (13.4) is satisfied by taking $a_1(t)$ and $a_2(t)$ as constant in time. We substitute (13.3) into the last equation, multiply both sides by u_2^*, and integrate. Using the orthonormality relations

$$\langle 1|2\rangle = 0, \qquad \langle 1|1\rangle = \langle 2|2\rangle = 1 \tag{13.5}$$

as well as

$\mu_{11} = \mu_{22} = 0$ (This will be the case if u_1 and u_2 have definite parity since, then, according to the discussion of Section 6.3, $y_{11} = y_{22} = 0$.)

we obtain

$$i\hbar \frac{da_2}{dt} = -\frac{\mu_{21} E_0}{2} a_1 \left(e^{i(\omega_2 - \omega_1 + \omega)t} + e^{i(\omega_2 - \omega_1 - \omega)t}\right) \tag{13.6}$$

We limit our treatment to the case where the applied frequency is nearly equal to that of the transition, so that $\omega_2 - \omega_1 \simeq \omega$. In this case the first exponential on the right side of (13.6) oscillates at $\sim 2\omega$ and will average to zero over intervals long compared with the oscillation period $2\pi/\omega$. We thus keep only the second "slow" term $\exp[i(\omega_2 - \omega_1 - \omega)t]$ so that approximately

$$i\hbar \frac{da_2}{dt} = \frac{\mu E_0}{2} a_1 e^{-i\Delta t} \tag{13.7}$$

and

$$i\hbar \frac{da_1}{dt} = \frac{\mu E_0}{2} a_2 e^{i\Delta t}$$

where

$$\mu_{12} = \mu_{21}^* \equiv -\mu$$

$$\Delta \equiv \omega - (\omega_2 - \omega_1) \quad (\Delta = \text{"frequency offset"})$$

These equations can now be solved to yield $a_1(t)$ and $a_2(t)$. We indicate some of the main steps.

Assuming solutions in the form

$$a_{1,2}(t) = A_{1,2} e^{i(S_{1,2})t} \tag{13.8}$$

(where the subscripts 1 and 2 are to be taken one at a time) in (13.7), we find that in order that all terms have the same time dependence, we must have

$$S_2 = S_1 - \Delta \tag{13.9}$$

The two resulting homogeneous equations in A_1 and A_2 possess nontrivial solutions only if the determinant of their coefficients vanishes. This happens when

$$S_1^{(\pm)} = \frac{\Delta \pm \sqrt{\Delta^2 + 4\Omega^2}}{2} = \tfrac{1}{2}(\Delta \pm \omega')$$

so that according to (13.9)

$$S_2^{(\pm)} = \frac{-\Delta \pm \sqrt{\Delta^2 + 4\Omega^2}}{2} = \tfrac{1}{2}(-\Delta \pm \omega') \tag{13.10}$$

where

$$\Omega = \frac{\mu E_0}{2\hbar}, \qquad \omega' = \sqrt{\Delta^2 + 4\Omega^2} \tag{13.11}$$

The total solution for $a_1(t)$ can then be taken as

$$a_1(t) = A_1^+ e^{iS_1^+ t} + A_1^- e^{iS_1^- t} \tag{13.12}$$

A_1^+ and A_1^- remain to be determined.

If at $t=0$ the system is in the lower state, then, according to (13.3), $a_1(0)=1$ and $a_2(0)=0$. It follows from (13.12) that

$$A_1^+ + A_1^- = 1 \tag{13.13}$$

Next we use (13.7) and (13.12) to solve for $a_2(t)$ in terms of A_1^+ and A_1^-. Imposing the initial condition $a_2(0)=0$ leads to

$$\frac{A_1^+}{\Delta - \omega'} + \frac{A_1^-}{\Delta + \omega'} = 0 \tag{13.14}$$

From the last two equations we obtain

$$A_1^+ = \frac{\omega' - \Delta}{2\omega'}, \qquad A_1^- = \frac{\omega' + \Delta}{2\omega'} \tag{13.15}$$

which, when combined with (13.10) and (13.12), results in

$$a_1(t) = \left[\cos\left(\frac{\omega' t}{2}\right) - i\left(\frac{\Delta}{\omega'}\right)\sin\left(\frac{\omega' t}{2}\right)\right] e^{i\Delta t/2} \tag{13.16}$$

$$a_2(t) = -\left(i\frac{\Omega}{\omega'}\right)\sin\left(\frac{\omega' t}{2}\right) e^{-i\Delta t/2} \tag{13.17}$$

These are the final solutions.

For the case of exact resonance, $\omega = \omega_2 - \omega_1$, so that $\Delta = 0$ and $\Omega = \omega'$. In this case $a_1(t)$ and $a_2(t)$ become

$$a_1(t) = \cos(\Omega t), \qquad a_2 = -i\sin(\Omega t) \tag{13.18}$$

so that the occupation probabilities $|a_1(t)|^2$ and $|a_2(t)|^2$ oscillate between zero and unity at a frequency

$$\Omega = \frac{\mu E_0}{2\hbar}$$

The plot of $|a_1(t)|^2$ and $|a_2(t)|^2$ is shown in Fig. 13.1. This oscillatory behavior that on resonance involves complete periodic oscillation of the atom between states 2 and 1, is characteristic of any two linearly coupled oscillators. This behavior is modified strongly if the interaction is interrupted by elastic and inelastic collisions. This point is considered in Section 13.4.

Having solved for $\psi(t)$ in the presence of a perturbing radiation field, we may now obtain the expectation value of different physical observables. For problems involving radiation or absorption by the atoms, we need to determine the expectation value of the dipole matrix element $\int \psi^*(t)\hat{\mu}\psi(t)\,dv$. From (13.3) and the fact that $\langle 1|\hat{\mu}|1\rangle = \langle 2|\hat{\mu}|2\rangle = 0$, we obtain

$$\langle \hat{\mu} \rangle = \mu a_1^*(t) a_2(t) e^{i(\omega_1 - \omega_2)t} + \text{c.c.} \tag{13.19}$$

If the perturbing field $E_0 \cos \omega t$ is applied to a sample containing N atoms, all initially in the ground state u_1, then the resulting dipole moment is

$$\begin{aligned} p(t) &= N\langle \hat{\mu} \rangle \\ &= N\mu a_1^*(t) a_2(t) e^{i(\omega_1 - \omega_2)t} + \text{c.c.} \end{aligned} \tag{13.20}$$

This "giant" dipole will act as a multi-element radiating antenna, and for sufficiently large N, its emitted radiation (at $\omega_2 - \omega_1$) can be readily detected. As a matter of fact, if the perturbing field is turned off when $\Omega t = \pi/2$, then according to (13.18) $a_1(\Omega t = \pi/2) = (2)^{-1/2}$, $a_2(t) = -i(2)^{-1/2}$, and the total radiating dipole has its *maximum* possible value:

$$\begin{aligned} p(t) &= -\frac{iN\mu}{2} e^{i(\omega_1 - \omega_2)t} + \text{c.c.} \\ &= N\mu \sin(\omega_1 - \omega_2)t \end{aligned} \tag{13.21}$$

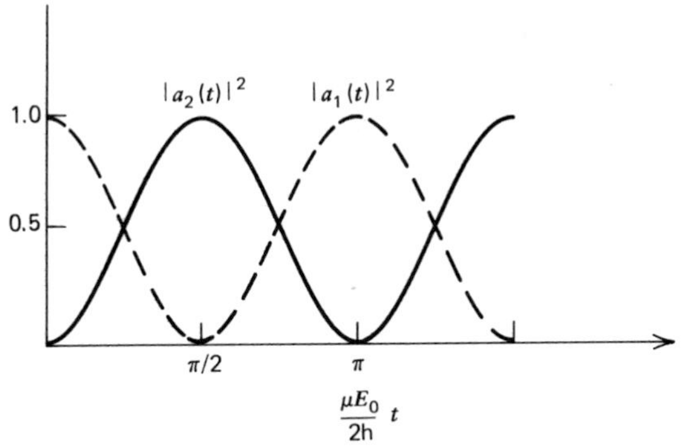

Figure 13.1 The probability $|a_1(t)|^2$ of finding an atom in the lower state and that of the upper state $|a_2(t)|^2$ when the applied field $E_0 \cos \omega t$ is exactly on resonance ($\omega = \omega_2 - \omega_1$).

This state of affairs in which N atoms are "prepared" by a radiation pulse so as to acquire phase coherent dipole moments was termed the superradiant state by Dicke.[2]

Before we close this section, it is important to remind ourselves that the formalism just developed applies only to a collisionless system. This situation is approached in cases where the field intensity is sufficiently high so that the atom can oscillate between the states 1 and 2 a number of times before it collides and "loses" its phase memory. This condition can be written as $\mu E_0/\hbar \gg 1/\tau_{\text{collision}}$. When the reverse is true, that is, when $\mu E_0/\hbar \ll \tau_{\text{collision}}^{-1}$, then according to Section 11.2 we can use the results of time perturbation theory. This collision-dominated regime applies in most experimental situations including that of laser oscillators.

13.2 ABSORPTION AND AMPLIFICATION IN ATOMIC SYSTEMS—COLLISION DOMINATED REGIME

We have shown in Section 12.4 that under the influence of an applied electromagnetic field an atom will be induced to undergo transitions between its eigenstates. If an atom happens to be initially at some eigenstate, which for specificity we shall call state 2, then the induced rate for transitions into state 1 with $E_2 > E_1$ was found in (12.52*a*) to be given by

$$W_i = \frac{\lambda^3 g(\nu)}{8\pi h c n^2 t_{\text{spont}}} I_\nu \tag{13.22}$$

where λ is the wavelength of the radiation, t_{spont} is the spontaneous lifetime for $2 \to 1$ transitions, I_ν is the wave intensity, and $g(\nu)$ is the lineshape function. It must be emphasized that the notion of a lineshape function $g(\nu)$ applies only in the collision regime as does, consequently, the analysis of the remainder of this section. Consider next what happens when a plane electromagnetic wave of intensity I_ν (watts/m^2) at frequency ν passes through a medium with atomic population densities (atoms/m^3) of N_2 and N_1 in levels 2 and 1, respectively. The number of downward ($2 \to 1$) transitions per unit time per unit volume is

$$N_{2\to 1} = N_2 W_i$$

while that in the opposite direction is given by

$$N_{1\to 2} = N_1 W_i$$

The excess of upward over downward transitions per unit time per unit volume is thus $W_i(N_1 - N_2)$. Each $1 \to 2$ transition involves a loss of one photon by the wave, while a $2 \to 1$ transition adds a photon to the wave. The net loss of power by the wave per unit volume is thus

$$\frac{P}{\text{volume}} = (N_1 - N_2) W_i h\nu \tag{13.23}$$

---

[2] R. H. Dicke, Coherence in spontaneous radiation processes, *Phys. Rev.* **93**, 99 (1954).

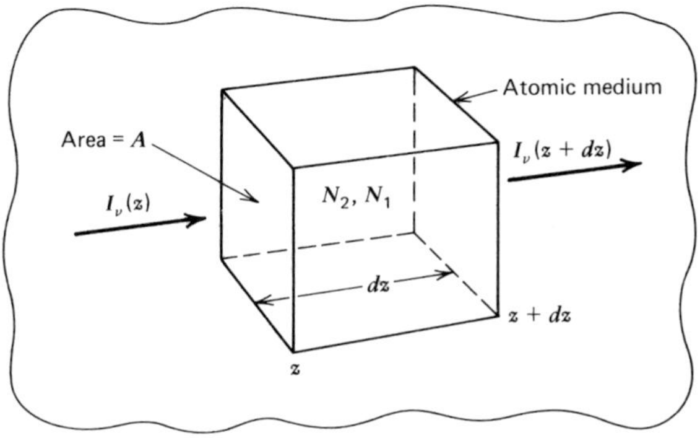

Figure 13.2 A differential volume power "bookkeeping." The power induced by the radiation field $(N_2 - N_1)W_i h\nu A\,dz$ must be equal to the increase in the power carried by the wave. If $N_2 < N_1$, then power is absorbed and the wave must attenuate in its passage through the medium.

This loss of power must be accompanied by a decrease of the wave intensity as shown in Fig. 13.2. In the figure we consider a differential volume $A\,dz$ of an atomic medium with population densities N_2 and N_1. A wave with an intensity $I_\nu(z)$ is incident normally on the face at z and emerges from the face at $z + dz$ with an intensity $I_\nu(z + dz)$. At steady state $(\partial/\partial t = 0)$ the net change in power flow between the two faces is

$$\left[I_\nu(z + dz) - I_\nu(z)\right]A = \frac{dI_\nu}{dz}A\,dz$$

This change must be equal to the power $-(N_1 - N_2)W_i h\nu A\,dz$, which is generated by the atoms within $A\,dz$, whence

$$\frac{dI_\nu}{dz} = -(N_1 - N_2)W_i h\nu \tag{13.24}$$

Using expression (12.52*b*) for W_i in (13.24), we obtain

$$\frac{dI_\nu}{dz} = -(N_1 - N_2)\frac{\lambda^2 g(\nu)}{8\pi n^2 t_{\text{spont}}}I_\nu \tag{13.25}$$

so that the radiation decays according to

$$I_\nu(z) = I_\nu(0)e^{-\alpha z} \tag{13.26}$$

$$\alpha = (N_1 - N_2)\frac{\lambda^2 g(\nu)}{8\pi n^2 t_{\text{spont}}} \tag{13.27a}$$

In media in thermal equilibrium $N_1 = N_2 e^{(E_2 - E_1)/kT}$ so that $N_1 > N_2$ and $\alpha > 0$. Such media will absorb incident radiation with an absorption coefficient given by (13.27*a*). It is possible, however, in many cases to disturb, by "pumping" atoms into level 2, the equilibrium conditions to the point where

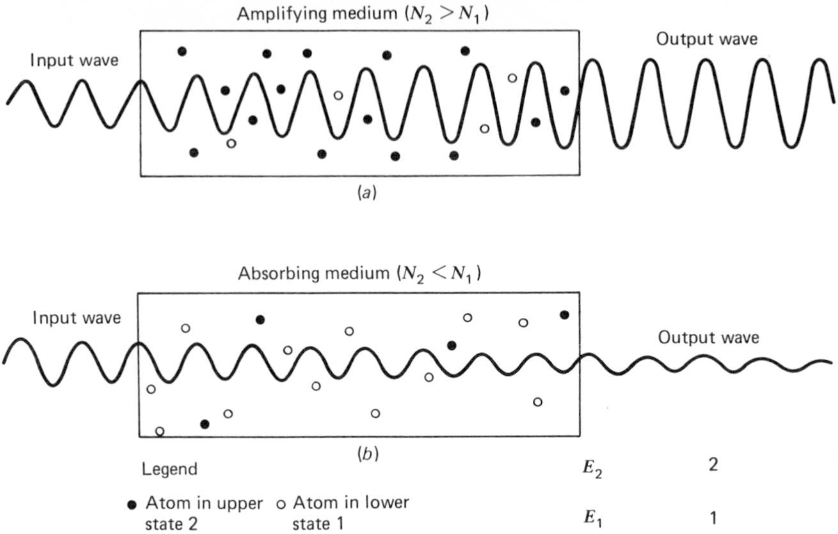

Figure 13.3 Amplification of a traveling electromagnetic wave in (*a*) an inverted population ($N_2 > N_1$), and (*b*) its attenuation in an absorbing ($N_2 < N_1$) medium.

$N_2 > N_1$. This causes a reversal in the sign of α so that, according to (13.26), the incident wave is amplified in its passage through the medium. This is the basis of laser amplification,[3] and the condition $N_2 > N_1$ is referred to as population inversion. The resulting gain constant is thus given by

$$\gamma = -\alpha = (N_2 - N_1)\frac{\lambda^2 g(\nu)}{8\pi n^2 t_{\text{spont}}} \tag{13.27b}$$

The difference in the behavior of absorbing ($N_1 > N_2$) and amplifying media is shown schematically in Fig. 13.3.

13.3 ELECTRIC POLARIZATION, SUSCEPTIBILITY, AND THE DIELECTRIC CONSTANT

In the last section we have shown, using time-dependent perturbation theory, how electromagnetic radiation can be absorbed (or amplified) in its passage through an atomic medium. Another consequence of this interaction is the modification of the phase velocity of the waves. We account, numerically, for this fact by taking the velocity of light in a medium as c/n, where c is the velocity of light in vacuum, $n = \sqrt{\varepsilon/\varepsilon_0}$ is the index of refraction of the medium (ε is its dielectric constant). In this section we derive an expression for ε using a detailed atomic model.

[3]Suggested, in its optical version, by A. L. Schawlow and C. H. Townes, Infrared and optical lasers, *Phys. Rev.* **112**, 1940 (1958).

The physical mechanism responsible for the slowing down of electromagnetic waves is the following: The alternating electric field of a traveling wave exerts a force on the electrons and nuclei in the medium, thus setting them into oscillation at the frequency of the incident wave. The oscillating charges radiate at the same frequency as the incident field so that the total field is a coherent superposition of both constituents. This superposition is such that the wave phase velocity is modified from that in vacuum.

To account for the effect of atomic response on electromagnetic wave propagation we start with Maxwell's equations in a charge free medium

$$\nabla \times \mathbf{E} = -\frac{\partial \mathbf{B}}{\partial t}$$

$$\nabla \times \mathbf{H} = \frac{\partial \mathbf{D}}{\partial t}$$

and take the electric displacement vector **D** as

$$\mathbf{D} = \varepsilon_0 \mathbf{E} + \mathbf{P} = \varepsilon \mathbf{E} \tag{13.28}$$

where **P** is the electric polarization (dipole moment per unit volume) due to the atoms and ε is the dielectric constant. Classically, the polarization is given by

$$\mathbf{P}(t) = \lim_{V \to 0} \frac{1}{V} \sum_{i=1}^{n} \boldsymbol{\mu}_i$$

$\boldsymbol{\mu}_i$ is the dipole moment of atom i. The summation includes all the atoms in V. Quantum mechanically we will use the form

$$\mathbf{P}(t) = N\langle \hat{\boldsymbol{\mu}}(t) \rangle \tag{13.29}$$

where N is the number of atoms per unit volume and $\hat{\boldsymbol{\mu}}(t)$ is the electric dipole operator per atom. In the case, say, of a single electron atom, $\hat{\boldsymbol{\mu}}(t) = -e\mathbf{r}$, where $\mathbf{r}$ is the radius vector from the nucleus to the electron. The dispersive effects due to magnetic moments will not be considered here.

In the linear treatment of absorption and dispersion, an electric field in the form

$$\begin{aligned} E(t) &= E_0 \cos \omega t \\ &= \mathrm{Re}\left(E_0 e^{i\omega t}\right) \end{aligned} \tag{13.30}$$

(here we use a simple scalar notation) will be shown in Section 13.4 to induce in each atom an oscillating electric dipole moment with an expectation value

$$\langle \hat{\boldsymbol{\mu}}(t) \rangle = \mu_0 \cos(\omega t - \phi) \tag{13.31}$$

so that according to (13.29) the polarization (dipole moment per unit volume) is given by

$$\begin{aligned} P(t) &= N\langle \hat{\boldsymbol{\mu}}(t) \rangle \\ &= N\mu_0 \cos(\omega t - \phi) \\ &= \mathrm{Re}\left(P_0 e^{i\omega t}\right) \end{aligned} \tag{13.32}$$

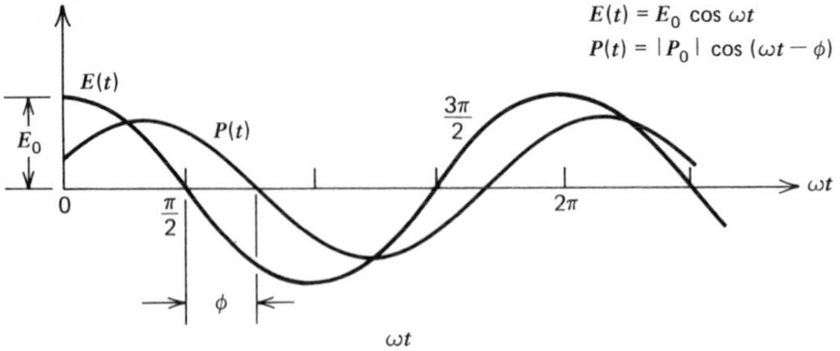

Figure 13.4 The electric field $E(t)$ and the induced polarization in an atomic medium. Note the (relative) lag of $P(t)$ with respect to $E(t)$.

where

$$P_0 = N\mu_0 e^{-i\phi} \tag{13.33}$$

E_0 and P_0 are, respectively, the complex amplitudes of the electric field and the induced polarization. We chose the origin of the time variable such as to render E_0 real. P_0 will, in general, be complex. The phase angle $(-\phi)$ associated with P_0 tells us that $P(t)$ lags $E(t)$ by a time delay ϕ/ω. The situation is depicted in Fig. 13.4.

We will show in Sec. 13.4 that the induced dipole moment per atom μ_0 is proportional to the electric field E_0. We can use this fact to define a physical constant $\chi_a(\omega)$ through the relation

$$P_0 = \varepsilon_0 \chi_a E_0 \tag{13.34}$$

(or)

$$P(t) = \text{Re}\left(\varepsilon_0 \chi_a E_0 e^{i\omega t}\right) \tag{13.35}$$

The complex constant χ_a—the electric susceptibility—is then a measure of the polarizability of the atomic medium. If we express χ_a by

$$\chi_a = \chi_a' - i\chi_a'' \quad (\chi_a', \chi_a'' \text{ real}) \tag{13.36}$$

then from (13.35)

$$P(t) = \varepsilon_0 \chi_a' E_0 \cos \omega t + \varepsilon_0 \chi_a'' E_0 \sin \omega t \tag{13.37}$$

Since the electric field was taken as $E(t) = E_0 \cos \omega t$, a real value of χ_a (i.e., $\chi_a'' = 0$) signifies that $E(t)$ and $P(t)$ are in phase. The complex susceptibility χ_a can be related to the complex dielectric constant (permittivity) ε by using the constitutive relations

$$\begin{aligned} D_0 &= \varepsilon_0 E_0 + P_0 = \varepsilon_0 (1 + \chi) E_0 \\ &= \varepsilon_0 (1 + \chi_b + \chi_a) E_0 = \varepsilon E_0 \end{aligned} \tag{13.38}$$

where χ_b accounts for the residual (background) polarization in the medium, while χ_a represents the contribution from the specific atomic transition under study.

The total dielectric constant of the medium is obtained from (13.28) and (13.38) as

$$\varepsilon = \varepsilon_b\left(1 + \frac{\varepsilon_0}{\varepsilon_b}\chi_a\right) \tag{13.39}$$

where $\varepsilon_b = \varepsilon_0(1+\chi_b)$ is thus the background dielectric constant which accounts for all sources of polarization except the specific atomic transition labeled by the subscript a.

The Significance of $\chi_a(\omega)$

We have shown in Section 12.1 that a plane monochromatic electromagnetic wave propagating inside a medium with a dielectric constant ε has the form of

$$\begin{aligned} E(z,t) &= E_0\cos(\omega t - k'z) \\ &= \mathrm{Re}\left[E_0 e^{i(\omega t - k'z)}\right] \end{aligned} \tag{13.40}$$

where

$$k' = \omega\sqrt{\mu\varepsilon} \tag{13.41}$$

Using (13.39) we obtain

$$\begin{aligned} k' &= \omega\sqrt{\mu\varepsilon} \\ &= \omega\sqrt{\mu\varepsilon_b\left[1+(\varepsilon_0/\varepsilon_b)\chi_a(\omega)\right]} \\ &\simeq \omega\sqrt{\mu\varepsilon_b}\left(1+\frac{\varepsilon_0}{2\varepsilon_b}\chi_a(\omega)\right) \end{aligned} \tag{13.42}$$

where the last approximation took advantage of the fact that usually $(\varepsilon_0|\chi_a|/2\varepsilon_b) \ll 1$. Defining $\omega\sqrt{\mu\varepsilon_b} \equiv k$ and expressing $\chi_a(\omega)$ in terms of χ'_a and χ''_a, as in (13.36), we obtain

$$k' = k\left(1+\frac{\chi'_a(\omega)}{2n^2}\right) - ik\frac{\chi''_a(\omega)}{2n^2} \tag{13.43}$$

where $n = (\varepsilon_b/\varepsilon_0)^{1/2}$ is the background index of refraction (physically, the index of refraction experienced by waves whose frequency is far from resonance with the specific atomic transition considered). The effect of the atomic polarization on wave propagation is best appreciated by substituting (13.43) in (13.40) so that the wave propagation is described by

$$E(z,t) = \mathrm{Re}\left\{E_0 \exp i\left[\omega t - k\left(1+\frac{\chi'_a(\omega)}{2n^2}\right)z\right] - k\frac{\chi''_a(\omega)}{2n^2}z\right\} \tag{13.44}$$

The effect of χ'_a is thus merely to modify the propagation constant k by the factor $(1+\chi'_a/2n^2)$, or equivalently to lower the wave velocity by this factor. The effect of $\chi''_a(\omega)$ is to give rise to an attenuation [when $\chi''_a(\omega) > 0$] or amplification [when $\chi''_a(\omega) < 0$] of the electromagnetic wave.

All that remains is to derive the expression for the complex susceptibility $\chi_a(\omega)$.

13.4 DENSITY MATRIX DERIVATION OF THE ATOMIC SUSCEPTIBILITY

In this section we apply the density matrix formalism developed in Section 11.3 to derive an expression for the susceptibility of an ensemble of atoms (or spins, ions, etc.) interacting with a time-harmonic electromagnetic field. The assumption is made that *only two levels*, with energies E_1 and E_2, are involved in the interaction as shown in Fig. 13.5. This assumption is justified when the angular frequency ω of the field satisfies $\omega \sim (E_2 - E_1)/\hbar$. As a result the density matrix is reduced to a 2×2 matrix with elements $\rho_{11}, \rho_{12}, \rho_{21}, \rho_{22}$.

We assume that the interaction Hamiltonian $\hat{\mathcal{H}}'(t)$ is of the dipole type and can be written as[4]

$$\hat{\mathcal{H}}'(t) = -\hat{\mu}E(t) \tag{13.45}$$

where $\hat{\mu}$ is the projection of the dipole operator along the direction of field $\mathbf{E}(t)$. In our initial analysis field $E(t)$ will not be quantized, and will be considered as a classical variable. The diagonal matrix elements of $\hat{\mu}$ are taken as zero

$$\mu_{11} = \mu_{22} = 0 \tag{13.46}$$

as appropriate to electric dipole transitions between states of definite parity. The phases of the eigenfunctions $|2\rangle$ and $|1\rangle$ are taken, without loss of generality, such that

$$\mu_{21} = \mu_{12} \equiv \mu \tag{13.47}$$

The total Hamiltonian of the two-level system is taken

$$\hat{\mathcal{H}} = \hat{\mathcal{H}}_0 + \hat{\mathcal{H}}' \tag{13.48}$$

where $\hat{\mathcal{H}}_0$ is the Hamiltonian of the system in the absence of any field $[E(t)=0]$, and $\hat{\mathcal{H}}'$ is the atom-field Hamiltonian. The effect of relaxation and collisions is not included in $\hat{\mathcal{H}}$ and will be considered separately.

Our task consists of calculating the ensemble average $\langle\hat{\mu}\rangle$ of the dipole moment of the atom that is induced by field $E(t)$. For the matrix representation of $\hat{\mu}$ and $\hat{\rho}$ we choose the (energy) eigenfunctions u_n of $\hat{\mathcal{H}}_0$ satisfying $\hat{\mathcal{H}}_0 u_n = E_n u_n$. Using (11.45) and (13.47) we obtain

$$\langle\hat{\mu}\rangle = \mathrm{tr}(\hat{\rho}\hat{\mu})$$

$$= \mathrm{tr}\left\{\begin{pmatrix} \rho_{11} & \rho_{12} \\ \rho_{21} & \rho_{22} \end{pmatrix} \begin{pmatrix} 0 & \mu \\ \mu & 0 \end{pmatrix}\right\} \tag{13.49}$$

$$= \mu(\rho_{12} + \rho_{21}) \tag{13.50}$$

The matrix elements of ρ are obtained using (11.49):

$$\frac{d\hat{\rho}}{dt} = -\frac{i}{\hbar}\left[(\hat{\mathcal{H}}_0 + \hat{\mathcal{H}}'), \hat{\rho}\right] \tag{13.51}$$

---

[4]The form (13.45) of $\hat{\mathcal{H}}'(t)$ can be shown to be valid in the limit where the wavelength of the radiation is long compared to atomic dimensions.

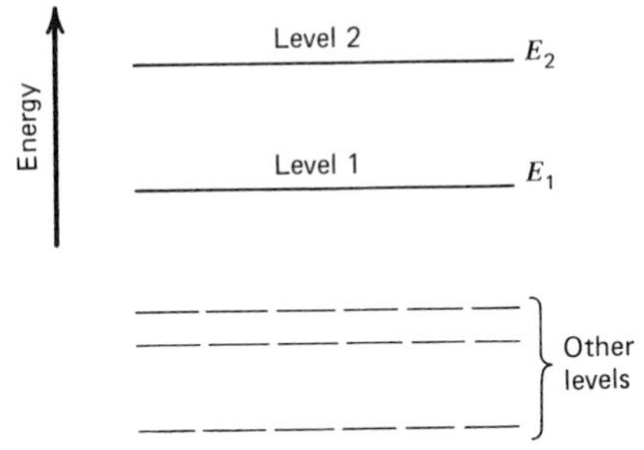

Figure 13.5 A two-level atomic system interacting with a radiation field whose frequency ω is nearly resonant with the transition $2 \leftrightarrow 1$, that is, $\hbar\omega \simeq E_2 - E_1$. Other levels that do not participate in the interaction are shown by dashed lines and are not involved explicitly in the analysis.

In the u_n representation the matrices corresponding to $\hat{\mathcal{H}}_0\hat{\rho}$ and $\hat{\mathcal{H}}'\hat{\rho}$ are

$$\hat{\mathcal{H}}_0\hat{\rho} = \begin{pmatrix} E_1 & 0 \\ 0 & E_2 \end{pmatrix} \begin{pmatrix} \rho_{11} & \rho_{12} \\ \rho_{21} & \rho_{22} \end{pmatrix}$$

$$\hat{\mathcal{H}}'\hat{\rho} = -E(t)\begin{pmatrix} 0 & \mu \\ \mu & 0 \end{pmatrix} \begin{pmatrix} \rho_{11} & \rho_{12} \\ \rho_{21} & \rho_{22} \end{pmatrix} \tag{13.52}$$

so that

$$\frac{d\rho_{21}}{dt} = -i\omega_0\rho_{21} + i\frac{\mu E(t)}{\hbar}(\rho_{11} - \rho_{22})$$

$$\frac{d}{dt}(\rho_{11} - \rho_{22}) = i\frac{2\mu E(t)}{\hbar}(\rho_{21} - \rho_{21}^*) \tag{13.53}$$

where we defined

$$\omega_0 \equiv (E_2 - E_1)/\hbar$$

and used the Hermiticity of $\hat{\rho}$ to write $\rho_{21}^* = \rho_{12}$. From the basic ensemble definition (11.43)

$$\rho_{nm} = \overline{C_n C_m^*}$$

$$= \frac{1}{N}\sum_{i=1}^{N} C_n^{(i)} C_m^{(i)*} \tag{13.54}$$

where the summation is over the N atoms of the ensemble. It follows that when $E(t)$ is turned off the off-diagonal elements $\rho_{nm}\, n \neq m$ relax back to zero. This is due to the fact that $C_n^{(i)}C_m^{(i)*}$ of the ith atom is a complex number whose phase will vary from one atom to another, and which changes randomly after each atomic collision. The sum of (13.54) will thus tend toward zero in a few collisions.

The modification of the phase of the wavefunction by an elastic collision can be understood qualitatively on the basis of the material we have already digested. Imagine a gas atom in the quantum state n with a wave function $u_n(\mathbf{r})\exp(-iE_n t/\hbar)$. Let this atom be exposed during a "collision" to the Coulomb forces of another atom moving close by. Let us assume that the collision lasts exactly τ seconds and that the interaction leads to a change ΔE_n in the state energy, where ΔE_n can be calculated using the results of Section 11.1. During the "collision" the wavefunction is given approximately by

$$u_n(\mathbf{r})\exp\left(-i\frac{E_n+\Delta E_n}{\hbar}t\right)$$

so that after the collision is "turned-off" it becomes

$$u_n(r)\exp\left(-i\frac{E_n}{\hbar}t-i\phi\right)$$

where

$$\phi=\frac{\Delta E_n\tau}{\hbar}$$

The effect of collisions is thus to modify the wavefunction by the phase factor $\exp(-i\phi)$. Since the collisions are random, it follows that the ensemble average of (13.54) with $n\neq m$ would tend to zero.

Using similar arguments we expect that the diagonal elements ρ_{nn} will relax toward their equilibrium values $\rho_{nn}^{(0)}$.

The above requirements are satisfied mathematically by adding to (13.53) and (13.54) phenomenological relaxation terms according to

$$\frac{d\rho_{21}}{dt}=-i\omega_0\rho_{21}+i\frac{\mu E(t)}{\hbar}(\rho_{11}-\rho_{22})-\frac{\rho_{21}}{T_2} \tag{13.55}$$

$$\frac{d}{dt}(\rho_{11}-\rho_{22})=i\frac{2\mu E(t)}{\hbar}(\rho_{21}-\rho_{21}^*) - \frac{(\rho_{11}-\rho_{22})-(\rho_{11}^{(0)}-\rho_{22}^{(0)})}{\tau} \tag{13.56}$$

(Here we follow a symbol convention dating back to F. Bloch.[5]) T_2 is the so-called "transverse relaxation time." It is the time constant for loss of phase coherence between individual atoms of the ensemble when $E(t)$ is turned off. The diagonal elements ρ_{11} and ρ_{22} will regain their equilibrium values with a time constant τ.

In the case of a sinusoidal field

$$E(t)=\frac{E_0}{2}(e^{i\omega t}+e^{-i\omega t})$$

it is convenient to transform to a set of variables σ_{21}, σ_{12} defined by

$$\begin{aligned}\sigma_{21}(t)&=\rho_{21}(t)e^{i\omega t}\\ \sigma_{12}(t)&=\sigma_{21}^*(t)=\rho_{12}(t)e^{-i\omega t}\end{aligned} \tag{13.57}$$

---

[5]F. Bloch, W. W. Hansen, and H. M. Packard, Nuclear induction, *Phys. Rev.* **70**, 960 (1946). The population relaxation time, τ in our notation, is designated by T_1.

so that (13.53) becomes

$$\frac{d}{dt}\left[\sigma_{21}e^{-i\omega t}\right]=-i\omega_0\sigma_{21}e^{-i\omega t}+i\frac{\mu E_0}{2\hbar}\left(e^{i\omega t}+e^{-i\omega t}\right)(\rho_{11}-\rho_{22})-\frac{\sigma_{21}}{T_2}e^{-i\omega t} \tag{13.58}$$

Since σ_{ij} and ρ_{ii} are "slowly" varying in comparison to $\exp(\pm i\omega t)$, the term involving $\exp(i\omega t)$ on the right side is nonsynchronous (i.e., oscillates with a different frequency) with all the other terms, so that its effect averages out to zero over time intervals large compared with $2\pi/\omega$. Neglecting it, we write

$$\frac{d\sigma_{21}}{dt}=-i(\omega_0-\omega)\sigma_{21}+i\frac{\mu E_0}{2\hbar}(\rho_{11}-\rho_{22})-\frac{\sigma_{21}}{T_2} \tag{13.59}$$

and, from (13.56),

$$\frac{d}{dt}(\rho_{11}-\rho_{22})=i\frac{\mu E_0}{\hbar}(\sigma_{21}-\sigma_{21}^*)-\frac{(\rho_{11}-\rho_{22})-\left(\rho_{11}^{(0)}-\rho_{22}^{(0)}\right)}{\tau} \tag{13.60}$$

In (13.60) we kept only the terms with no time dependence in the exponents.

At steady state we can set all the time derivatives in (13.59) and (13.60) equal to zero. By first adding (13.59) and its complex conjugate, and then subtracting them, we obtain

$$-\frac{1}{T_2}(\sigma_{21}-\sigma_{21}^*)-i(\omega_0-\omega)(\sigma_{21}+\sigma_{21}^*)+i\frac{\mu E_0}{\hbar}(\rho_{11}-\rho_{22})=0$$

$$-i(\omega_0-\omega)(\sigma_{21}-\sigma_{21}^*)-\frac{1}{T_2}(\sigma_{21}+\sigma_{21}^*)=0 \tag{13.61a}$$

while (13.60) becomes

$$i\frac{\mu E_0}{\hbar}(\sigma_{21}-\sigma_{21}^*)-\frac{1}{\tau}(\rho_{11}-\rho_{22})=-\frac{\rho_{11}^{(0)}-\rho_{22}^{(0)}}{\tau} \tag{13.61b}$$

The three algebraic equations for $(\sigma_{21}-\sigma_{21}^*)$, $(\sigma_{21}+\sigma_{21}^*)$, and $(\rho_{11}-\rho_{22})$ can be solved straightforwardly. As an example:

$$\sigma_{21}-\sigma_{21}^*=2i\,\mathrm{Im}\,\sigma_{21}$$

$$=\frac{\begin{vmatrix} -\frac{1}{\tau}\left(\rho_{11}^{(0)}-\rho_{22}^{(0)}\right) & 0 & \tau^{-1} \\ 0 & -i(\omega_0-\omega) & i2\Omega \\ 0 & -T_2^{-1} & 0 \end{vmatrix}}{\begin{vmatrix} i2\Omega & 0 & -\tau^{-1} \\ -T_2^{-1} & -i(\omega_0-\omega) & i2\Omega \\ -i(\omega_0-\omega) & -T_2^{-1} & 0 \end{vmatrix}}$$

so that

$$\mathrm{Im}\,\sigma_{21}=\frac{\Omega T_2(\rho_{11}-\rho_{22})_0}{1+(\omega-\omega_0)^2T_2^2+4\Omega^2T_2\tau} \tag{13.62}$$

Similarly,

$$\mathrm{Re}\,\sigma_{21}=\frac{(\omega_0-\omega)T_2^2\Omega\left(\rho_{11}^{(0)}-\rho_{22}^{(0)}\right)}{1+(\omega-\omega_0)^2T_2^2+4\Omega^2T_2\tau}$$

$$\Omega\equiv\mu E_0/2\hbar \tag{13.63}$$

Returning to our basic definition (13.50),

$$\begin{aligned}\langle\hat{\mu}(t)\rangle&=\mu(\rho_{12}+\rho_{21})\\&=\mu\left(\sigma_{21}e^{-i\omega t}+\sigma_{21}^{*}e^{i\omega t}\right)\\&=2\mu\,\mathrm{Re}\left(\sigma_{21}e^{-i\omega t}\right)\\&=2\mu\left[(\mathrm{Re}\,\sigma_{21})\cos\omega t+(\mathrm{Im}\,\sigma_{21})\sin\omega t\right]\end{aligned} \tag{13.64}$$

The polarization-dipole moment per unit volume is

$$\begin{aligned}P(t)&=N\langle\mu(t)\rangle\\&=\frac{\mu^2\left(N_1^{(0)}-N_2^{(0)}\right)T_2}{\hbar}\left(\frac{(\omega_0-\omega)T_2E_0\cos\omega t+E_0\sin\omega t}{1+(\omega-\omega_0)^2T_2^2+4\Omega^2T_2\tau}\right)\end{aligned} \tag{13.65}$$

where the population densities are

$$N_1^{(0)}=N\rho_{11}^{(0)},\qquad N_2^{(0)}=N\rho_{22}^{(0)}$$

Comparing (13.65) with (13.37) we obtain our final result:

$$\chi_a''(\omega)=\frac{\mu^2T_2\left(N_1^{(0)}-N_2^{(0)}\right)}{\varepsilon_0\hbar}\;\frac{1}{1+(\omega-\omega_0)^2T_2^2+4\Omega^2T_2\tau}$$

$$\chi_a'(\omega)=\frac{\mu^2T_2\left(N_1^{(0)}-N_2^{(0)}\right)}{\varepsilon_0\hbar}\;\frac{(\omega_0-\omega)T_2}{1+(\omega-\omega_0)^2T_2^2+4\Omega^2T_2\tau} \tag{13.66a}$$

$\chi_a''(\omega)$ and $\chi_a'(\omega)$ are plotted in Fig. 13.6 for the weak field case. The functional dependence of $\chi_a''(\omega)\propto[1+(\omega-\omega_0)^2T_2^2]^{-1}$ is called *Lorentzian.* The width of the atomic response curve is taken conventionally as the separation between the two frequencies, where $\chi''(\omega)/\chi''(\omega_0)=0.5$. This width is

$$\Delta\omega=2/T_2 \tag{13.66b}$$

From Eq. (13.44) we concluded that the induced dipole moment is associated with an absorption of the inducing radiation with an absorption coefficient

$$\alpha(\omega)=\frac{k\chi_a''(\omega)}{2n^2} \tag{13.66c}$$

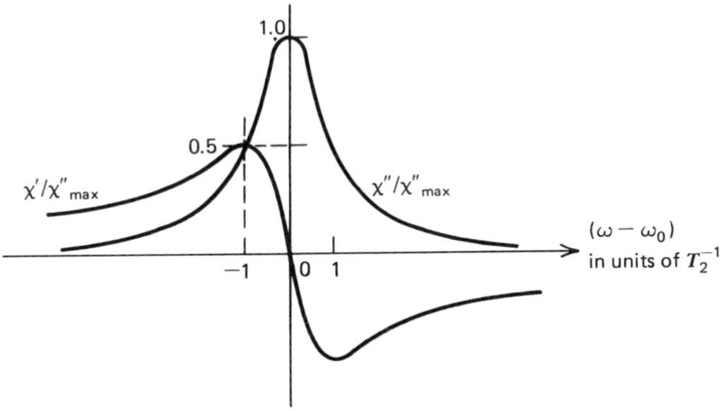

Figure 13.6 A plot of the real (χ') and imaginary (χ'') parts of the susceptibility for negligible saturation ($\mu^2 E_0^2 T_2 \tau/\hbar^2 \ll 1$).

which, using (13.66), becomes

$$\alpha(\omega) = \frac{\mu^2 T_2\left(N_1^{(0)} - N_2^{(0)}\right)}{\varepsilon_0 \hbar\left[1 + (\omega - \omega_0)^2 T_2^2\right]}, \qquad \Omega^2 T_2 \tau \ll 1 \tag{13.67}$$

At this point we have two expressions for the atomic absorption coefficient. The first, (13.27), was derived from Fermi's golden rule and assumed a general lineshape function $g(\nu)$ for the atomic transition. The second expression, (13.67), is derived from a model that incorporates "collision" time constants T_2 and τ. Since the two expressions describe the same physical phenomenon, we may equate them to each other. We also substitute for t_{spont} in (13.27) its value according to (12.71):

$$\begin{aligned} t_{\text{spont}}^{-1} &= \frac{2n^3 e^2 |y_{12}|^2 \omega^3}{\varepsilon h c^3} \\ &= \frac{2n^3 \mu^2 \omega^3}{\varepsilon h c^3} \qquad \left(\mu^2 \equiv e^2 |y_{12}|^2\right) \end{aligned}$$

The result is an explicit expression for the collision dominated lineshape function

$$\begin{aligned} g(\nu) &= \frac{2T_2}{1 + 4\pi^2(\nu - \nu_0)^2 T_2^2} \\ &= \frac{\Delta\nu/2\pi}{(\nu - \nu_0)^2 + (\Delta\nu/2)^2} \end{aligned} \tag{13.68}$$

The width of $g(\nu)$, the atomic response function, which we designate as $\Delta\nu$ is thus related to the atomic "collision" time T_2 by

$$\Delta\nu = 1/\pi T_2 \tag{13.69}$$

The theoretical tools we have just developed can be used to treat a very large number of experimental situations involving the interaction of light and atoms. In the next chapter we apply them to the special, but important, case of laser oscillators.

The Significance of $\chi'(\omega)$

The effect of the interaction of a traveling electromagnetic wave with an atomic transition was shown in Section 13.3 to cause the propagation to take the form

$$E(z,t)=E_0\exp\left[i\omega t-ik\left(1+\frac{\chi_a'(\omega)}{2n^2}\right)z-k\frac{\chi_a''(\omega)}{2n^2}z\right] \qquad (13.70)$$

where n_0 is the index of refraction of the medium at frequencies far from resonance and $k=\omega n_0/c$. The phase velocity of the wave is thus

$$\begin{aligned} v_p &= \frac{\omega}{k\left[1+\chi_a'(\omega)/2n^2\right]} \\ &= \frac{c}{n\left(1+\chi_a'(\omega)/2n^2\right)} \\ &= c/n(\omega) \end{aligned} \qquad (13.71)$$

where

$$n(\omega)=n\left(1+\chi_a'(\omega)/2n^2\right) \qquad (13.72)$$

In transitions where $N_1>N_2$, the effect of the interaction is thus to slow the wave down at frequencies $\omega<\omega_0$, where $\chi_a'(\omega)>0$. At $\omega>\omega_0$ the effect of dispersion is to increase the phase velocity. In inverted atomic media, $N_1<N_2$, the sign of $\chi_a'(\omega)$ is reversed and the phase velocity exceeds c/n at $\omega<\omega_0$, while at $\omega>\omega_0$ the wave is slowed down.

PROBLEMS

1. Compare the formalism leading to Eq. (13.66) with that used by Bloch to describe absorption and dispersion in magnetic resonance (see the cited reference).
2. Derive Eq. (13.66*a*).
3. Show that if we neglect saturation, that is, $\Omega^2T_2\tau\ll 1$, then the tip of the vector $\chi_a'(\omega)+i\chi_a''(\omega)$ describes a circle as ω is varied.
4. In the collision-free two-level description of atomic transition (see Section 13.1), the exchange of energy between an atom and an electromagnetic field is periodic so that the time-averaged exchange is zero. In the collision dominated regime [see (13.23)] energy is either absorbed or emitted, depending on the sign of N_1-N_2. Can you explain the difference? (The answer is not mathematical unless you are willing to go to immensely sophisticated and complicated lengths. Try simple intuitive reasoning.)

CHAPTER FOURTEEN

Laser Oscillation

The possibility of using an inverted atomic population to amplify and generate electromagnetic waves was suggested during 1953–1955 by a number of groups in the U.S. and U.S.S.R.[1] Following a proposal by Schawlow and Townes,[2] Maiman[3] in 1960 demonstrated a laser (acronym for "light amplification by stimulated emission of radiation") oscillator, using a ruby crystal. This marked a genuine revolution in optics, since all the light sources previously available were incoherent—that is, not monochromatic. The highly coherent nature of the laser has made it possible to use and manipulate light today in much the same way that radio waves and microwaves are handled.

14.1 LASER OSCILLATION

We have already shown in Chapter 13 that an inverted atomic population—that is, one wherein a high quantum state E_2 has a higher population than a lower state E_1—is capable of amplifying, by stimulated emission, electromagnetic radiation at frequencies near $(E_2 - E_1)/h$. Now if such a medium is placed inside a structure that supports an electromagnetic mode—a resonator—and if the amount of power that is stimulated exceeds the losses in the resonator, a steady oscillation of the electromagnetic field can result.

Consider the basic configuration sketched in Fig. 14.1. The inverted atomic medium is placed inside a Fabry–Perot etalon, which is an optical

[1] J. Weber, Amplification of microwave radiation by substances not in thermal equilibrium, *IRE Trans. Prof. Group on Electron Devices* **3**, 1 (1953); J. P. Gordon, H. J. Zeiger, and C. H. Townes, Molecular microwave oscillator and new hyperfine structure in the microwave spectrum of NH_3, *Phys. Rev.* **95**, 282 (1954); The maser—new type of microwave amplifier, frequency standard, and spectrometer, *ibid.* **99**, 1264 (1955); N. G. Basov, and A. M. Prokhorov, Application of molecular beams to the radio spectroscopic study of the rotation spectrum of molecules, *J. Expt. Theoret. Phys. (USSR)* **27**, 431 (1954); On the possible methods of producing active molecules for a molecular generator, *ibid.* **28**, 249 (1955).

[2] A. L. Schawlow and C. H. Townes, Infrared and optical masers, *Phys. Rev.* **112**, 1940 (1958).

[3] T. H. Maiman, Stimulated optical radiation in ruby, *Nature* **187**, 493 (1960).

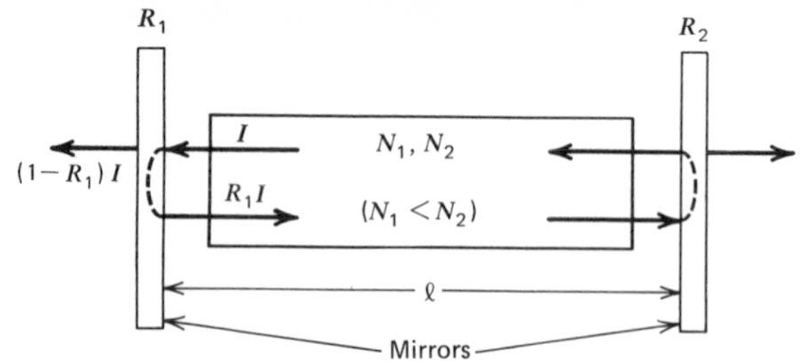

Figure 14.1 A laser oscillator consisting of an atomic medium with an inverted population ($N_2 > N_1$) placed between two plane parallel mirrors with reflectivities R_1 and R_2.

resonator formed by two opposing plane parallel (or slightly curved) mirrors with power reflectivities R. The exponential gain constant of an electromagnetic wave at frequency ν due to the inverted medium is given by (13.27) as

$$\gamma(\nu) = (N_2 - N_1)\frac{\lambda^2 g(\nu)}{8\pi n^2 t_{\text{spont}}} \tag{14.1}$$

where $g(\nu)$ is the normalized lineshape function of the $2 \to 1$ transition, λ is its vacuum wavelength, and N_2 and N_1 are the population densities of the upper and lower levels, respectively. Let a wave with intensity I be launched inside the medium in a direction normal to the reflectors. The wave intensity after a complete round trip is

$$IR_1 R_2 e^{2(\gamma - \alpha)l} \tag{14.2}$$

where α accounts for residual optical losses in the medium, while the factor $R_1 R_2$ represents the reduction in intensity due to the two reflections from the end mirrors. For sustained oscillation it is necessary that the intensity $IR_1 R_2 \exp[2(\gamma - \alpha)l]$ after one round trip equal the starting intensity I, that is, at oscillation,

$$R_1 R_2 e^{2(\gamma - \alpha)l} = 1$$

or, solving for the oscillation threshold gain,

$$\gamma = \alpha - \frac{1}{2l} \ln R_1 R_2 \tag{14.3}$$

The Fabry–Perot Laser

A more rigorous approach to the derivation of the laser oscillation condition is to consider what happens to the electric field rather than the intensity of a

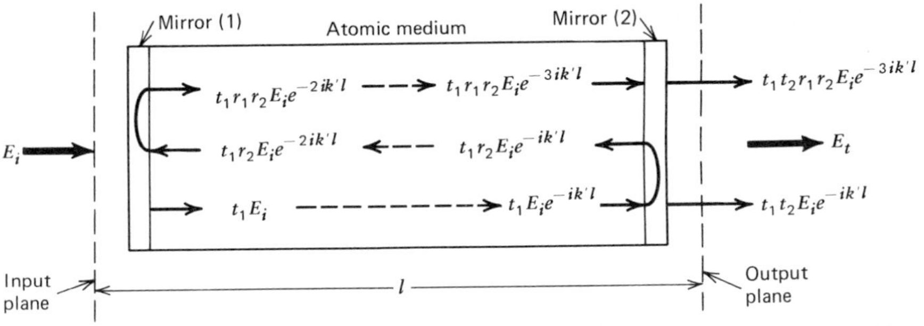

Figure 14.2 Model used to analyze a laser oscillator. A laser medium (i.e., one with an inverted atomic population) with a complex propagation constant

$$k'(\omega)=k\left(1+\frac{\chi'(\omega)}{2n^2}\right)-ik\frac{\chi''(\omega)}{2n^2}-i\frac{\alpha}{2}$$

is placed between two reflecting mirrors.

wave bouncing between two reflectors (Fabry–Perot étalon[4]) that contains an inverted population such as shown in Fig. 14.2. We do so formally by letting a traveling plane electromagnetic wave[5]

$$E(z,t)=E_i e^{i(\omega t-kz)}$$

be incident from the left on the resonator. The propagation constant of the electromagnetic wave inside the medium filling the resonator is taken from (13.43) as

$$k'(\omega)=k+k\frac{\chi_a'(\omega)}{2n^2}-ik\frac{\chi_a''(\omega)}{2n^2}-\frac{i\alpha}{2} \tag{14.4}$$

where α is the intensity absorption coefficient of the medium due to all possible mechanisms but excluding $2\leftrightarrow$transitions, which are represented by χ_a'', $k=\omega n/c$ is the propagation constant in the medium at frequencies well removed from that of the laser transition. $\chi_a(\omega)=\chi_a'(\omega)-i\chi_a''(\omega)$ is the complex dielectric susceptibility due to the laser transition.

The ratio of transmitted to incident fields at the left mirror is taken as t_1 and that at the right mirror as t_2. The ratios of reflected to incident fields inside the laser medium at the left and right reflectors are r_1 and r_2, respectively.

The propagation factor corresponding to a single transit is $\exp(-ik'l)$ where k' is given by (14.4) and l is the length of the etalon.

---

[4]The étalon (standard in French) is an optical resonator formed by opposing two plane parallel mirrors named after its inventors. [See C. Fabry and A. Perot, *Ann. Chim. Phys.* **16**, 115 (1899).]

[5]The real space- and time-dependent electric field should be written as $E(z,t)=\mathrm{Re}[E_i e^{i(\omega t-kz)}]$. We merely omit the Re sign to simplify the notation.

Adding the partial waves at the output plane to get the total outgoing wave E_t, we obtain

$$E_t = t_1 t_2 E_i e^{-ik'l}\left(1 + r_1 r_2 e^{-i2k'l} + r_1^2 r_2^2 e^{-i4k'l} + \cdots\right) \tag{14.5}$$

which is a geometric progression with a sum (if $|r_1 r_2 e^{-i2k'l}| \leq 1$)

$$\begin{aligned} E_t &= E_i\left(\frac{t_1 t_2 e^{-ik'l}}{1 - r_1 r_2 e^{-i2k'l}}\right) \\ &= E_i\left(\frac{t_1 t_2 e^{-i(k+\Delta k)l} e^{(\gamma-\alpha)l/2}}{1 - r_1 r_2 e^{-2i(k+\Delta k)l} e^{(\gamma-\alpha)l}}\right) \end{aligned} \tag{14.6}$$

where we used (14.4) $k' = k + \Delta k + i(\gamma - \alpha)/2$ with

$$\Delta k = k\frac{\chi_a'(\omega)}{2n^2} \tag{14.7}$$

$$\begin{aligned} \gamma &= -k\frac{\chi_a''(\omega)}{n^2} \\ &= (N_2 - N_1)\frac{\lambda^2}{8\pi n^2 t_{\text{spont}}} g(\nu) \end{aligned} \tag{14.8}$$

The transmission characteristics $|E_t/E_i|$ of the Fabry–Perot étalon are plotted in Fig. 14.3. If the medium between the reflectors is nonabsorbing ($\alpha = 0$) and nonamplifying ($\gamma = 0$), then $|E_t/E_i|$ is equal to unity at an infinite number of equispaced frequencies that are separated by $\Delta\nu = c/2nl$. These unity transmission peaks occur whenever $kl[=(\omega/c)nl] = m\pi$ where m is an integer. These frequencies are called the longitudinal Fabry–Perot resonances. If the medium possesses net gain, $\gamma - \alpha > 0$, then the transmission coefficient at these resonances exceeds unity, that is, $|E_t| > |E_i|$, and the device acts as a *coherent electromagnetic amplifier*. If the excess gain $(\gamma - \alpha)$ is sufficiently large so that

$$r_1 r_2 e^{-2i[k+\Delta k(\omega)]l} e^{[\gamma(\omega)-\alpha]l} = 1 \tag{14.9}$$

then the denominator of (14.6) vanishes and the transmission coefficient $|E_t/E_i|$ is infinite. This singularity may be interpreted as resulting in a finite outgoing wave E_t with a vanishingly small input wave E_i—that is, *laser oscillation*. These singularities would appear as a series of zero width infinite height spikes in Fig. 14.3. The frequencies at which these singularities occur are the oscillation frequencies of the laser.

The oscillation condition (14.9) is merely a statement of the fact that after a complete round trip inside the resonator a wave returns to its starting plane with the same amplitude and, except for some multiple integer of 2π, with the same phase. This last condition leads to constructive interference and hence to a buildup of the radiation field.

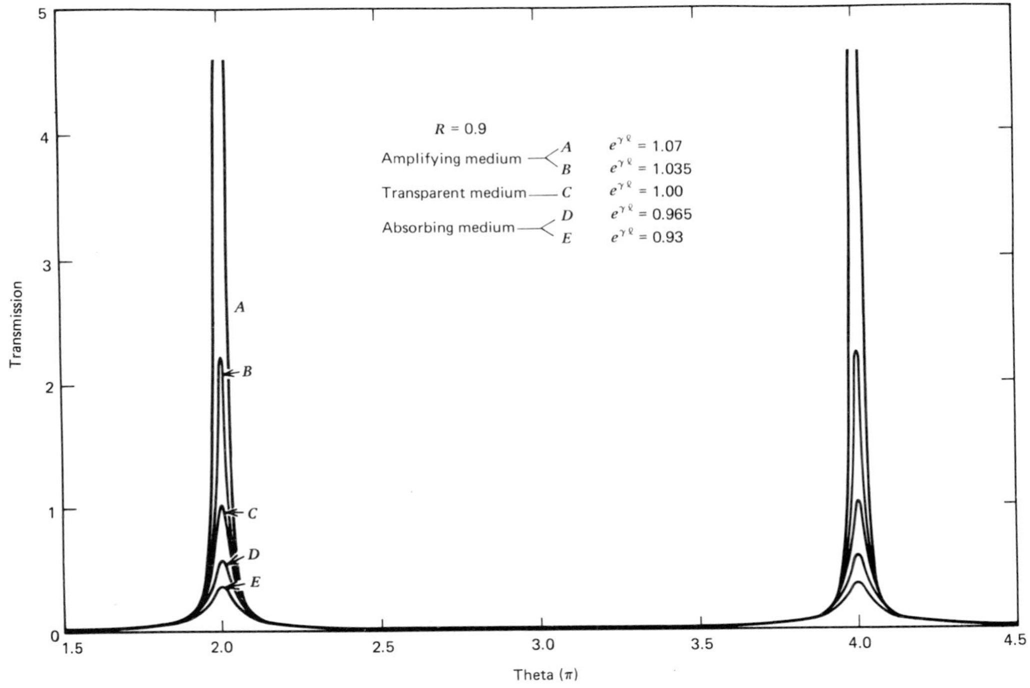

Figure 14.3 (*a*) A plot of the transmission gain $|E_t/E_i|^2$ of a Fabry–Perot resonator [see (14.6)] as a function of the incremental phase shift angle per round trip $\boldsymbol{\theta}$. $\boldsymbol{\theta} = 2(kl - m\pi)$, where m is some integer. $R \equiv r_1 r_2 = 0.9$. The intensity amplification factor per pass is $\exp(\gamma l)$ and is thus equivalent to the term $\exp(\gamma - \alpha)$ in (14.6). Plots D and E, where $\exp(\gamma l) > 1$, thus correspond to a lossy medium, curve Y [$\exp(\gamma l) = 1$], to a transparent medium, while A and B correspond [$\exp(\gamma l) > 1$] to amplifying media. We note that for $\exp(\gamma l) > 1$ the peak transmission gain exceeds unity, while for $\exp(\gamma l) < 1$ it is less than unity. This is true for all values of R. When $\exp(\gamma l)R = 1$—that is, when the one-pass amplification factor just makes up for the mirror transmission—the gain becomes infinite and oscillation results.

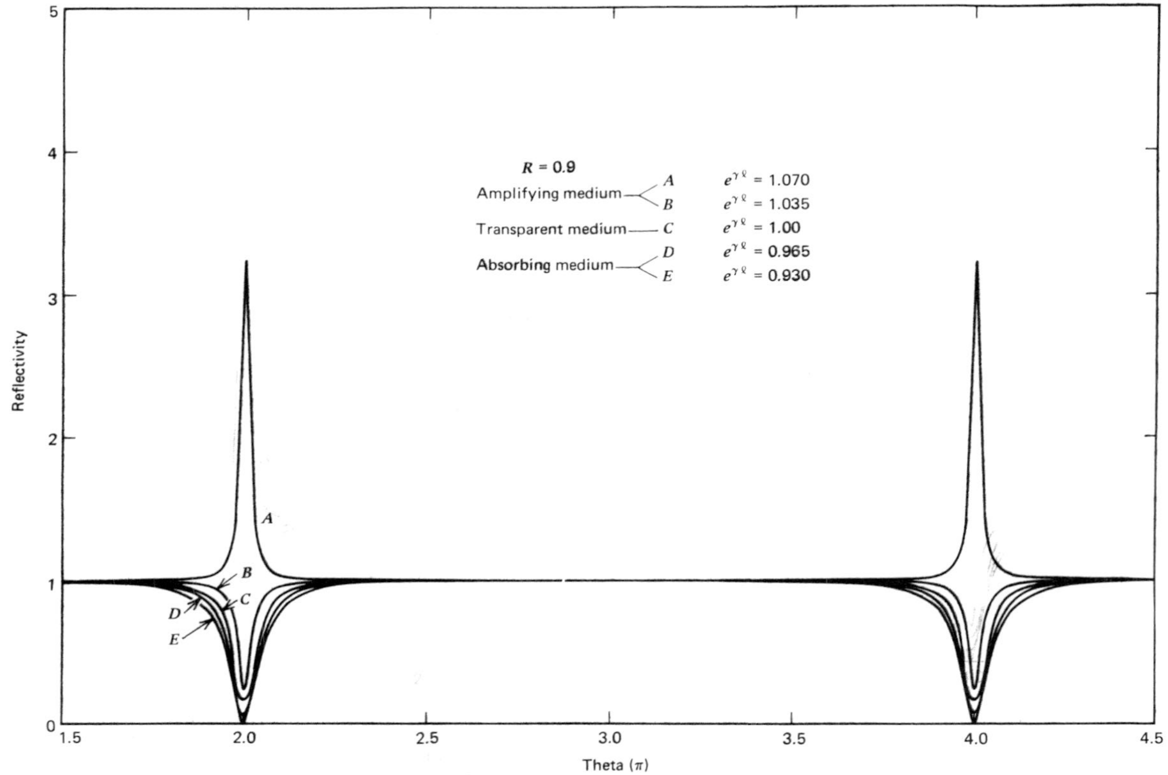

Figure 14.3 (*b*) Plots of the reflection coefficient $|E_r/E_i|^2$ of the Fabry–Perot amplifier. Note that when $\exp(\gamma l) > 1$ the peaks of $|E_r/E_i|^2$ exceed unity—that is, net reflection gain results. The reflection gain becomes infinite when $\exp(\gamma l)R = 1$. (Note curve C is the one that reaches to the abcissa.)

Separating the oscillation condition (14.9) into amplitude and phase parts gives

$$r_1 r_2 e^{[\gamma(\omega)-\alpha]l} = 1 \tag{14.10}$$

for the threshold gain constant $\gamma(\omega)$ and

$$2[k+\Delta k(\omega)]l = 2\pi m \quad (m=1,2,3,\ldots) \tag{14.11}$$

for the phase condition. The amplitude condition (14.10) can be written as

$$\gamma(\omega) = \alpha - \frac{1}{l}\ln r_1 r_2 \tag{14.12}$$

and is identical to that given by (14.3) if we put $r_1 r_2 = \sqrt{R_1 R_2}$. Using (14.1) we can write the threshold gain condition in terms of the critical (threshold) atomic inversion density needed to start oscillation:

$$N_t \equiv (N_2 - N_1)_t = \frac{8\pi n^2 t_{\text{spont}}}{g(\nu)\lambda^2}\left(\alpha - \frac{1}{l}\ln r_1 r_2\right) \tag{14.13}$$

The Laser Oscillation Frequencies

The phase part of the oscillation condition, Eq. (14.11), is satisfied at an infinite set of frequencies ν_m, which are obtained by substituting (14.7) for Δk

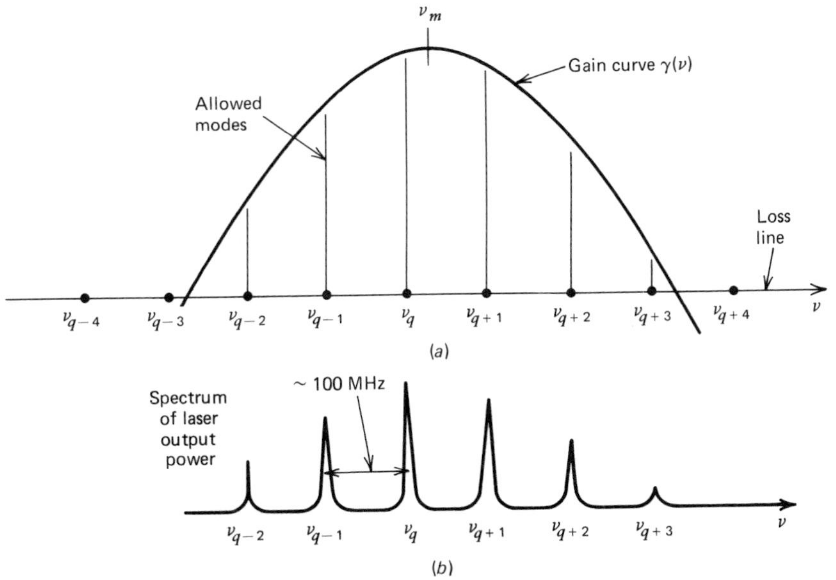

Figure 14.4 (*a*) Inhomogeneously broadened Doppler gain curve of the 6328 Å Ne transition and position of allowed longitudinal-mode frequencies. (*b*) Intensity versus frequency profile of an oscillating He–Ne laser. Six modes have sufficient gain to oscillate [after R. L. Fork, D. R. Herriott, and H. Kogelnik, *Appl. Opt.* 3, 1471 (1964)].

and recalling that ($k=(2\pi n/c)\nu$,

$$\frac{2\pi\nu_m n}{c}\left(1+\frac{\chi_a'(\nu_m)}{2n^2}\right)=m\pi \tag{14.14}$$

where we used $k=2\pi\nu n/c$. For $|\chi_a'(\nu)|\ll 1$ the oscillation frequencies are given by

$$\nu_m \cong m\frac{c}{2nl} \tag{14.15}$$

The laser will oscillate at (or very near) the frequencies ν_m provided the laser gain at this frequency satisfies the threshold condition (14.12). A plot of the gain function $\gamma(\nu)$ and the oscillation power as a function of frequency of the popular red 0.6328 μm He-Ne laser is shown in Fig. 14.4.

The process of oscillation results in a buildup of optical energy in the resonator. This causes the population inversion to decrease (saturate) due to the induced transitions. If new atoms are not "pumped" into the upper laser level 2, the oscillation will cease once the inversion drops below the critical value as given by (14.3). If atoms are pumped continuously into level 2, continuous wave (CW) oscillation can be maintained. An example of a pulsed laser is that of the ruby system.

The Ruby Laser

The first material in which laser action was demonstrated[6] was the ruby. The atoms undergoing the laser transition are Cr^{3+}, which are present as impurities in the host Al_2O_3 crystal. A typical ruby laser setup is shown in Fig. 14.5, while the pertinent energy level diagram is shown in Fig. 14.6. The pumping is performed by subjecting the crystal to an intense burst of light (quite similar to that used in flash photography). Cr^{3+} ions absorbing this light undergo a transition to the $^4F_2$ or $^4F_1$ bands.[7] From these bands they pass very quickly (within $\sim 10^{-7}$ s) to the upper laser level ^2E. Since the (spontaneous) lifetime of that level is relatively long, $\sim 3\times 10^{-3}$ s, a considerable fraction ($\gtrsim 0.5$) of the ground-state population is transferred by the pump and is stored in the ^2E upper laser level. Oscillation starts as soon as the gain due to the population difference $N_{^2\mathrm{E}}-\mathrm{N}_{^4A_2}$ is large enough to overcome the resonator losses.

Lasers are now used in an ever increasing number of scientific and technological applications. Their extreme monochromaticity makes them an invaluable tool in precise determinations of atomic and molecular transition energies (spectroscopy) and as length and time standards. This property also makes them a suitable choice as carrier waves for transmitting communication. As a matter of fact, transmission channels based on information modulated light beams propagating inside hair-thin silica fibers are beginning to

---

[6] T. H. Maiman, Stimulated optical radiation in ruby, *Nature* **187**, 493 (1960).

[7] The level designations ($^4F_2, ^4F_1\, ^2E, ^4A_2$) follow the conventional usage.

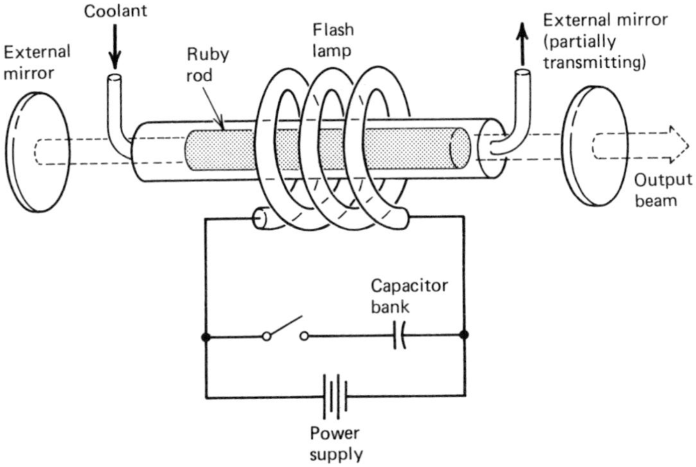

Figure 14.5 Typical setup of a pulsed ruby laser using flashlamp pumping and external mirrors.

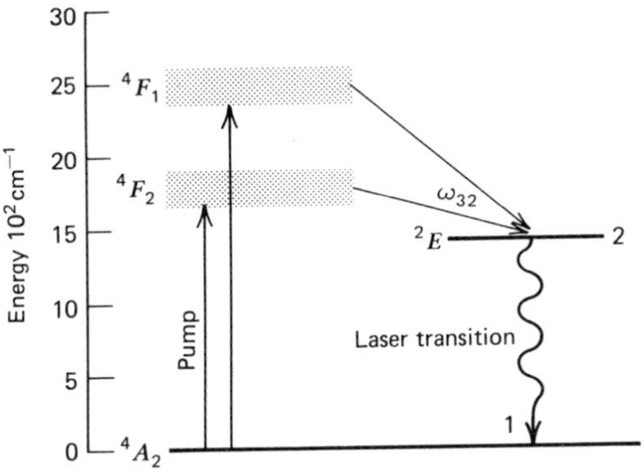

Figure 14.6 Energy levels pertinent to the operation of a ruby laser.

play an important role in the communication business. A schematic representation of such a system is shown in Fig. 14.7.

The spatial coherence of lasers—that is, the fact that the waves can have near ideal planar wavefronts—makes it possible to focus them down to volumes of the order of λ^3 (wavelength cubed) and thus achieve huge intensities. These intensities are useful in the generation of extremely hot plasmas, hot enough to undergo fusion reaction, and in producing higher harmonics $(2,3,\ldots,9)$ of the incident radiation by forcing the valence electrons into highly nonlinear regimes of oscillation. The latter phenomenon is now part of a new field known as nonlinear optics. Another type of a laser oscillator, the cw semiconductor diode laser (See Fig. 14.7), is playing a

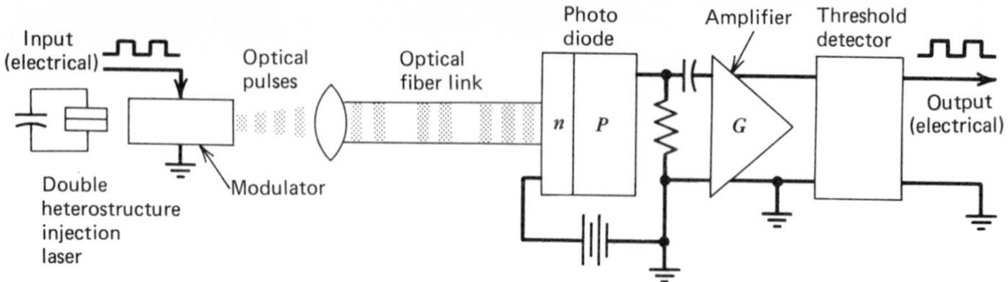

Figure 14.7 An optical fiber communication link. The information to be transmitted has been coded onto a stream of electrical binary pulses. These pulses are used to modulate the continuous wave (cw) output of a semiconductor injection laser. The pulses are thus replicated as the power envelope of the light that is fed into a fiber and carried to the receiving point. An optical detector converts the light pulses to electrical pulses, which are then decoded to recover the original information. [See, for example, S. Ramo, J. R. Whinnery, and T. Van Duzer in *Fields and Waves in Communication Devices* (J. Wiley & Sons, New York, 1965), p. 474.]

central role in the optical fiber communication technology. It will be described in detail in Chapter 21.

PROBLEMS

1. Obtain an expression for the ratio of the reflected to incident wave of a Fabry–Perot étalon.

2. Plot the reflection coefficient $|E_r/E_i|$ of a Fabry–Perot étalon as a function of frequency. Let the frequency interval be big enough to include two minima (or maxima). Assume the laser gain to be independent of ω. Assume $r_1^2 = r_2^2 \equiv 0.8$ and take (a) $e^{(\gamma-\alpha)l} = 1.0$, and (b) $e^{(\gamma-\alpha)l} = 1.1$.

3. Obtain an expression for the FWHM (full width at half maximum) of the transmission plot $|E_t/E_i|$ vs. ν of a Fabry–Perot etalon. Assume that $r_1 = r_2 \equiv \sqrt{R}$ and that the FWHM is small compared with the intermode spacing $c/2$ ml (i.e., $R \lesssim 1$).

   Answer: $$\text{FWHM} = \frac{c}{2nlF}$$

   $$F = \frac{\pi\sqrt{R}}{1-R} = \text{“Finesse”}$$

4. Plot the transmission of a passive Fabry–Perot etalon ($\gamma = \alpha = \Delta k = 0$) as a function of frequency. Use the fact that

   $$t_2 t_1 + r_2 r_1 = 1$$

   What is the qualitative difference between the case where $tt_1 \ll 1$ (high "finesse") and the case where $t_2 t_1$ is nearly equal to unity.

5. (a) Solve for the complex reflection coefficient E_r/E_i of the Fabry–Perot etalon shown in Fig. 14.1.
   (b) Show that the laser oscillation condition (14.9) results from insisting that $E_r/E_i=\infty$ [while keeping $\gamma(\omega)l$ finite].
6. What is the qualitative change in the expression (14.15) for the laser oscillation frequency if the term $[\chi_a'(\nu)]/2n^2$ in (14.14) is not neglected. You may use the Lorentzian form (13.66) for $\chi_a'(\nu)$. Show that the oscillation frequency is "pulled" toward the atomic resonance frequency ω_0.
7. Discuss qualitatively how the amount of frequency "pulling" depends on the width $\delta\nu\equiv 1/T_2$ of the Lorentzian response $\chi_a'(\nu)$.

CHAPTER FIFTEEN

Quantum Statistics

The need for a statistical treatment in describing quantum mechanical systems is motivated by the same reasons that prompted the development of classical statistical mechanics. Given a system with a large number N of constituent particles, it is possible in principle to follow the evolution of the system by solving the Schrödinger equation

$$H(p_1, q_1, \ldots, p_N, q_N, t)\psi(q_1, q_2, \ldots, q_N, t) = i\hbar\frac{\partial\psi}{\partial t} \tag{15.1}$$

given the initial state $\psi(q_1, q_2, \ldots, q_N, 0)$. This procedure, however, may not be practical even when the number of particles N is small, since the initial conditions are not available with sufficient accuracy. In cases of large N[1] it is simply impractical to attempt to compute ψ.

Under these circumstances we find it profitable to consider the statistical properties of a large ensemble of similar systems and assume that *the ensemble statistics are a reliable predictor of the expected behavior of a single system.*

15.1 THE THREE TYPES OF QUANTUM PARTICLES

A *single* quantum mechanical system in our model is taken to consist of N weakly interacting "particles" constrained to some volume V. These "particles" may be atoms, electrons, nuclei, protons, harmonic oscillators, electromagnetic radiation modes, normalized lattice vibrations, photons, and mixtures thereof. It will be assumed that the particle density is low enough so that, to first order, the energy of the system can be considered as the *sum* of the individual particle energies.

A given *microscopic* energy eigenstate of the system is constructed by assigning to each particle s a quantum state $\psi_i^{(s)}$, where

$$\hat{\mathcal{H}}(p_s, q_s)\psi_i^{(s)} = \varepsilon_i\psi_i^{(s)} \tag{15.2}$$

[1]We may appreciate the possible enormity of N in cases of interest by recalling that the density of conduction band electrons in a metal, for example, is $\sim 10^{23}\ \text{cm}^{-3}$.

where $\hat{\mathcal{H}}(p_s, q_s)$ is the Hamiltonian of particle s when occupying the volume V all by itself, so that ψ_i is the *single-particle* energy eigenfunction of the particle s in the ith quantum state with an energy eigen value ε_i. The total energy of the system is thus

$$E = \sum_{i=1}^{\infty} n_i \varepsilon_i \tag{15.3}$$

where n_i is the number of particles with energy ε_i.

There is, in general, more than one way of distributing the N particles among the energy levels ε_i so as to result in a given total system energy of E. (More precisely, we should say $E \pm \delta E$, where δE is the resolution of our energy measuring apparatus.) Each such individual arrangement that is describable, in principle, by a *unique* system eigenfunction, is called a *distinguishable microscopic arrangement*. It is a fundamental postulate of quantum statistics that in equilibrium *different (distinguishable) microscopic arrangements of the system that correspond (within δE) to the same total energy are equally likely to occur.*

The aim of this chapter is to determine the most probable distribution $(n_1, n_2, \ldots, n_s, \ldots)$ of finding n_1 particles out of a total of N with energy ε_1, n_2 with energy ε_2, and so on, such that the total energy of the system as given by (15.3) is a constant E. Since, in general, there is more than one microscopically distinguishable arrangement corresponding to a given distribution $(n_1, n_2, \ldots, n_s, \ldots)$, it follows from the basic postulate of quantum statistics described above that the probability of finding $(n_1, n_2, \ldots, n_s, \ldots)$ is proportional to the number of such microscopic arrangements, each resulting in a total energy E. We conclude that *the most probable distribution $(n_1, n_2, \ldots, n_s, \ldots)$ is the one associated with the largest number of microscopically distinguishable arrangements.*

The plan of attack would be to derive an expression for $P(n_1, n_2, \ldots, n_s, \ldots)$, the total number of microscopically distinct arrangements corresponding to a given arbitrary sequence $(n_1, n_2, \ldots, n_s, \ldots)$ and then find the particular sequence that maximizes P subject to the conditions that the total number of particles and the total energy be constant. The corresponding distribution of particles among the energy levels is the most likely to occur.

The particles of concern to us fall into one of three categories.

(A) *Identical but distinguishable particles.* Two examples of such particles are: (1) very heavy point masses that are confined inside a large box, and (2) a collection of harmonic oscillators.

(B) *Identical indistinguishable particles of half-odd-integral spin—Fermions (total spin angular momentum of $\hbar/2, 3\hbar/2, 5\hbar/2$, etc.).* Examples: electrons, protons. These particles were shown in Section 8.1 to obey the Pauli exclusion principle, according to which no two particles can occupy the same quantum state, or alternatively, that the total N-particle eigenfunction be antisymmetric (change sign) upon the permutation of any two particles. This constraint will be found to have a profound effect on the particle statistics.

(C) *Identical indistinguishable particles of integral spin—Bosons.* Examples: photons, phonons, nuclei with an even number of nucleons, and atoms with an even number of elementary particles. The Pauli exclusion principle does not apply to systems made up of bosons, so that more than one particle in the system can possess the same set of quantum numbers. The eigenfunction is symmetric upon permutation of two particles.

The most important point to understand before we proceed to derive the particle distribution probabilities is what does or does not constitute a microscopically distinguishable arrangement. The answer depends on whether the particles involved are of type (A), (B), or (C). To illustrate the point, consider a system of two identical particles with two allowed eigenstates, as shown in Fig. 15.1.

In case (A) of Fig. 15.1 we may label the particles by numbers 1 and 2, since, although identical, they are distinguishable (by their unique positions, for example). All four arrangements shown are thus microscopically distinguishable. An equivalent statement would be to say that there are four different energy eigenstates with eigenfunctions as shown. In case (B)—Fermions—only arrangement (2) is possible. Arrangements (1) and (4) are forbidden by the Pauli exclusion principle, since they involve the assignment of two identical Fermions to the same quantum state. Arrangement (3) is excluded since it cannot be distinguished from that of (2) (they are both described by the same eigenfunction, which is the antisymmetric combination of the two) so

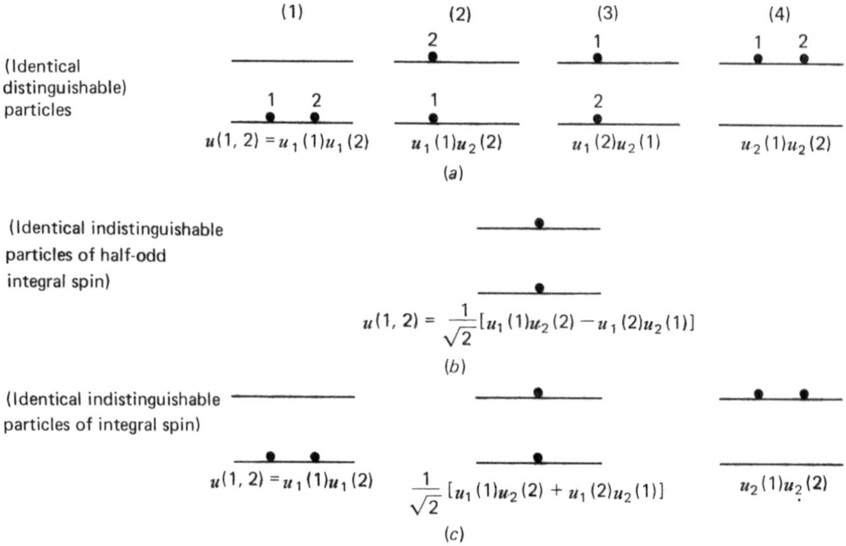

Figure 15.1 The various possible arrangements and corresponding eigenfunctions of two particles in two eigenstates.

that both count as one arrangement. All we can say is that one particle is in the upper energy state and one in the lower state. We cannot, however, specify which particle occupies which state. Arrangements (1) and (4) are allowed in case (C), since bosons do not obey the Pauli exclusion principle. In case (c), arrangements (2) and (3) give one state—the symmetric combination.

15.2 THE COUNTING ALGEBRA FOR QUANTUM SYSTEMS

We are now ready to solve for the most probable distribution of the different classes of particles. Our model system is taken to contain N identical "particles," each with a spectrum of allowed energy levels (derived by solving the time-independent Schrödinger equation for a single particle)

$$\varepsilon_1, \varepsilon_2, \ldots, \varepsilon_s, \ldots$$

The energies may belong to degenerate states so that in general an energy level ε_s may be associated with more than one elementary (i.e., single-particle) quantum state.

In deriving the distribution laws we employ the following device: We divide the single-particle energy spectrum between $-\infty$ and ∞ into energy "bins." Bin s, as an example, represents all the elementary quantum states whose energies lie within some arbitrarily chosen interval $\Delta\varepsilon_s$ centered about ε_s ($\Delta\varepsilon_s$ can be made as small as allowed by our energy resolution limit $\delta\varepsilon$). The number of quantum states in bin s is denoted by g_s. A *macroscopic* state $(n_1, n_2, \ldots, n_s, \ldots)$ of the N-particle system is specified by giving the number of particles $n_1, n_2, \ldots, n_s, \ldots$ in each bin. There will be, in general, many microscopic ways of arranging the N particles so as to lead to a given macroscopic distribution. An individual microscopic arrangement is considered distinguishable if it has a unique N-particle eigenfunction.

According to the fundamental postulate of quantum statistics cited above, in equilibrium any two microscopically distinguishable arrangements of a system with the same total energy (within δE) are equally likely. The probability of finding the system with n_1 particles in bin 1, n_2 in bin 2, and so on, is consequently proportional to the number of microscopically distinguishable arrangements that correspond to the same macroscopic distribution $(n_1, n_2, \ldots, n_s, \ldots)$. We call this number $P(n_1, n_2, \ldots, n_s, \ldots)$ or $P(n_s)$.

Our next task will be to evaluate, one at a time, $P(n_s)$ for the three basic types of particles.

Identical but Distinguishable Particles

We seek to determine the number of microscopically distinguishable arrangements of N particles such that bin 1 contains n_1 particles, bin 2 n_2 particles, and so on. We determine first the number of ways in which we can choose n_1 particles out of a total of N to be placed in bin 1, ignoring for the moment the manner in which these particles are distributed between the quantum states of the bin. The first particle can be chosen from among a total of N particles, the second from the remaining $N-1$, and so on, so that the total number of distinguishable choices is

$$N(N-1)(N-2),\ldots,(N-n_1+1)=\frac{N!}{(N-n_1)!} \tag{15.4}$$

In arriving at the last result, we counted as distinct all permutations among a given group of n_1 particles. The sequence of particles 1,4,7 and 7,4,1, for example, was counted as two distinct arrangements, since we agreed to ignore temporarily the internal ordering. We need to divide the last equation by $n_1!$, the number of permutations of n_1 particles among themselves. The result is

$$\frac{N!}{n_1!(N-n_1)!} \tag{15.5}$$

We now address ourselves to the internal arrangement of the n_1 particles in bin 1. Let the number of quantum states combined in bin 1 be g_1. We can place particle 1 into any of the g_1 states of the bin. The second particle can, likewise, also be put into any of the g_1 states. (Since the particles are distinguishable we can assign more than one particle to any one state.) The total number of distinguishable *microscopic* arrangements of the n_1 particles is thus $g_1^{n_1}$. This result, multiplied by (15.5), gives

$$P_1=\frac{N!g_1^{n_1}}{n_1!(N-n_1)!} \tag{15.6}$$

for the total number of microscopic distinguishable ways of putting n_1 particles out of a total N into the g_1 states of bin 1. The number of distinct microscopic arrangements of putting n_2 of the remaining $(N-n_1)$ particles into bin 2 is, in analogy with (15.6),

$$P_2=\frac{(N-n_1)!g_2^{n_2}}{n_2!(N-n_1-n_2)!} \tag{15.7}$$

The total number of distinct microscopic arrangements in which n_1 particles are in bin 1, n_2 in bin 2, and so on, is thus

$$\begin{aligned} P(n_s) &\equiv P(n_1,n_2,\ldots,n_s,\ldots)\\ &=P_1P_2,\ldots,P_s,\ldots\\ &=N!\prod_{s=1}^{\infty}\frac{g_s^{n_s}}{n_s!} \qquad \text{Case (A)} \end{aligned} \tag{15.8}$$

Identical Indistinguishable Particles of Half-Odd-Integral Spin—Fermions

Since the particles are indistinguishable, we cannot tell which particle occupies a particular state. All we can do is determine whether a given state in a bin is occupied or not. Since each state is associated with a one-particle energy eigenfunction, the Pauli exclusion principle (which applies to particles with half-odd-integral spin) prevents us from putting more than one particle in any one state. Let us pretend, for a moment, that the particles are distinguishable. The task of fitting n_s particles into g_s states is begun by putting one particle into any one state in bin s. There are g_s ways of doing so. Since no two particles can occupy the same state, the second particle can be placed into any of the remaining (g_s-1) states, so that altogether there are

$$g_s(g_s-1)\cdots(g_s-n_s+1)$$

configurations (because of the Pauli exclusion principle $g_s \geq n_s$). This last expression includes all possible permutations of the n_s particles among themselves. The number of these permutations is $n_s!$. Since the particles are indistinguishable, these permutations do not lead to distinguishable arrangements (they correspond to the same eigenfunction) so that the total number of distinct microscopic arrangements with n_s particles in bin s is

$$P_s = \frac{g_s(g_s-1)\cdots(g_s-n_s+1)}{n_s!}$$

$$= \frac{g_s!}{(g_s-n_s)!n_s!} \tag{15.9}$$

and the total number of configurations with n_1 particles in bin 1, n_2 in bin 2, and so on, is

$$P(n_s) = P(n_1, n_2, \ldots, n_s, \ldots) = \prod_{s=1}^{\infty} P_s$$

$$= \prod_{s=1}^{\infty} \frac{g_s!}{(g_s-n_s)!n_s!} \qquad \text{Case (B)} \tag{15.10}$$

Identical, Indistinguishable Particles of Integral Spin—Bosons

In this case, as in (B), the indistinguishability of the particles prevents us from stating which particle is placed in any particular bin or, within a given bin, which particle occupies any given elementary quantum state. There is, however, no restriction on the number of particles occupying a given state. Stated quantum mechanically, more than one particle can have the same single-particle eigenfunction. The number of distinct arrangements of n_s particles in the g_s states of bin s can be determined by means of a graphic device[2] as shown in Fig. 15.2. Each dot in the figure represents a particle and each black line a partition. Each space between two adjacent partitions, as well as between the two extreme partitions and the walls, represents one state. The g_s-1 partitions thus account for the g_s states in bin s. The number of white dots is n_s and is equal to the number of particles in the bin. In the

[2] R. C. Tolman, *The Principles of Statistical Mechanics* (Oxford University Press, Oxford, 1959), Chapter X.

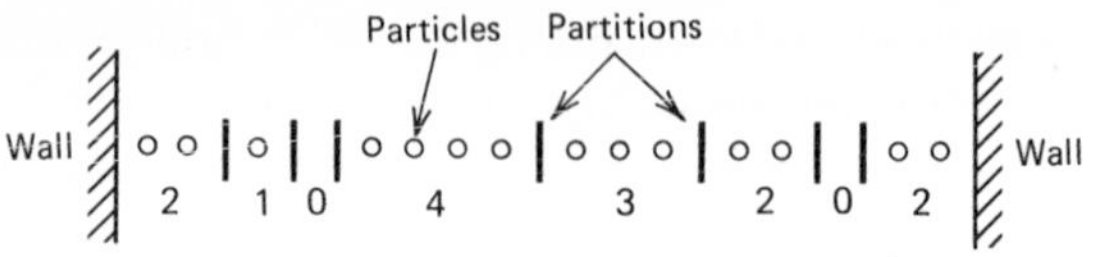

Figure 15.2 A graphical device for finding the number of distinct ways of arranging n_s indistinguishable particles (white dots) in the g_s states (each state is bounded by two partitions or by a partition and a wall) of bins. In the figure $n_s = 14$, $g_s = 8$.

example of Fig. 15.2, $g_s = 8$ and $n_s = 14$. If the n_s white dots and the $g_s - 1$ black lines were all distinguishable, the number of distinct arrangements of white dots and black lines would be equal to the number of permutations of $(n_s + g_s - 1)$ objects, which is $(n_s + g_2 - 1)!$. Since the particles and partitions are in fact indistinguishable, the $(g_s - 1)!$ permutations of the black lines among themselves or the $n_s!$ permutations of the white dots do not lead to distinguishable arrangements. The number of distinguishable arrangements is thus reduced to

$$P_s = \frac{(n_s + g_s - 1)!}{n_s!(g_s - 1)!} \tag{15.11}$$

so that the total number of distinct arrangements involving $n_1, n_2, \ldots, n_s$ is

$$\begin{aligned} P(n_s) &= P(n_1, n_2, \ldots, n_s, \ldots) \\ &= \prod_{s=1}^{\infty} \frac{(n_s + g_s - 1)!}{n_s!(g_s - 1)!} \qquad \text{Case (C)} \end{aligned} \tag{15.12}$$

15.3 THE MAXWELL–BOLTZMANN, FERMI–DIRAC, AND BOSE–EINSTEIN STATISTICS

In the last section we obtained expressions for the number of microscopically distinguishable ways of putting n_1 particles into bin 1, n_2 into bin 2, and so on. The resulting expression

$$P(n_s) = P(n_1, n_2, \ldots, n_s, \ldots)$$

was obtained for (a) identical and distinguishable particles as well as for (b) Fermions, and (c) Bosons. We postulated at the outset that, in equilibrium, all microscopically distinguishable distributions with a fixed total number of particles and the *same* total energy are *equally* likely. *It follows that the most probable (likely) macroscopic distribution* $(n_1, n_2, \ldots, n_s, \ldots)$ *is that for which the number of microscopically distinguishable arrangements* $P(n_1, n_2, \ldots, n_s, \ldots)$ *is a maximum.*

Our next task, then, consists of finding the distribution $n_1, n_2, \ldots, n_s$ for which $P(n_s)$ is a maximum, subject to the auxiliary conditions

$$\sum_{s=1}^{\infty} n_s = N = \text{const} \quad \text{(constancy of the number of particles)} \tag{15.13}$$

$$\sum_{s=1}^{\infty} \varepsilon_s n_s = E = \text{const} \quad \text{(constancy of energy)} \tag{15.14}$$

We will find it more convenient to maximize $\ln P$, subject to conditions (15.13) and (15.14), rather than P itself. The problem of maximizing a general function $G(n_1, n_2, \ldots)$ subject to auxiliary conditions $f(n_1, n_2, \ldots) = 0$ and $g(n_1, n_2, \ldots) = 0$ is treated by the Lagrange method of undetermined multipliers. In applying this method, we define

$$F(n_1, n_2, \ldots, \alpha, \beta) = G(n_1, n_2, \ldots) - \alpha f(n_1, n_2, \ldots) - \beta g(n_1, n_2, \ldots) \tag{15.15}$$

We solve for the values of $n_1, n_2, \ldots$ for which G is a maximum as well as for the undetermined coefficients α and β by requiring that

$$\frac{\partial F}{\partial n_s} = 0 \quad \text{for all } s \tag{15.16}$$

and

$$\frac{\partial F}{\partial \alpha} = 0, \qquad \frac{\partial F}{\partial \beta} = 0 \tag{15.17}$$

Applying this method to the problem at hand we take

$$G = \ln P(n_s)$$

$$f = \left(\sum_{s=1}^{\infty} n_s \right) - N$$

$$g = \left(\sum_{s=1}^{\infty} \varepsilon_s n_s \right) - E$$

so that

$$F = \ln P - \alpha \left[\left(\sum_{s=1}^{\infty} n_s \right) - N \right] - \beta \left[\left(\sum_{s=1}^{\infty} \varepsilon_s n_s \right) - E \right] \tag{15.18}$$

Case (A) Identical Distinguishable Particles

Here we take P as given by (15.8). Equation (15.18) becomes

$$\begin{aligned} F &= \ln \left(N! \prod_{s=1}^{\infty} \frac{g_s^{n_s}}{n_s!} \right) - \alpha \left[\left(\sum_{s=1}^{\infty} n_s \right) - N \right] - \beta \left[\left(\sum_{s=1}^{\infty} \varepsilon_s n_s \right) - E \right] \\ &= \ln N! + \sum_{s=1}^{\infty} (n_s \ln g_s - \ln n_s!) - \alpha \left[\left(\sum_{s=1}^{\infty} n_s \right) - N \right] - \beta \left[\left(\sum_{s=1}^{\infty} \varepsilon_s n_s \right) - E \right] \end{aligned} \tag{15.19}$$

We now assume that $n_s \gg 1$; then using Stirling's formula[3] we obtain

$$\frac{\partial F}{\partial n_s} = \ln g_s - \ln n_s - \alpha - \beta \varepsilon_s = 0 \quad \text{(for all } s\text{)} \tag{15.20}$$

while the application of (15.17) merely reproduces the auxiliary conditions (15.13) and (15.14).

Case (B) Fermions

From (15.10),

$$\ln P = \sum_{s=1}^{\infty} \{\ln g_s! - [\ln n_s! + \ln(g_s - n_s)!]\}$$

Using Stirling's formula, the last result can be approximated by

$$\begin{aligned}\ln P &= \sum_{s=1}^{\infty} \{\ln g_s! - (n_s \ln n_s - n_s) - [(g_s - n_s)\ln(g_s - n_s) - (g_s - n_s)]\} \\ &= \sum_{s=1}^{\infty} \{\ln g_s! - n_s \ln n_s - (g_s - n_s)\ln(g_s - n_s) + g_s\}\end{aligned}$$

Defining

$$F = \ln P - \alpha\left[\left(\sum_{s=1}^{\infty} n_s\right) - N\right] - \beta\left[\left(\sum_{s=1}^{\infty} \varepsilon_s n_s\right) - E\right]$$

we obtain

$$\text{(B)} \quad \frac{\partial F}{\partial n_s} = -\ln n_s + \ln(g_s - n_s) - \alpha - \beta\varepsilon_s = 0 \tag{15.21}$$

for all s. Here we assumed $g_s - n_s \gg 1$.

Case (C) Bosons

We repeat the same procedure as that leading to (15.21), except that P is given by (15.12). The result is:

$$\text{(C)} \quad \ln(n_s + g_s) - \ln n_s - \alpha - \beta\varepsilon_s = 0 \tag{15.22}$$

From Eqs. (15.20), (15.21), and (15.22) we obtain directly the distribution laws for thermal equilibrium:

$$\text{(A)} \quad n_s = \frac{g_s}{e^{\alpha + \beta\varepsilon_s}} \quad \text{(Maxwell–Boltzmann)} \tag{15.23}$$

$$\text{(B)} \quad n_s = \frac{g_s}{e^{\alpha + \beta\varepsilon_s} + 1} \quad \text{(Fermi–Dirac)} \tag{15.24}$$

$$\text{(C)} \quad n_s = \frac{g_s}{e^{\alpha + \beta\varepsilon_s} - 1} \quad \text{(Bose–Einstein)} \tag{15.25}$$

---

[3] Stirling's formula is $\ln n! \cong n \ln n - n$ $(n \gg 1)$.

15.4 SYSTEMS WITH MORE THAN ONE CONSTITUENT

The extension of the above ideas to systems containing more than one species of particles is straightforward. Specifically, consider a system consisting of N identical particles of type (A) and M particles of type (B). The particles in each of these two groups may consist of bosons, fermions, or particles obeying Maxwell–Boltzmann statistics. The number of distinguishable microscopic arrangements of the N particles of group (A) is $P^{(1)}(n_1,\ldots,n_s,\ldots)$, while that of group (B) is $P^{(2)}(m_1,\ldots,m_s,\ldots)$. The expressions for P_1 and P_2 are the same as those obtained above [see (15.8), (15.10), and (15.12)] and depend, of course, on the type of particle involved.

The total number of distinguishable arrangements that involve $(n_1, n_2,\ldots,n_s,\ldots)$ particles of type (A) and $(m_1, m_2,\ldots,m_k,\ldots)$ of type (B) is

$$P(n_s, m_k) = P^{(1)}(n_1,\ldots,n_s,\ldots)P^{(2)}(m_1,\ldots,m_k,\ldots)$$

To find the most probable distribution, we maximize $\ln P(n_s, m_k)$ subject to the auxiliary conditions

$$\sum_{s=1}^{\infty} n_s = N \tag{15.26}$$

$$\sum_{\lambda=1}^{\infty} m_\lambda = M \tag{15.27}$$

$$\sum_{s=1}^{\infty} n_s\varepsilon_s + \sum_{\lambda=1}^{\infty} m_\lambda\varepsilon_\lambda = E \tag{15.28}$$

where ε_s are the single-particle eigenvalues ("bin" energies in our model) of system (A) and ε_λ of system (B).

The application of Lagrange's method of undetermined multipliers involves a procedure identical to that leading to (15.20). We define

$$F = \ln P^{(1)}(n_1, n_2,\ldots) + \ln P^{(2)}(m_1, m_2,\ldots) - \alpha\left[\left(\sum_{s=1}^{\infty} n_s\right) - N\right]$$
$$- \alpha_2\left[\left(\sum_{\lambda=1}^{\infty} m_\lambda\right) - M\right] - \beta\left[\sum_{s=1}^{\infty} n_s\varepsilon_s + \sum_{\lambda=1}^{\infty} m_\lambda\varepsilon_\lambda - E\right] \tag{15.29}$$

and proceed to require that

$$\frac{\partial F}{\partial n_s} = 0 \quad \text{for all } s \tag{15.30}$$

$$\frac{\partial F}{\partial m_\lambda} = 0 \quad \text{for all } \lambda \tag{15.31}$$

$$\frac{\partial F}{\partial \alpha_1} = 0, \quad \frac{\partial F}{\partial \alpha_2} = 0, \quad \frac{\partial F}{\partial \beta} = 0 \tag{15.32}$$

From (15.30) we obtain

$$\frac{\partial F}{\partial n_s} = \frac{\partial \ln P_1}{\partial n_s} - \alpha_1 - \beta\varepsilon_s = 0 \tag{15.33}$$

From (6),

$$\frac{\partial F}{\partial m_\lambda} = \frac{\partial \ln P_2}{\partial m_\lambda} - \alpha_2 - \beta\varepsilon_\lambda = 0 \tag{15.34}$$

Expressions (15.33) and (15.34) are *identical* in *form* to those used in the case of a single constituent and thus lead to distribution laws *identical* to those given by (15.23), (15.24), and (15.25). The only difference is that the particles in (A) involve the undetermined parameter α_1, while particles in (B) involve α_2. Both distributions, however, involve the *same* β. (The parameter β will be shown subsequently to be related to temperature.)

To illustrate this point, assume for example that group (A) consists of particles obeying Maxwell–Boltzmann statistics, while group (B) consists of fermions. The most probable distribution of particles in an equilibrated mixture of the two constituents is thus

$$n_s = \frac{g_s}{e^{\alpha_1 + \beta\varepsilon_s}} \quad \text{for particles of group (A) (identical, distinguishable)} \tag{15.35}$$

$$m_\lambda = \frac{g_\lambda}{e^{\alpha_2 + \beta\varepsilon_\lambda} + 1} \quad \text{for particles of group (B) (fermions)} \tag{15.36}$$

15.5 EVALUATING THE PARAMETER β IN THE DISTRIBUTION LAWS

In order to be able to use the distribution laws (15.23), (15.24), and (15.25), we need to determine the values of the constants α and β, which appear in these laws. This is done by enforcing the auxiliary conditions (15.13) and (15.14), which causes (15.17) to be satisfied. We have shown in the last section that β is of a more fundamental nature than α, since a single value of β is used in the distribution laws, even when the system contains more than one type of particle, while the α parameters are different for the different types. Consequently, we start by evaluating the parameter β.

To determine β we consider a gas made up of N massive atoms of zero spin. The large mass of these particles, which renders them relatively immobile, and the large volume V in which they are enclosed, allow us to consider them as distinguishable so that their distribution is governed by the Maxwell–Boltzmann law (15.23)

$$n_s = \frac{g_s}{e^{\alpha_1 + \beta\varepsilon_s}} \tag{15.37}$$

Let us define bin s as comprising the energy eigenstates whose energies fall within a small interval $\Delta\varepsilon_s$ centered on ε_s. The number of states in the bin is

$$g_s = \frac{4\pi V m (2m)^{1/2}}{h^3}\varepsilon_s^{1/2}\Delta\varepsilon_s \qquad (15.38)$$

where V is the volume of the physical space to which the particles are confined and m is their mass.

Derivation of g_s (15.38)

The energy eigenfunctions u of a free particle of mass m in a rectangular box with impenetrable walls need satisfy the Schrödinger eigenvalue equation

$$-\frac{\hbar^2}{2m}\left(\frac{\partial^2}{\partial x^2}+\frac{\partial^2}{\partial y^2}+\frac{\partial^2}{\partial z^2}\right)u = \varepsilon u \qquad (15.39)$$

subject to the requirement that u vanish at the walls. The appropriate solutions are of the form

$$\begin{aligned} u_{lmn} &\propto \sin l\frac{\pi x}{a}\sin m\frac{\pi y}{b}\sin n\frac{\pi z}{c} \\ &= \sin(k_x x)\sin(k_y y)\sin(k_z z) \qquad (15.40) \\ k_x &= l\pi/a, \quad k_y = m\pi/b, \quad k_z = n\pi/c \end{aligned}$$

where a, b, c are the sides of the box and l, m, n are any positive integers.[4] We may thus associate with each state a vector:

$$\mathbf{k} = \mathbf{i}k_x + \mathbf{j}k_y + \mathbf{k}k_z \quad (\mathbf{x},\mathbf{y},\mathbf{z} = \text{unit Cartesian vectors})$$

The state energy is from (15.39):

$$\begin{aligned} \varepsilon_{l,m,n} &= \frac{\hbar^2}{2m}\left(k_x^2+k_y^2+k_z^2\right)_{l,m,n} \\ &= \frac{\hbar^2 k_{l,m,n}^2}{2m} \qquad (15.41) \end{aligned}$$

Since a change of any of l or m or n by unity results in a new eigenstate u_{lmn}, we may associate a volume

$$d^3\mathbf{k} = \left(\frac{\pi}{a}\right)\left(\frac{\pi}{b}\right)\left(\frac{\pi}{c}\right) = \frac{\pi^3}{V}$$

in k_x, k_y, k_z space with each eigenstate. The total number of eigenstates whose

[4] Negative integers do not generate linearly independent solutions.

$k=\sqrt{k_x^2+k_y^2+k_z^2}$ parameter falls between zero and k is thus

$$N(k)=\frac{\left(\frac{1}{8}\right)\left(\frac{4\pi k^3}{3}\right)}{\pi^3/V}$$

$$=\frac{k^3 V}{6\pi^2} \tag{15.42}$$

where $4\pi k^3/3$ is the volume in k_x, k_y, k_z space occupied by the states' $\mathbf{k}$ vectors. The $\left(\frac{1}{8}\right)$ factor accounts for the fact that only positive values of k_x, k_y, and k_z are counted. π^3/V is the volume per eigenstate.

The number of states with k values between k and $k+dk$ is

$$\frac{dN(k)}{dk}dk=\frac{3k^2V}{6\pi^2}dk \tag{15.43}$$

and using $\varepsilon_k=\hbar^2k^2/2m$, $dk=md\varepsilon_k/\hbar^2k$, we obtain

$$g_s=\frac{4\pi Vm(2m)^{1/2}}{h^3}\varepsilon_s^{1/2}\Delta\varepsilon_s \tag{15.44}$$

for the number of states in $\Delta\varepsilon_s$. This result was stated in (15.38).

If the density of particles is large, we can use (15.38) to replace the summation

$$\sum_{s=1}^{\infty} n_s=\sum_{s=1}^{\infty}\frac{g_s}{e^{\alpha_1+\beta\varepsilon_s}}=N \tag{15.45}$$

by

$$N=\frac{4\pi Vm(2m)^{1/2}}{h^3}e^{-\alpha_1}\int_0^{\infty}\varepsilon^{1/2}e^{-\beta\varepsilon}\,d\varepsilon$$

$$=\frac{4\pi Vm(2m)^{1/2}}{h^3}e^{-\alpha_1}\beta^{-3/2}\Gamma\left(\tfrac{3}{2}\right) \tag{15.46}$$

where we used the definition of the gamma function

$$\Gamma(x)=\int_0^{\infty}e^{-u}u^{x-1}\,du \tag{15.47}$$

The total energy of the N particles is given by

$$E=\sum_{s=1}^{\infty}n_s\varepsilon_s$$

$$=\frac{4\pi Vm(2m)^{1/2}}{h^3}e^{-\alpha_1}\int_0^{\infty}\varepsilon^{3/2}e^{-\beta\varepsilon}\,d\varepsilon$$

$$=\frac{4\pi Vm(2m)^{1/2}}{h^3}e^{-\alpha_1}\beta^{-5/2}\Gamma\left(\tfrac{5}{2}\right) \tag{15.48}$$

Dividing (15.46) by (15.48)

$$\frac{N}{E}=\beta\frac{\Gamma(\frac{3}{2})}{\Gamma(\frac{5}{2})}=\tfrac{2}{3}\beta$$

so that

$$\frac{1}{\beta}=\frac{2}{3}\frac{E}{N} \tag{15.49}$$

The N particles were chosen as sufficiently heavy so as to make the particles distinguishable. In this case the results of classical statistical mechanics apply. The average energy per particle in thermal equilibrium is thus[5]

$$E/N=\tfrac{3}{2}kT \tag{15.50}$$

where T is the temperature, so that

$$1/\beta=kT \tag{15.51}$$

where k is the Boltzmann constant.

The value of β in the case of fermions is determined using the following conceptual device. Let us assume that the system contains, in addition to the fermions, N_1 massive particles described by the Maxwell–Boltzmann distribution law. In equilibrium the energy of the subsystem of heavy particles is E_1, while that of the fermions is E, so that the total energy of the system is $E+E_1$.

In Section 15.4 we determined that in the case of a mixture of more than one type of particle, each type is characterized by its distribution law [of the form (15.23), (15.24), or (15.25), depending on the type of particle] with a *common* parameter β.

A calculation of the β parameter of the subsystem of heavy particles obeying the Maxwell–Boltzmann statistics will be identical to that leading to (15.51), so that $\beta=(kT)^{-1}$. Since both types of particles must have the same β, the β parameter of the fermion system is also equal to $(kT)^{-1}$.

The same argument may be applied to a mixture of heavy particles and bosons so that the parameter β of any equilibrium mixture of particles of *any type* is given by $(kT)^{-1}$.

To complete the derivation of the quantum statistical distribution laws, we need to solve for the α parameters in (15.23), (15.24), and (15.25). This requires that we be more specific about the system of particles. Some important examples are considered in Chapter 16.

PROBLEMS

1. Complete the derivation leading to (15.22).
2. Derive Eqs. (15.23), (15.24), and (15.25).
3. Use any standard text of classical statistical mechanics to outline the derivation of Eq. (15.50) $E/N=\frac{3}{2}kT$.

---

[5]See, for example, C. Kittel, *Elementary Statistical Physics* (John Wiley and Sons, New York, 1967).

CHAPTER SIXTEEN

Some Specific Applications of the Statistical Distribution Laws

In the last chapter we derived expressions for the statistical distribution laws of quantum particles. These expressions contain a parameter α that must be determined in each specific case. The general approach for doing it is discussed in this chapter. The important cases of a Fermion gas and a photon gas are emphasized.

16.1 THE MAXWELL–BOLTZMANN DISTRIBUTION

In the case of particles obeying Maxwell–Boltzmann statistics we obtain directly from (15.46) after using $\beta=(kT)^{-1}$, $\Gamma(\frac{3}{2})=\sqrt{\pi}/2$:

$$\alpha=\ln\left(\frac{2\pi^{3/2}Vm(2m)^{1/2}}{h^3N}(kT)^{3/2}\right) \tag{16.1}$$

where N is the number of particles, m is their mass, and V is the volume to which they are confined.

The fraction of the total number of atoms with energy ε_i (or more exactly, which are in the energy "bin" i) is obtained from (15.23):

$$f_i = \frac{n_i}{N}$$

$$= \frac{g_i e^{-\alpha-\varepsilon_i/kT}}{\sum_s g_s e^{-\alpha-\varepsilon_s/kT}}$$

$$= \frac{g_i e^{-\varepsilon_i/kT}}{\sum_s g_s e^{-\varepsilon_s/kT}} \tag{16.2}$$

and is independent of α. The ratio of the populations of any two levels, say i and j, is thus

$$\frac{f_i}{f_j} = \frac{g_i}{g_j} e^{-(\varepsilon_i-\varepsilon_j)/kT} \tag{16.3}$$

and is known as the Boltzmann ratio.

16.2 FERMI–DIRAC DISTRIBUTION

The distribution function of identical, indistinguishable particles of half-odd-integral spin (Fermions) is given in (15.24) as

$$n_s = \frac{g_s}{e^{\alpha+\beta\varepsilon_s}+1} \tag{16.4}$$

Using $\beta = (kT)^{-1}$ and defining

$$\mu \equiv -\alpha kT$$

we may rewrite (15.4) as

$$n_s = \frac{g_s}{e^{(\varepsilon_s-\mu)/kT}+1} \tag{16.5}$$

If we use the label s to denote a bin containing but a single state (rather than the set of g_s degenerate states having an energy ε_s), then $g_s = 1$ and (16.5) becomes

$$n(\varepsilon) \equiv f(\varepsilon) = \frac{1}{e^{[\varepsilon-\mu(T)]/kT}+1} \tag{16.6}$$

The quantity μ is called the chemical potential of the system of particles. The function $f(\varepsilon)$, the Fermi–Dirac distribution law, gives the probability that the quantum state ε is occupied. It follows that $f(\varepsilon) \leq 1$.

Figure 16.1 shows a plot of $f(\varepsilon)$ at $T=0$ and also at some finite temperature. At $T=0$, $f(\varepsilon)=1$ for $\varepsilon<\mu$ and $f(\varepsilon)=0$ for $\varepsilon>\mu$. In words: At zero temperature the states whose energies lie below the chemical potential $\mu(0)$ are occupied, while those above it are empty. At finite temperatures the

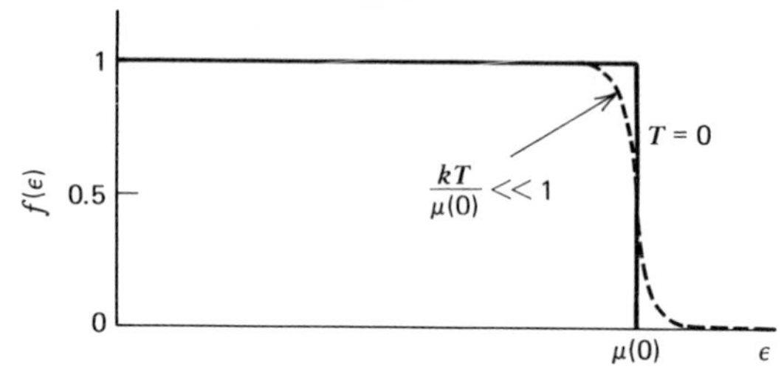

Figure 16.1 Sketch of the Fermi–Dirac distribution function at absolute zero and at a low temperature. The width of the transition region between $f(\varepsilon) \simeq 1$ and $f(\varepsilon) \simeq 0$ is of the order of kT.

transition from occupied states is "fuzzed" over a region of width $\sim kT$, as shown in the figure.

At energies ε such that $(\varepsilon - \mu) \gg kT$, $f(\varepsilon) \sim e^{-(\varepsilon - \mu(T)/kT}$. This exponential dependence of the occupation probability is the same as that of the Boltzmann law (16.3) so that in this limit the statistical behavior of fermions and of classical particles is similar.

To evaluate the chemical potential $\mu(0)$, we consider the case of a free electron gas that is confined to a volume V. The number of available electron states in $d\varepsilon$ is obtained from (15.44):

$$g(\varepsilon)\, d\varepsilon = \frac{8\pi V m (2m)^{1/2}}{h^3} \varepsilon^{1/2}\, d\varepsilon \tag{16.7}$$

[The factor of 2 difference between (16.7) and (15.44) is due to the fact that, in the case of electrons, each spatial wavefunction is associated with two associated quantum states, one with $m_s = \frac{1}{2}$ "spin up" and one with $m_s = -\frac{1}{2}$.] The number of occupied states in the energy interval $d\varepsilon$ is given by the product of the number of such states $g(\varepsilon)\, d\varepsilon$ and $f(\varepsilon)$, the occupation probability. Since the total number of electrons N is fixed, it follows that

$$\int_0^\infty g(\varepsilon) f(\varepsilon)\, d\varepsilon = N$$

which, using (16.6) and (16.7), becomes

$$N = \frac{8\pi V m (2m)^{1/2}}{h^3} \int_0^\infty \varepsilon^{1/2} \frac{1}{e^{[\varepsilon - \mu(T)]/kT} + 1}\, d\varepsilon \tag{16.8}$$

At zero temperature $f(\varepsilon) = 1$ for $\varepsilon < \mu$, and $f(\varepsilon) = 0$ for $\varepsilon > \mu$, so that the last integral becomes

$$\lim_{T \to 0} \int_0^\infty \varepsilon^{1/2} \frac{d\varepsilon}{e^{[\varepsilon - \mu(0)]/kT} + 1} = \int_0^{\mu(0)} \varepsilon^{1/2}\, d\varepsilon = \tfrac{2}{3} \mu^{3/2}(0)$$

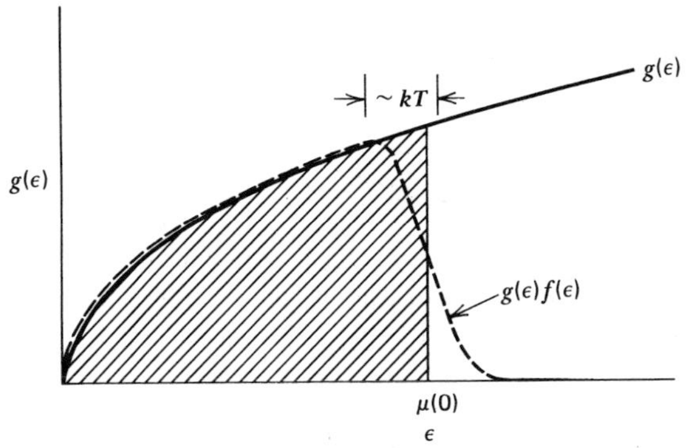

Figure 16.2 Plot of density of states $g(\varepsilon)$ and the density of electrons $g(\varepsilon)f(\varepsilon)$, as a function of energy. At absolute zero the states up to $\mu(0)$ are filled. The dotted curve indicates the density of filled states at a temperature $T \ll \mu(0)/k$.

and after substituting in (16.8)

$$\mu(0) = \frac{\hbar^2}{2m}\left[3\pi^2\left(\frac{N}{V}\right)\right]^{2/3} \tag{16.9}$$

The chemical potential μ, often called the Fermi energy, plays a key role in the theories of metals and semiconductors. The behavior of many metals can often be explained to a good approximation using a model of a free electron gas. According to this model at very low temperatures all the electron states with energies up to $\varepsilon = \mu$ are filled, and those above it are empty. At higher temperatures, levels within $\sim kT$ of μ are only partially filled. This situation is depicted in Fig. 16.2.

As an example we calculate the chemical potential (Fermi energy) of a metal with 10^{29} electrons/m^3. Using this result in (16.9) gives

$$\begin{aligned}\mu(0) &= 1.257 \times 10^{-18} \text{ joules} \\ &= 7.85 \text{ eV}\end{aligned}$$

(It is useful to recall here that it is the Pauli exclusion principle that causes the energy levels in the metal, even at $T=0$, to be filled up to $\varepsilon \simeq 7.85$ eV. In a system of bosons, for example, where a given state may be occupied by any number of particles at zero temperature, the particles will all "condense" to $\varepsilon = 0$). Electrons near the top of the occupied states move with velocities v_F such that at $T=0$

$$\tfrac{1}{2} m v_F^2 = \mu(0)$$

In the above example, $v_F \simeq 1.67 \times 10^8$ cm/s. It is important to note that these high velocities persist even at zero temperature, and the explanation of their existence is *purely quantum mechanical.*

A very simple yet important significance of the chemical potential $\mu(T)$ is that it corresponds to the *increase* of the *total energy of the system when one particle is added to it*, that is,

$$\mu(T)=\frac{\partial}{\partial N}\int_0^\infty \varepsilon g(\varepsilon)f(\varepsilon)\,d\varepsilon \tag{16.10}$$

This statement follows directly from the Fermi–Dirac distribution law (16.6). Since the states up to $\mu(T)$ are occupied [here we ignore the small transition region of width $\sim kT$ near $\mu(T)$], any additional particle must be placed at an energy $\mu(T)$. It will be assigned as an exercise to show that at $T=0$

$$\frac{\partial E}{\partial N}=\frac{\partial E}{\partial \mu(0)}\frac{\partial \mu(0)}{\partial N}=\mu(0) \tag{16.11}$$

where $E=\int_0^{\mu(0)}\varepsilon g(\varepsilon)\,d\varepsilon$ is the energy of the N-particle system (at zero temperature). It follows that if two materials with different chemical potentials are brought into contact with one another, particles will flow from the region of high μ to that of low μ. Equilibrium will be reestablished when the chemical potential has a single uniform value throughout the sample, since otherwise the total energy may be lowered by transferring particles from regions of high chemical potential to those where it is lower.

The general dependence of $\mu(T)$ on temperature can be obtained using (16.8) by taking $T\neq 0$. We will not derive the result (the interested reader can consult any basic text in statistical mechanics or solid state physics[1]), but merely state it:

$$\mu(T)\cong\mu(0)\left[1-\left(\frac{\pi^2}{12}\right)\left(\frac{kT}{\mu(0)}\right)^2\right] \tag{16.12}$$

Another important consequence of the Pauli exclusion principle manifests itself in the heat capacity of an electron gas. Viewed as a classical particle, each electron would have a total energy $3kT/2$, so that the energy per unit volume in a sample with an electron density (N/V) is $U=\frac{3}{2}(N/V)kT$ and the heat capacity per unit volume is $C=\partial U/\partial T=\frac{3}{2}(N/V)k$. This result is bigger by some two orders of magnitude than the measured values of the heat capacity at room temperature.

We can understand the nature of this discrepancy by referring to Fig. 16.2. The total energy of the electron system is given by

$$E=\int_0^\infty g(\varepsilon)f(\varepsilon)\varepsilon\,d\varepsilon$$

Most of the value of this integral comes from the fully occupied states $[f(\varepsilon)=1]$ below the chemical potential. Their contribution to U is thus a

---

[1]See, for example, C. Kittel, *Introduction to Solid State Physics*, 5th ed. (J. Wiley & Sons, New York, 1976).

constant that does not depend on temperature, so that their heat capacity

$$C = \partial E / \partial T$$

is zero. The main contribution to C is thus due to the fraction $\sim kT/\mu$ of the electrons with energies within kT of the chemical potential μ. These may be considered approximately as classical particles with an energy per particle of $\frac{3}{2}kT$. The contribution to the total energy due to these electrons is, per unit volume,

$$E' \sim \left(\frac{N}{V}\right)\left(\tfrac{3}{2}kT\right)\frac{kT}{\mu(0)}$$

so that the heat capacity becomes

$$C = \frac{\partial E}{\partial T} \simeq \frac{\partial E'}{\partial T} = 3\left(\frac{N}{V}\right)k\frac{kT}{\mu(0)}$$

An exact analysis[2] yields

$$C = \frac{\pi^2}{2}\left(\frac{N}{V}\right)k\frac{kT}{\mu(0)} \tag{16.13}$$

In the case of the numerical example considered above and for $T = 300°\text{K}$, $kT/\mu(0) \sim \frac{1}{300}$. This factor is of the order of magnitude of the discrepancy between the experimental data and the prediction of classical theory.

16.3 THE BOSE–EINSTEIN DISTRIBUTION

We conclude this chapter by considering the case of the Bose–Einstein distribution law (15.25). To determine α in this case, one must invoke, as in the case of the other distributions, the constancy of the number of particles in the system. This involves a good deal of specialized calculation. There exists, however, a very important yet simple case we can use to demonstrate the basic principles. This is the case of electromagnetic radiation at thermal equilibrium. We have already considered this case in Chapter 12. In that treatment, the elementary particles of the system were the radiation modes (oscillators) that, due to their distinguishability, obey the Boltzmann distribution law (16.2).

We may, alternatively, consider the photons, rather than the oscillators, as the elementary particles of the radiation fields. The one-particle states are taken as the electromagnetic modes

$$\psi(\mathbf{r}) \propto e^{i\mathbf{k}\cdot\mathbf{r}} \tag{16.14}$$

which were considered in Section 12.2. A "particle" (photon) in state $\mathbf{k}$ thus has an energy $h\nu$ ($\nu = kc/2\pi$). The photon is characterized, in addition to the

[2] Kittel, *ibid.*

propagation vector **k**, as well as by its "spin" state $s=1$, $m_s=\pm 1$. The quantum numbers $m_s=\pm 1$ describe the two senses of circular polarization associated with a given direction of propagation **k**. The photons being indistinguishable and possessing an integral ($s=1$) spin obey the Bose–Einstein distribution law (15.25):

$$n_s = \frac{g_s}{e^{\alpha+\varepsilon_s/kT}-1} \tag{16.15}$$

where we used $\beta=(kT)^{-1}$. We recall that the parameter α was introduced into the formalism [see for example (15.18)], in the process of insuring that the number of particles N in the system be a constant. The total number of photons (particles), however, is not restricted.

(This is a very subtle point and deserves amplification. If the field energy is increased, then, using the radiation modes as the basic "particles" of the system, we need merely increase the energy of the modes, keeping their number a *constant*. If we choose instead to take the field particles as photons, then an increase in the excitation of the system results in an increase of the number of photons per mode. The number of photons is thus not conserved.)

Since the number of particles is not fixed, the auxiliary condition (15.13) that gave rise to α in the distribution laws is now meaningless, so that in (16.15) $\alpha=0$ and

$$n_s = \frac{g_s}{e^{h\nu_s/kT}-1} \qquad \text{(for photons)} \tag{16.16}$$

Let g_s in (16.16) be the number of states **k** whose frequencies lie within an interval $d\nu$ centered on ν. The number of such states was given by (12.35) as

$$g_s = \frac{8\pi\nu^2 n^3 V}{c^3}\,d\nu$$

The number of particles (photons) in this range is from (16.16):

$$n_s = \frac{8\pi\nu^2 n^3 V}{c^3\left(e^{h\nu/kT}-1\right)}\,d\nu$$

Multiplying the last expression by the energy $h\nu$ per photon, we obtain (after dividing by the enclosure volume V) an expression for the energy density of a black body radiation field due to frequencies between ν and $\nu + d\nu$:

$$\rho(\nu)\,d\nu = \frac{8\pi h\nu^3 n^3}{c^3\left(e^{h\nu/kT}-1\right)}\,d\nu \tag{16.17}$$

in agreement with (12.37).

PROBLEMS

1. Show that at equilibrium the "temperature" of a mixture of three types of particles is the same.
2. Derive the Fermi energy of a two-dimensional electron gas confined to an area of $a \times b$.
3. Show that the two orthogonally plane-polarized states of an electromagnetic wave propagating along some arbitrary direction can be considered, equivalently, as two waves with opposite senses of circular polarization.

   Hint: Show that we can express each representation as a linear superposition of the modes of the other.

4. Prove Eq. (16.11).
5. Derive the black body energy density formula (16.17)

$$\rho(\nu) = \frac{8\pi h\nu^3 n^3}{c^3(e^{h\nu/kT} - 1)}$$

   considering the electromagnetic modes, rather than the photons, as the "particles."

   Clue: Are the modes distinguishable?

CHAPTER SEVENTEEN

The Band Theory of Electrons in Crystals

We have considered in early chapters of this book the quantum properties of single particles and of some two-particle systems. These included the hydrogen atom (Chapter 7), a particle in a potential well (Chapter 4), and as an example of a two-particle system, the helium atom (Section 8.2). In the last chapter we discussed the statistical properties of a many-electron gas.

In this chapter we consider the problem of an electron (or particle) in a spatially periodic potential field. The analysis is relevant to the electronic properties of crystals.

We encounter here, for the first time, some concepts of central importance in solid state physics. These include the ideas of forbidden energy gaps and of Brillouin zones. These concepts reflect the lattice periodicity and arise in other branches of physics that involve wave propagation in periodic media. Some examples are: The propagation of acoustic waves in crystals[1] and of electromagnetic waves in periodic waveguides.

17.1 THE KRONIG–PENNEY MODEL

The basic features of the wavefunctions of electrons in crystals may be demonstrated with the aid of the one-dimensional Kronig–Penney model. The potential field used in this model is shown in Fig. 17.1. The plot is that of the potential energy profile $V(x)$ experienced by a single electron. The wavefunction $\psi(x)$ of an electron in this model obeys the Schrödinger

[1]C. Kittel, *Introduction to Solid State Physics*, 5th ed. (John Wiley & Sons, New York, 1976).

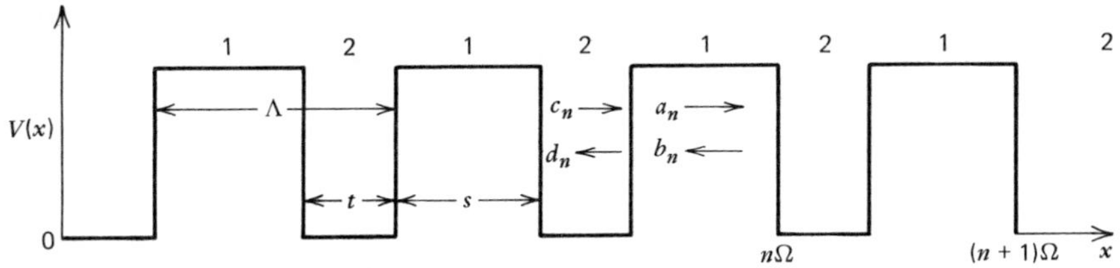

Figure 17.1 A one-dimensional periodic potential.

equation:

$$-\frac{\hbar^2}{2m}\frac{d^2\psi}{dx^2}=[E-V(x)]\psi(x) \tag{17.1}$$

Before proceeding to solve for $\psi(x)$, we consider some general properties of wavefunctions in periodic media. According to the Floquet theorem in mathematics (also known as the Bloch theorem in solid state physics[2]), the solution of the wave equation in a periodic medium is of the form

$$\psi_{\mathbf{K}}(\mathbf{r})=u_{\mathbf{K}}(\mathbf{r})e^{i\mathbf{K}\cdot\mathbf{r}} \tag{17.2}$$

where $u_{\mathbf{K}}(\mathbf{r})$ has the same periodicity as the medium. In one dimension (17.2) becomes

$$\psi_K(x)=u_K(x)e^{iKx} \tag{17.3}$$

where

$$u_K(x+\Lambda)=u_K(x) \tag{17.4}$$

Λ is the periodic distance.

To prove that the wavefunction $\psi(x)$ has the Bloch form (17.3), we introduce the translation operation $\hat{T}_\Lambda$ defined by

$$\hat{T}_\Lambda f(x)=f(x+\Lambda) \tag{17.5}$$

where $f(x)$ is any arbitrary function. We will first prove that $\hat{T}_\Lambda$ commutes with the Hamiltonian $\hat{\mathcal{H}}$. Since $\psi(x)$ is an eigenfunction of $\hat{\mathcal{H}}$, $\hat{\mathcal{H}}\psi(x)=E\psi(x)$, where E is the energy of the electron. Using (17.5) we write

$$\begin{aligned}\hat{T}_\Lambda\hat{\mathcal{H}}\psi(x)&=E\hat{T}_\Lambda\psi(x)\\&=E\psi(x+\Lambda)\\\hat{\mathcal{H}}\hat{T}_\Lambda\psi(x)&=\hat{\mathcal{H}}\psi(x+\Lambda)\\&=E\psi(x+\Lambda)\end{aligned}$$

so that

$$\hat{\mathcal{H}}\hat{T}_\Lambda-\hat{T}_\Lambda\hat{\mathcal{H}}=0$$

[2] F. Bloch, *Z. Physik* **52**, pp. 555–560 (1928).

and $\hat{\mathcal{H}}$ and $\hat{T}_\Lambda$ commute. Two commuting operators have common eigenfunctions so that $\psi(x)$ can be chosen to be an eigenfunction of $\hat{T}_\Lambda$ as well as of $\hat{\mathcal{H}}$ (see the discussion in Chapter 3). Let the corresponding eigenvalue be c so that

$$\begin{aligned}\hat{T}_\Lambda\psi(x)&=\psi(x+\Lambda)\\&=c\psi(x)\end{aligned}\tag{17.6a}$$

where c is some constant. Then

$$\psi(x+g\Lambda)=c^g\psi(x)\tag{17.6b}$$

To determine c we need to impose boundary conditions on $\psi(x)$. The one commonly used[3] is that $\psi(x)=\psi(x+L)$, where $L=N\Lambda$ is the length of our one-dimensional crystal:

$$\begin{aligned}\psi(x+N\Lambda)&=c^N\psi(x)\\&=\psi(x)\end{aligned}\tag{17.7}$$

so that

$$C^N=1,\qquad C=e^{i(2\pi g/N)}\,(g=0,1,2,\ldots,N-1)\tag{17.8}$$

A solution satisfying (17.5) and (17.7) can be taken as

$$\psi_g(x)=e^{i2\pi g(x/N\Lambda)}u_g(x)\tag{17.9}$$

where $u_g(x)$ *is periodic in* Λ. Defining

$$K=\frac{2\pi g}{N\Lambda}=\frac{2\pi}{L}g(g=0,\pm1,\pm2,\ldots)\tag{17.10}$$

the total eigenfunction can thus be written as

$$\psi_K(x)=u_K(x)e^{iKx}\tag{17.11}$$

which is the Bloch form.

We are now ready to proceed with the solution of the Schrödinger equation:

$$\frac{d^2\psi}{dx^2}+\frac{2m}{\hbar^2}[E-V(x)]\psi=0\tag{17.12}$$

Since $V(x)$ has a constant value of V_0 in regions 1 and is zero in regions 2, the solution of (17.12) in each region is a superposition of the two linearly independent exponential solutions:

$$\psi(x)=a_ne^{k_1(x-n\Lambda)}+b_ne^{-k_1(x-n\Lambda)}\quad\text{in regions 1}\tag{17.13}$$

$$\psi(x)=c_ne^{ik_2(x-n\Lambda)}+d_ne^{-ik_2(x-n\Lambda)}\quad\text{in regions 2}\tag{17.14}$$

with

$$k_1=\frac{\sqrt{2m(V_0-E)}}{\hbar}\tag{17.15}$$

$$k_2=\frac{\sqrt{2mE}}{\hbar}\tag{17.16}$$

---

[3]C. Kittel, *Introduction to Solid State Physics*, 5th ed. (John Wiley & Sons, New York, 1975).

Our next task is that of solving for the coefficients a_n, b_n, and d_n that uniquely determine $\psi(x)$. Requiring that $\psi(x)$ and $d\psi/dx$ be continuous at $x=n\Lambda$ yields

$$\begin{aligned} a_n + b_n &= c_{n+1}e^{-ik_2\Lambda} + d_{n+1}e^{ik_2\Lambda} \\ k_1 a_n - k_1 b_n &= ik_2 c_{n+1}e^{-ik_2\Lambda} - ik_2 d_{n+1}e^{ik_2\Lambda} \end{aligned} \tag{17.17}$$

while the same boundary conditions applied at $x=n\Lambda+t=(n+1)\Lambda-s$ gives

$$\begin{aligned} c_{n+1}e^{-ik_2 s} + d_{n+1}e^{ik_2 s} &= a_{n+1}e^{-k_1 s} + b_{n+1}e^{k_1 s} \\ ik_2 c_{n+1}e^{-ik_2 s} - ik_2 e^{ik_2 s} d_{n+1} &= k_1 a_{n+1}e^{-k_1 s} - k_1 b_{n+1}e^{ik_1 s} \end{aligned} \tag{17.18}$$

Equations (17.17) and (17.18) can be expressed using matrix notation as

$$\begin{pmatrix} 1 & 1 \\ 1 & -1 \end{pmatrix} \begin{pmatrix} a_n \\ b_n \end{pmatrix} = \begin{pmatrix} e^{-ik_2\Lambda} & e^{ik_2\Lambda} \\ i\dfrac{k_2}{k_1}e^{-ik_2\Lambda} & -i\dfrac{k_2}{k_1}e^{ik_2\Lambda} \end{pmatrix} \begin{pmatrix} c_{n+1} \\ d_{n+1} \end{pmatrix} \tag{17.19}$$

and

$$\begin{pmatrix} e^{-ik_2 s} & e^{ik_2 s} \\ e^{-ik_2 s} & -e^{ik_2 s} \end{pmatrix} \begin{pmatrix} c_{n+1} \\ d_{n+1} \end{pmatrix} = \begin{pmatrix} e^{-k_1 s} & e^{k_1 s} \\ -i\dfrac{k_1}{k_2}e^{-k_1 s} & i\dfrac{k_1}{k_2}e^{k_1 s} \end{pmatrix} \begin{pmatrix} a_{n+1} \\ b_{n+1} \end{pmatrix} \tag{17.20}$$

By obvious matrix manipulation (pre- and post-multiplication) we can relate (a_n, b_n) to (a_{n+1}, b_{n+1}):

$$\begin{vmatrix} a_n \\ b_n \end{vmatrix} = \begin{vmatrix} A & B \\ C & D \end{vmatrix} \begin{vmatrix} a_{n+1} \\ b_{n+1} \end{vmatrix} \tag{17.21}$$

where

$$\begin{aligned} A &= e^{-k_1 s}\left[\cos k_2 t + \frac{1}{2}\left(\frac{k_2}{k_1} - \frac{k_1}{k_2}\right)\sin k_2 t\right] \\ B &= e^{k_1 s}\left[\frac{1}{2}\left(\frac{k_2}{k_1} + \frac{k_1}{k_2}\right)\sin k_2 t\right] \\ C &= e^{-k_1 s}\left[-\frac{1}{2}\left(\frac{k_2}{k_1} + \frac{k_1}{k_2}\right)\sin k_2 t\right] \\ D &= e^{k_1 s}\left[\cos k_2 t - \frac{1}{2}\left(\frac{k_2}{k_1} - \frac{k_1}{k_2}\right)\sin k_2 t\right] \end{aligned} \tag{17.22a}$$

The matrix (A, B, C, D) is referred to as the unit cell transformation matrix. From (17.22a) it follows that

$$AD - BC = 1 \tag{17.22b}$$

that is, the transformation matrix is unimodular.

Using (17.21) we can obtain the coefficients a_n, b_n in any unit cell once their values in some cell are known. We can then solve—using (17.19)—for c_n, d_n, given a_{n-1}, b_{n-1}.

Consider the basic solution of $\psi(x)$ in region 1 of the nth unit cell, Eq. (17.13):

$$\psi(x)=a_n e^{k_1(x-n\Lambda)}+b_n e^{-k_1(x-n\Lambda)}$$

so that

$$\psi(x+\Lambda)=a_{n+1}e^{k_1[x+\Lambda-(n+1)\Lambda]}+b_{n+1}e^{-k_1[x+\Lambda-(n+1)\Lambda]}$$
$$=a_{n+1}e^{k_1(x-n\Lambda)}+b_{n+1}e^{-k_1(xn-\Lambda)}$$

The Bloch form of $\psi(x)$ imposed by the periodicity was given by (17.11) as

$$\psi(x)=\psi(x+\Lambda)e^{iK\Lambda}$$

The last two forms of $\psi(x)$ can be reconciled provided

$$\begin{vmatrix} a_n \\ b_n \end{vmatrix}=\begin{vmatrix} a_{n+1} \\ b_{n+1} \end{vmatrix}e^{-iK\Lambda} \tag{17.23}$$

which, using (17.21) and then letting $n+1\rightarrow n$, leads to

$$\begin{vmatrix} A & B \\ C & D \end{vmatrix}\begin{vmatrix} a_n \\ b_n \end{vmatrix}=e^{-iK\Lambda}\begin{vmatrix} a_n \\ b_n \end{vmatrix} \tag{17.24}$$

Equation (17.24) is in the form of the general operator eigenvalue problem

$$\hat{A}u_n=a_n u_n$$

where the matrix (A, B, C, D) can be considered as a matrix representation of the unit cell translation operator in a 2×2 function space, and the column matrix (a_n, b_n) is the eigenvector. The factor $\exp(-iK\Lambda)$ is thus the eigenvalue of the matrix. By subtracting the right side of (17.24) from the left side, we obtain

$$\begin{vmatrix} A-e^{-iK\Lambda} & B \\ C & D-e^{-iK\Lambda} \end{vmatrix}\begin{vmatrix} a_n \\ b_n \end{vmatrix}=0 \tag{17.25}$$

which is a set of two homogeneous equations for the unknowns a_n and b_n. The condition for the existence of nontrivial solutions is that the determinant vanish. After using (17.22a), this leads to

$$e^{-iK_{1,2}\Lambda}=\tfrac{1}{2}(A+D)\pm i\left\{1-\left[\tfrac{1}{2}(A+D)\right]^2\right\}^{1/2} \tag{17.26}$$

where the 1,2 subscripts correspond to $+$ and $-$ on the right side, respectively. The eigenvectors corresponding to the eigenvalues (17.26) are obtained by substituting (17.26) in (17.25), leading to

$$\begin{vmatrix} a_0 \\ b_0 \end{vmatrix}=N\begin{vmatrix} B \\ e^{-iK_{1,2}\Lambda}-1 \end{vmatrix} \tag{17.27}$$

where N is a normalization constant to ensure that $\int\psi\psi^*\,dV=1$. Given a_0 and b_0 we can use (17.13) to write the eigenfunction $\psi_K(x)$ in region 1 of the nth unit cell:

$$\psi_K(x)=N\Big[\Big(a_0e^{k_1(x-n\Lambda)}+b_0e^{-k_1(x-n\Lambda)}\Big)e^{-iK(x-n\Lambda)}\Big]e^{iKx} \qquad (17.28)$$

The expression for $\psi(x)$ has the required Bloch form (17.11), since the portion of the function within the square brackets is periodic in Λ. The propagation constant K is obtained from (17.26):

$$\cos(K\Lambda)=\tfrac{1}{2}(A+D) \qquad (17.29)$$

which is the basic dispersion relation for the propagation of the wavefunction. Using the specific form of A and D (17.22), we obtain

$$\cos(K\Lambda)=\cos(k_2t)\cosh(k_1s)+\sin(k_2t)\sinh(k_1s)\left(\frac{k_1^2-k_2^2}{2k_1k_2}\right) \qquad (17.30)$$

A comtemplation of the dispersion relation (17.30) is in order. The right side is, according to (17.15) and (17.16), a function of the particle energy $E=\hbar^2k_2^2/2m$. Given any value of E, we can thus use (17.30) to obtain $K\Lambda$ (to within a multiple of 2π). We distinguish between two regimes:

(a) Values of energy E such that

$$\tfrac{1}{2}(A+D)<1 \qquad (17.31)$$

It follows from (17.29) that in this case K is real and $\psi_K(x)$, according to (17.28), is a (modulated) propagating wave. These regions of E are referred to as "*allowed*."

(b) Values of E such that

$$\tfrac{1}{2}(A+D)>1 \qquad (17.32)$$

Here $\cos K\Lambda>1$ and K must be complex. Since $\frac{1}{2}(A+D)$ is real, it follows that

$$K=\frac{m\pi}{\Lambda}+iK_i \quad (m=\pm\pi,\pm2\pi,\ldots) \qquad (17.33)$$

[The simple proof of (17.33) is assigned as a problem.] In this regime the factor $\exp(iKx)$ in (17.28) becomes

$$\exp(iKx)=e^{i(m\pi/\Lambda)x}\exp(-|K_i|x)$$

and the $\psi_K(x)$ is an exponentially evanescent function. In the interior of a large crystal $\psi_K(x)$ is thus zero. (The solution corresponding to $\exp(|K_i|x)$ is ruled out, since it leads to unphysical solutions that increase exponentially without limit.

The energy intervals for which $\frac{1}{2}(A+D)>1$ so that K is complex are referred to as the *forbidden energy gaps*. No electrons with "forbidden" energies can "exist" in the interior of a large crystal. In practice, we can find such electrons only within a few times K_i^{-1} from an interface or some other discontinuity in the crystal structure.

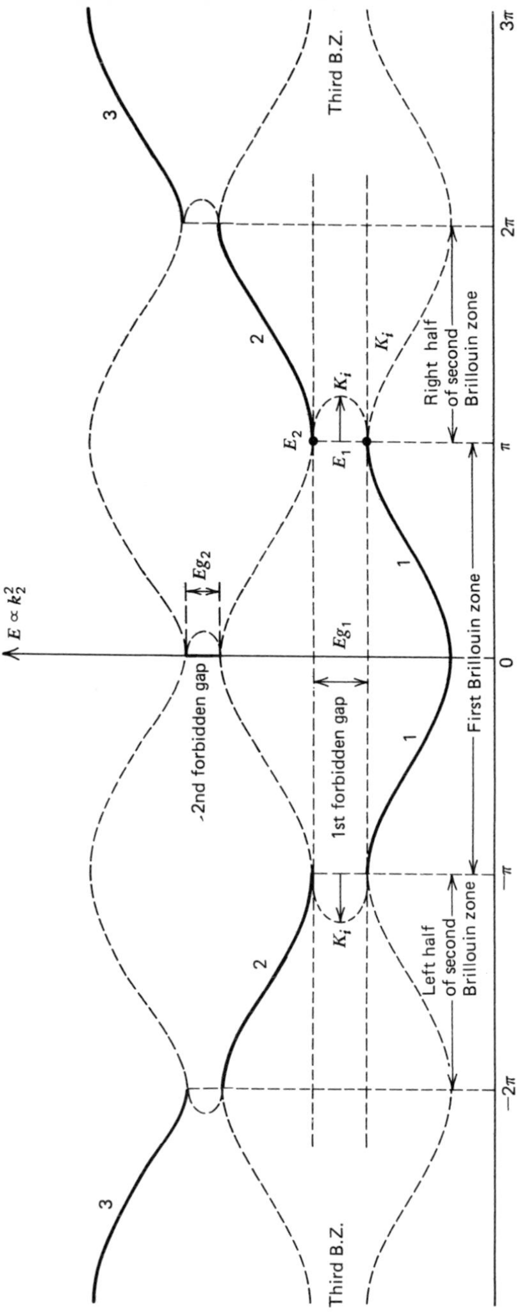

Figure 17.2 (a) The dispersion diagram E vs. $K\Lambda$ for electrons in the Kronig–Penney potential field.

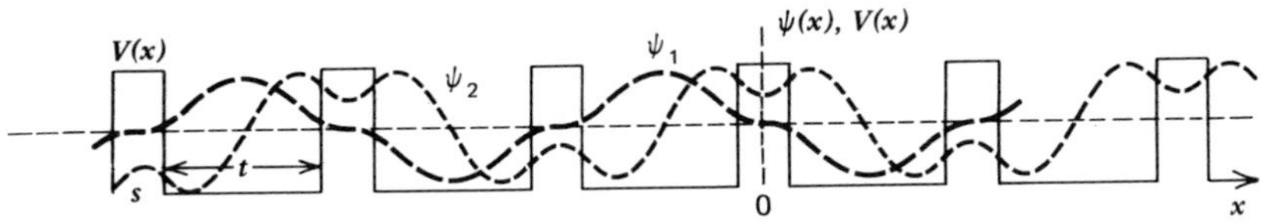

Figure 17.2 (*b*) A schematic representation of the eigenfunctions ψ_1 (odd symmetry) and ψ_2 at the edge of the forbidden gap, that is, $K\Lambda = m\pi$, $K_i = 0$. ψ_1 has its nodes inside the potential barriers, while ψ_2 has its maxima inside them. The eigenenergy E_1 of ψ_1 is thus lower than the energy E_2 of ψ_2.

A typical plot of E vs. K (and K_i) obtained from (17.30) is shown in Fig. 17.2*a*. We notice that the decay constant K_i is maximum at midgap.

The condition $K\Lambda = m\pi$ that, according to (17.33), marks the boundary of a forbidden gap, is formally equivalent to the one-dimensional Bragg condition for X rays.[4] When this condition is satisfied, the reflections of the electron wavefunction $\psi_K(x)$ from neighboring unit cells are in phase and reinforce each other since the round-trip phase delay is $2K\Lambda = 2n\pi$. Under these conditions $\psi_K(x)$ is strongly reflected and cannot "penetrate" into the bulk of the crystal, which results in the evanescent behavior of (17.33).

The perfect reflection of the particle wave that occurs when $K\Lambda = m\pi$ causes the wavefunction to behave like a standing wave rather than a running wave. (A standing sinusoidal wave, we recall, results from the interference of e^{iKx} and e^{-iKx}.) There are two independent standing wave solutions corresponding to the same value of $K = m\pi/\Lambda$—one even and one odd—since the eigenfunctions must possess definite parity. Since one of these two functions has its extrema inside the potential barriers, its eigenenergy E_2 is higher than that of the second function that has its nodes at the barriers. There are thus two eigenfunctions with the same value of $K = m\pi/\Lambda$ but with two different energies. These two energies E_1 and E_2 in Fig. 17.2 correspond to the top and bottom of the forbidden gap. The schematic behavior of the two eigenfunctions at the edge of the gap is illustrated in Fig. 17.2*b*.

The region $-\pi < K\Lambda \leq \pi$ is called the first Brillouin zone.[5] The two regions $\pi < K\Lambda < 2\pi$ and $-2\pi < K\Lambda < -\pi$ are designated collectively as the second Brillouin zone, and so on.

Since the dispersion relation (17.30) only determines $K\Lambda$ to within $2m\pi$, m being an integer, and since $\psi_K(x)$ as given by (17.28) is invariant when K is replaced by $K + m(2\pi/\Lambda)$, we may "collapse" the dispersion diagram (Fig. 17.2) to the interval $-\pi/\Lambda < K < \pi/\Lambda$ by shifting the dark portions of the curves horizontally by $(2\pi/\Lambda)l$ (l = some integer). The result, known as the reduced zone energy band diagram, is shown in Fig. 17.3.

---

[4] See Kittel, *ibid.*

[5] L. Brillouin is responsible for many of the basic concepts involving propagation in periodic media. See, for example, his book *Wave Propagation in Periodic Structures* (McGraw-Hill, New York, 1946).

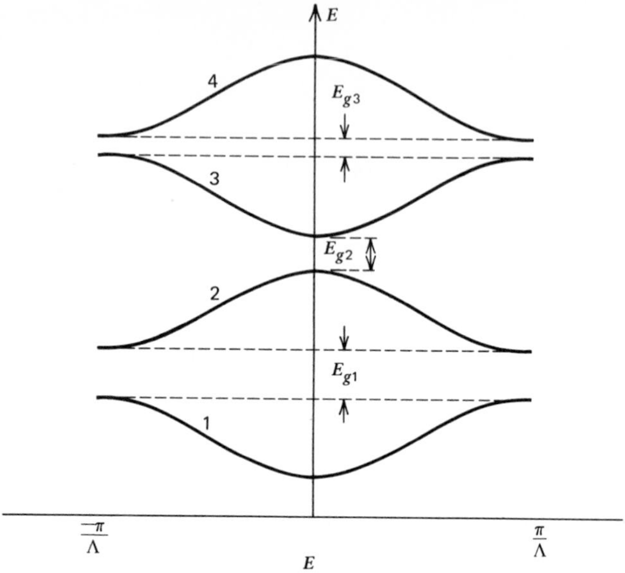

Figure 17.3 A reduced zone band diagram. The numerical designation corresponds to Fig. 17.2.

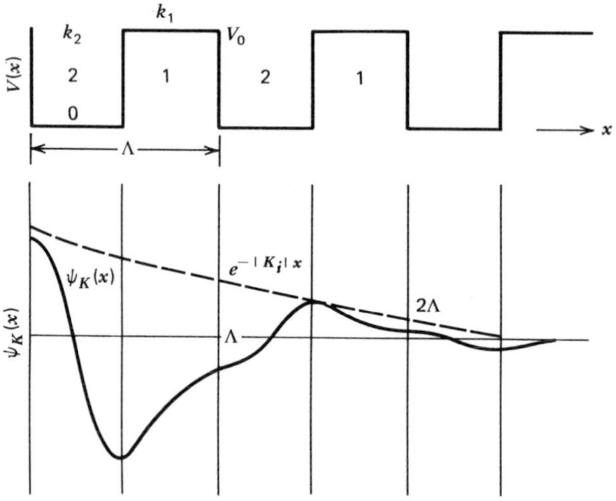

Figure 17.4 The behavior of the wavefunction of an electron whose energy is inside the first forbidden gap.

A schematic plot of the wavefunction $\psi_K(x)$ (17.28) for an energy E somewhere in the first forbidden gap ($K\Lambda = \pi + iK_i$) is shown in Fig. 17.4. Note the phase reversal [$\exp(iK\Lambda) = -1$] in each unit cell. The basic behavior of $\psi_K(x)$ is one of sinusoidal variation in region 2 and exponential in region 1. The whole pattern fits under an evanescent envelope $\exp(-K_i X)$ so that it decays quickly upon penetration into the bulk of the crystal and is thus associated with "surface electron states."

17.2 THE MULTIELECTRON CRYSTAL

The solution of the Schrödinger equation in a periodic potential field resulted in the one-particle eigenfunctions (17.28)

$$\psi_K(x) = u_K(x)e^{iKx} \tag{17.34}$$

and the associated energy E_K. If we neglect the interaction between electrons, we can treat the case of a real crystal with many electrons by associating with each allowed value of K one electron, until all the electrons are used up.

In the deliberations leading to (17.10), we showed that in a (one-dimensional) crystal *consisting of N unit cells*, the allowed set of K numbers is restricted to the set

$$K = \frac{2\pi g}{N\Lambda} = \frac{2\pi}{L}g \quad (g = 0, \pm 1, \pm 2, \ldots)$$

so that two adjacent K numbers are separated by $\Delta K = 2\pi/L$. These are shown as dots in Fig. 17.5. Since the length of a Brillouin zone is $2\pi/\Lambda$ there are

$$\frac{2\pi/\Lambda}{\Delta K} = \frac{2\pi/\Lambda}{2\pi/L} = \frac{L}{\Lambda} = N$$

allowed K values in each zone. Since with each K value we may associate 2 spin states ($m_s = \pm\frac{1}{2}$), each Brillouin zone can accommodate $2N$ electrons, where N, we recall, is the number of unit cells in the crystal. If the number of valence electrons per unit cell[6] is 2, then the first zone is exactly full. If the number is 1, the zone is only half full. In general, an even number of electrons per unit cell leads to fully occupied zones, while an odd number requires that the uppermost zone (band) is only half full.

When an electric field is applied to a crystal with an even number of valence electrons per unit cell so that the uppermost Brillouin zone is filled with no electrons left over for the next higher zone, as in Fig. 17.5, there is *no current flow.* This astounding fact follows from the symmetry of a filled band. For each electron moving to the right ($K>0$), there is a corresponding electron with ($K<0$) with an equal but oppositely directed velocity,[7] so that

---

[6]We assume that the inner atomic electrons are tightly bound and do not play a significant role in the electronic conduction process.

[7]The more precise explanation of this fact should use the concept of the electron group velocity, which is introduced in Section 17.3.

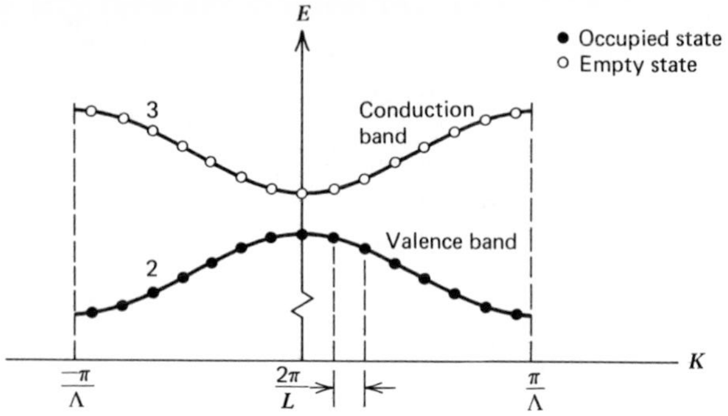

Figure 17.5 In an insulator (which includes the case of an intrinsic semiconductor at very low temperatures) all the quantum states in the valence band are occupied by electrons (black dots). All the energy bands lying above the valence band, of which only one (the conduction band) is shown, are empty.

the sum of the two velocities, hence the currents, cancel. The crystal is thus an insulator. This exact balancing on a one-to-one basis of electrons with positive K values by those with negative K numbers is disrupted by the application of an electric field to a crystal with a partially filled band, so that a net current flow results, that is, the crystal is metallic.

We thus find that the radically different conduction properties of crystals depend fundamentally on the crystal structure—that is, number of valence electrons per unit cell. As an example of metallic behavior we may take the case of sodium. The crystal is body centered cubic. Its Brillouin zone contains two states for each atom of the crystal. Each sodium atom, however, contributes but one $3s$ electron. The band is consequently only half full and the crystal is highly conductive.

Sodium chloride (NaC1) is a good example of an insulating crystal. The unit NaC1 associated with each unit cell has 28 electrons, which is an even number. It follows that the topmost occupied band is completely full.

In crystals with an even number of electrons per unit cell, one usually refers to the uppermost fully occupied band as the *valence band* and to the next higher band that is empty as the conduction band. The energy separating the extrema of these bands—that is, the smallest energy separation—is called the energy gap E_g.

Now our statement about the valence band being fully occupied while the conduction band is empty is strictly true only in the limit of zero temperature. At finite temperatures it follows from the Fermi–Dirac distribution law (16.6) that some electrons are to be found in the conduction band. Each such electron must leave an unoccupied state (*hole*) in the valence band. An application of an electric field will now cause a current flow, since the bands are not perfectly *full* or *empty*, so that crystals with sufficiently small values of E_g, where the excitation of carriers across the gap is appreciable, are called semiconductors. Some of the better known and widely used semiconductors are crystals of Si ($E_g = 1.1\,\text{eV}$), and GaAs ($E_g = 1.45\,\text{eV}$).

The conductivity of undoped semiconductors, unlike that of metals, depends on excitation of electrons across an energy gap and is thus a strong function of temperature, disappearing altogether at $T=0$, since according to (16.6) at zero temperature the valance band is completely full.

As mentioned above, the application of an electric field to a semiconductor or a conductor causes a net flow of charge that is due to the unbalanced motion of the electrons in the conduction band, in the case of semiconductors, and valence bands. Since the number of occupied states in the valence band is typically many orders of magnitude larger than that of the vacant states, it is a convenient matter of bookkeeping to consider the valence band as completely full with electrons and to add to it a number of positive charges equal to the actual number of vacant states so as to preserve the actual total electronic charge in the valence band. Since the fully occupied valence band does not contribute to the current flow, we can attribute it *solely* to the fictitious positive charges called "holes."

In the case of a pure ("intrinsic") semiconductor, the number of conduction band electrons and holes is equal, since each conduction band electron leaves behind a vacancy in the valence band. This situation is depicted in Fig. 17.6. The incorporation of impurity atoms into the crystal can change this balance and leads to crystals whose conductivity is dominated by conduction band electrons; that is, the number of electrons far exceeds that of the holes (N-type semiconductor), or by valence band holes (P-type).

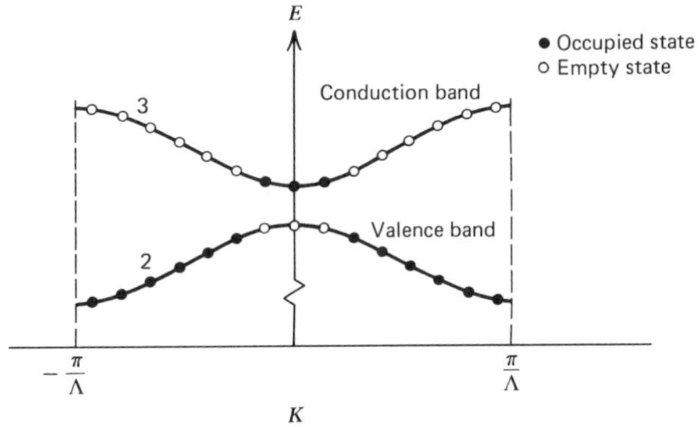

Figure 17.6 A semiconductor. The energy gap E_g is small enough so that thermal excitation elevates a sizeable number of electrons to the conduction band. Occupied states are shown as black dots, while empty ones are white. The conductivity is due to the electrons in the conduction band as well as to the missing electrons (holes) in the valence band.

The control of the conductivity type of a semiconductor (N or P) is the basis for the operation of the transistor, the semiconductor laser, and other devices currently changing the face of the electronic and optical fields. These topics are considered in Chapter 20.

17.3 THE MOTION OF ELECTRONS IN CRYSTALS

The classical motion of a pointlike electron is not directly compatible with the quantum mechanical behavior of an electron described by a Bloch eigenstate

$$\psi_K(x,t)=u_K(x)e^{i(Kx-E_Kt/\hbar)} \tag{17.35}$$

This is due to the fact that $\psi_K(t)$ is distributed over the whole volume of the crystal and is not localized. To pinpoint an electron we need to describe its wavefunction as a distribution of eigenstates $\psi_K(x,t)$.

Let the wavefunction at some time, which without loss of generality we take as $t=0$, be denoted by $\psi(x,0)$. Let $\psi(x,0)$ be limited (localized) to a characteristic distance δ as sketched in Fig. 17.7. Since the states $\psi_K(x,0)$ form a complete orthonormal set (see Section 2.6), we may expand $\psi(x,0)$ as

$$\psi(x,0)=\sum_K a_K u_K(x)e^{iKx} \tag{17.36}$$

$$a_K=\int_0^L \psi(x,0)u_K^*(x)e^{-iKx}\,dx \tag{17.37}$$

where L is the "crystal" length. The distribution of a_K is shown in Fig. 17.7*b*, and is centered about a mean value K_0. The wavefunction $\psi(x,t)$ $(t>t_0)$ must satisfy the time-dependent Schrödinger equation $\hat{\mathcal{H}}\psi=i\hbar(\partial\psi/\partial t)$ as well as the initial condition (17.36). This can be accomplished by taking

$$\psi(x,t>t_0)=\sum_K a_K u_K(x)e^{i(Kx-E_Kt/\hbar)} \tag{17.38}$$

Expanding E_K in the vicinity of K_0,

$$E_K\simeq E_{K_0}+\left.\frac{dE_K}{dK}\right|_{K_0}(K-K_0)$$

we rewrite (17.38) as

$$\begin{aligned}\psi(x,t)&=\sum_{\Delta K} A_{K_0+\Delta K}u_{K_0+\Delta K}(x)\\&\quad\times\exp\left\{i\left[(K_0+\Delta K)x-\frac{1}{\hbar}\left(E_{K_0}+\frac{dE_K}{dK}\Delta K\right)t\right]\right\}\\&=e^{i(K_0x-E_{K_0}t/\hbar)}\sum_{\Delta K} a_{K_0+\Delta K}u_{K_0+\Delta K}\\&\quad\times\exp\left[i\left(\Delta Kx-\frac{1}{\hbar}\frac{dE_K}{dK}\Delta Kt\right)\right]\end{aligned} \tag{17.39}$$

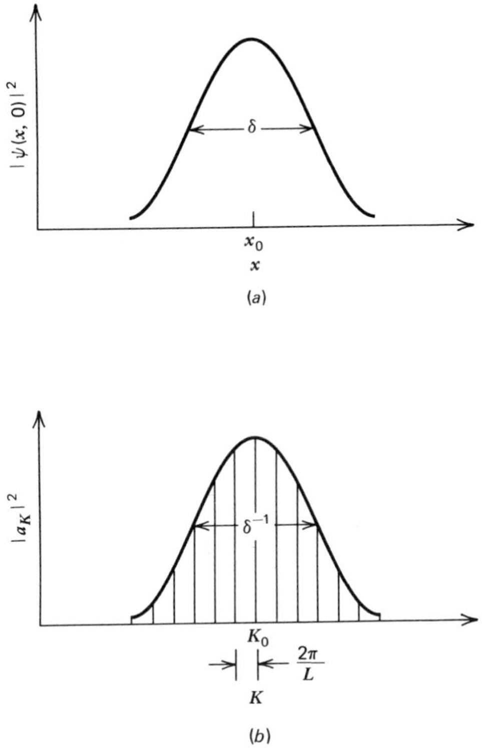

Figure 17.7 (a) The spatial probability distribution of a localized electron at $t=0$. (b) The corresponding "momentum" distribution function $|a_K|^2$.

where $\Delta K \equiv K - K_0$. An observer moving in such a way that the exponent

$$(\Delta K)x - \frac{1}{\hbar}\frac{dE_K}{dK}\Delta Kt \tag{17.40}$$

of eq. (17.39) remains a constant will "see" the same value of $|\psi(x,t)|$. This observer will have to travel at a velocity

$$\frac{dx}{dt} = \frac{1}{\hbar}\frac{dE_K}{dK} \equiv v_g \tag{17.41}$$

The quantity v_g is called the *group velocity* of the electron wave packet. It represents the velocity of the envelope $|\psi(x,t)|^2$ of Fig. 17.7a. For most practical purposes, *when one talks about the velocity of an electron in a crystal, one talks about* v_g.

Another important relation results from equating the work $-e\mathcal{E}v_g\,dt$, done on an electron in a time dt by an externally applied electric field $\mathcal{E}$, to the change in electron energy dE_K (we use the definition $e \equiv |e|$):

$$-e\mathcal{E}v_g\,dt = dE_K$$

$$= \frac{dE_K}{dK}\,dK \qquad (17.42)$$

Using $v_g = \hbar^{-1}dE_k/dK$ leads to

$$\frac{dK}{dt} = -\frac{e\mathcal{E}}{\hbar} \qquad (17.43)$$

or, in general,

$$\frac{d}{dt}(\hbar K) = \text{External Force} \qquad (17.44)$$

We thus find that as far as external forces are concerned the electron (i.e., the electron wave packet) behaves as if it possesses a momentum $\hbar K$. The quantity $\hbar K$ is called the *crystal momentum*. The acceleration of the electron wave packet in response to an external force is

$$\frac{dv_g}{dt} = \frac{1}{\hbar}\frac{d}{dt}\left(\frac{dE_K}{dK}\right)$$

$$= \frac{1}{\hbar}\frac{d^2E_K}{dK^2}\frac{dK}{dt} \qquad (17.45)$$

and using (17.43),

$$\frac{dv_g}{dt} = -e\mathcal{E}\left(\frac{1}{\hbar^2}\frac{d^2E_K}{dK^2}\right)$$

$$= \text{Force/Effective Mass} \qquad (17.46)$$

so that

$$m_e = \text{Effective Mass}$$

$$= \left(\frac{1}{\hbar^2}\frac{d^2E_K}{dK^2}\right)^{-1} \qquad (17.47)$$

The motion of an electron in a crystal in response to an external force is governed by an effective mass that is inversely proportional to the curvature of the dispersion (E_K vs. K) graph. The effective masses of carriers in some commonly used electronic semiconductor crystals are given in Table 19.1.

17.4 THE CONTROL OF CONDUCTIVITY OF SEMICONDUCTORS BY IMPURITIES

The occupation of electronic states in a semiconductor can be affected by the introduction of foreign atoms—a process referred to as impurity "doping." The impurity atoms often enter the lattice by displacing an original atom. In the process of bonding to the surrounding atoms the impurity atoms lose or gain electrons so as to end up with a number of valence electrons equal to that of the majority atoms. If, as an example, pentavalent arsenic is used to dope a crystal of Si (valence $=4$), then the extra, fifth electron is given up. Since all the states in the valence band of silicon are filled (see Fig. 17.5), the extra electron must be accommodated in the conduction band. This type of doping is called *N* (for negative) type.

If the impurity atom possesses fewer valence electrons, then the host atom—Zn in Si as an example—can complete its chemical bonding by removing an electron from the valence band. The vacancy left behind is called a "hole" and it takes part in the current transport, as discussed below. This type of doping is called *P* (for positive) type.

The main difference between this extrinsic conductivity and the intrinsic conductivity discussed earlier, which is due to excitation of electrons from the valence band to the conduction band in a pure (undoped) crystal, is that in the extrinsic case the number of holes and electrons is not equal. A more quantitative discussion of this case is given in Chapter 19.

Current Flow in Semiconductors

The process of current flow in an N doped semiconductor is illustrated in Fig. 17.8. The occupancy of the electron states in *K* space before the application of an electric field is shown in Fig. 17.8*a*. The physical location at $t=0$ of the single conduction band electron is shown in 17.8*b*. The application at $t=0$ of an electric field causes, according to (17.43), the *K* value of each electron to increase at a rate

$$\frac{dK}{dt} = -\frac{e\mathcal{E}}{\hbar}$$

The resulting distribution in *K* space after a time Δt is shown in Fig. 17.8*c*.

Since the valence band is full, it does not contribute to the current flow. This is due to the fact that for each electron with a positive group velocity $(1/\hbar)(dE/dK)$ there exists an electron (at $-K$) with an equal but oppositely directed group velocity. The net current is thus due to the conduction band electrons only. Since the single conduction electron of Fig. 17.8*a* moves in the positive *K* direction, it has a positive group velocity and its physical translation is to the right, as shown in 17.8*d*. The electron is thus accelerated in the direction of the force, indicating a positive mass. This is consistent with Eq. (17.47),

$$m_e = \left(\frac{1}{\hbar^2}\frac{d^2E}{dK^2}\right)^{-1}$$

which relates the effective electron mass to the curvature of the energy band $E(K)$ in *K* space.

The situation in a *P*-type semiconductor is depicted in Fig. 17.9. In our example a single state on top of the valence band is shown as unoccupied at

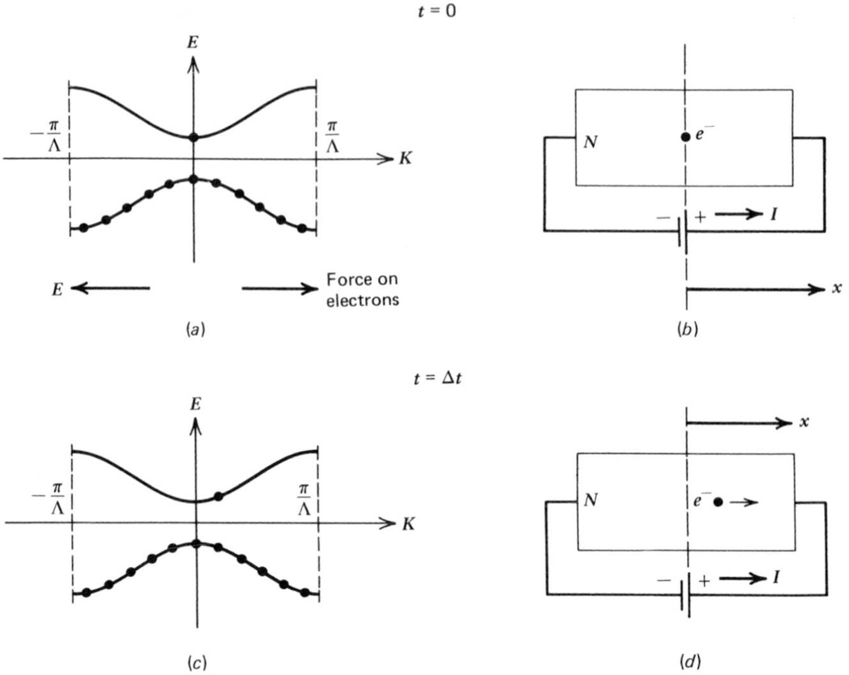

Figure 17.8 Current flow in an N-type semiconductor. At $t=0$: (a) the electron shown at the bottom ($K=0$) of the conduction band; (b) the electron is shown at $t=0$ as localized at $x=0$. At $t=\Delta t$: (c) the electron has moved in "K" space to the region of positive slope; hence, since $v_g=(1/\hbar)\,dE_K/dK>0$, it moved in physical space to the right. This corresponds to a positive effective electron mass ($m_e>0$).

$t=0$ (17.9a). The spatial and temporal bookkeeping of the electronic distribution will be unaltered if we fill the vacancy with an electron, thereby filling the valence band, provided we attach to the electron a positive charge that follows it everywhere, in *real space* as well as in *K space*. Since the full valence band does not contribute to the current flow, the latter can be accounted for completely by the motion of the fictitious positive charge—the "hole."

The motion of the hole in K space under the influence of an electronic field is shown in Fig. 17.9c. The added electron and its shadowing hole move according to (17.43) in the $+K$ direction. Since the slope dE/dK and hence the electron group velocity v_g is negative, the physical translation of the hole (which must accompany the electron) is to the left, in the direction of the electric field. Since the hole has positive charge, its acceleration is in the direction of the force $e\mathcal{E}$ so that its mass is positive. The mass of a hole is thus the negative of its associated electron, that is,

$$m_h=-\left(\frac{1}{\hbar^2}\frac{d^2E_K}{dK^2}\right)^{-1} \tag{17.48}$$

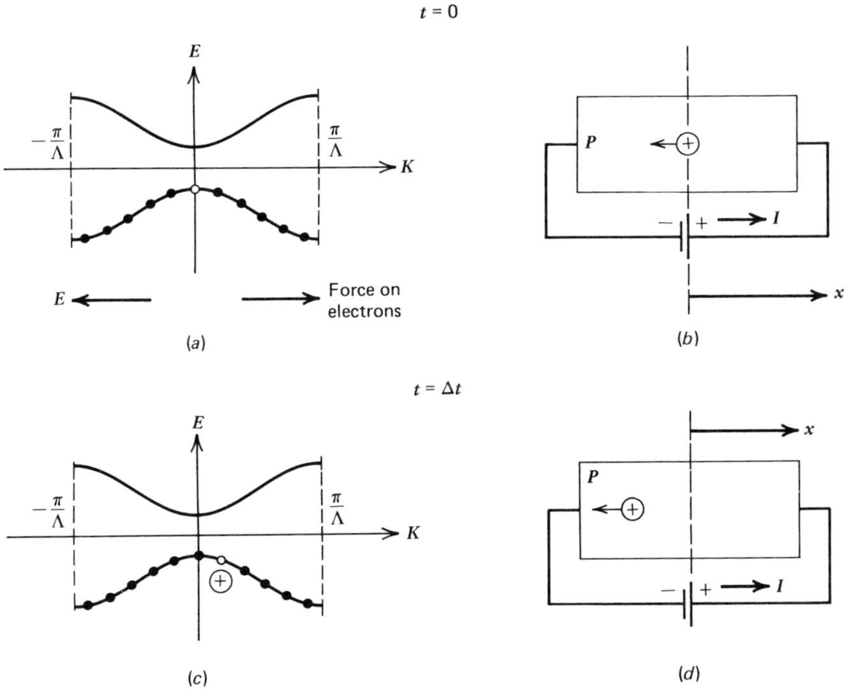

Figure 17.9 A P-doped semiconductor. (a) One electron state in the valence band is unoccupied at $t=0$. The corresponding excess mobile positive charge ("hole") is shown at $x=0$ [see (b)]. (c) Under the influence of the applied electric field the electron distribution moves to the right in K space, where $dE/dK \propto v_g < 0$. Since $v_g < 0$, the corresponding physical motion is to the left. This can be represented by the motion of a positive charge carrier ("hole") with a positive effective mass

$$m^* = \left(\frac{1}{\hbar^2} \frac{d^2E}{dK^2} \right)^{-1}$$

which both in K space and in real space is at the location of the electron vacancy.

The control of current flow in semiconductor crystals, which are doped selectively with N- and P-type impurities, is the basic principle behind the operation of transistors and of the technology of integrated electronics. This topic is discussed in some detail in Chapter 19.

PROBLEMS

1. Derive the expressions for the matrix elements A, B, C, D of (17.22).
2. Prove Eq. (17.33).
3. (Requires numerical analysis sophistication.) Solve numerically for the eigenvalues and eigenfunctions of an electron in a periodic Kronig–Penney

potential with $s = t = 4\,\text{Å}$, $V_0 = 5$ volts. Assume a value of $E = \hbar^2 k_2^2/2m$, then solve for the corresponding value of K from (17.30) and the eigenvector (a_0, b_0) as given by (17.27). Repeat for values of K spanning the interval 0 to π/Λ ($\Lambda = s + t = 8\,\text{Å}$).

4. If the electron momentum in a crystal is taken as $\hbar k_i$ show that the classical expression Power = Force × Velocity is still valid provided the electron velocity is taken as the group velocity.

CHAPTER EIGHTEEN

The Interaction of Electrons and Nuclei with Magnetic Fields, Magnetic Resonance, the Maser

Electrons and nuclei possess magnetic dipole moments. In electrons these moments are due to their orbital motion about the nuclei as well as to the intrinsic spin. In nuclei the moments are due to intrinsic spin alone. These moments interact with external magnetic fields and with each other, thus modifying the observable quantum energies.

The study of the response of these moments to external harmonic magnetic fields is known by the name of magnetic resonance—a field with a large following among both chemists and physicists.

A number of practical devices including microwave maser amplifiers and the hydrogen maser atomic clock utilize the interaction between atomic magnetic moments and external fields.

18.1 ORBITAL MAGNETIC MOMENTS

Experiments reveal that the frequency of the radiation emitted by excited hydrogen atoms in a magnetic field is shifted with respect to their zero field values. In addition, the magnetic field causes some of these lines to split. To explain these effects we need to modify the Schrödinger equation to include the effect of the external magnetic field. This will be done by adding a term to the Hamiltonian that accounts for the extra energy of the electron due to the presence of the magnetic field. The more rigorous derivation of the perturbation Hamiltonian is rather formal and is given below. We will start with a derivation that, though less formal and general, better displays the relationship between the perturbation Hamiltonian and the work done on the electron by the perturbing magnetic field.

Consider an electron in some closed orbit C. The work done on the electron (charge $-e$) in one revolution by an electric field $\mathbf{E}$ is

$$\begin{aligned}\frac{W}{\text{Rev}} &= -e\oint_C \mathbf{E}\cdot d\mathbf{s} \\ &= -e\oint_S (\nabla\times\mathbf{E})\cdot\mathbf{n}\,da \end{aligned} \tag{18.1}$$

In the last equality we employed the Stokes' theorem of vector calculus:

$$\int_S (\nabla\times\mathbf{A})\cdot\mathbf{n}\,da = \int_C \mathbf{A}\cdot d\mathbf{s} \tag{18.2}$$

which relates the contour integral (over C) of some arbitrary vector $\mathbf{A}$ to the surface integral of $\nabla\times\mathbf{A}$. The surface S is bounded by C. $\mathbf{n}$ is the unit vector normal to S, while da is the differential surface area. $d\mathbf{s}$ is the differential distance along C. Using the Maxwell equation

$$\nabla\times\mathbf{E}(t) = -\frac{\partial\mathbf{B}(t)}{\partial t} \tag{18.3}$$

in (18.1) leads to

$$\begin{aligned}\frac{W}{\text{Rev}} &= e\frac{\partial}{\partial t}\int \mathbf{B}\cdot\mathbf{n}\,da \\ &= \frac{\partial}{\partial t}\left[e\pi r^2 B(t)\right]\end{aligned} \tag{18.4}$$

In the last equality we assumed a circular electron orbit of radius r and a uniform magnetic field with a component B normal to the plane of the orbit. An electron with momentum p and mass m will execute one revolution in a period

$$T = \frac{2\pi r}{p/m}$$

so that the work done on the electron in a time dt is equal to the work per revolution (18.4) multiplied by the number of revolutions in dt, which is dt/T:

$$\begin{aligned} dW &= \left(e\pi r^2\frac{\partial B}{\partial t}\right)\times\frac{p}{2\pi rm}dt \\ &= \frac{erp}{2m}\frac{\partial B}{\partial t}dt \\ &= \frac{erp}{2m}dB \end{aligned} \tag{18.5}$$

where $dB = (\partial B/\partial t)\,dt$ is the change of B during dt.

To find the change in energy of an electron due to a magnetic field B, we integrate (18.5) from 0 to B:

$$\Delta W = \frac{erp}{2m} B \tag{18.6}$$

Since $\mathbf{B}$ is the field component normal to the orbit, we may rewrite the Hamiltonian operator corresponding to ΔW as

$$\begin{aligned} \Delta \hat{W} &= \frac{e}{2m} (\mathbf{r} \times \hat{\mathbf{p}}) \cdot \mathbf{B} \\ &= \frac{e}{2m} \hat{\mathbf{l}} \cdot \mathbf{B} = -\hat{\boldsymbol{\mu}}_l \cdot \mathbf{B} \end{aligned} \tag{18.7}$$

where $\hat{\mathbf{l}} = \mathbf{r} \times \hat{\mathbf{p}}$ is the orbital angular momentum operator. Since the interaction energy of a magnetic dipole $\boldsymbol{\mu}$ in a magnetic induction $\mathbf{B}$ is $-\boldsymbol{\mu} \cdot \mathbf{B}$, we identify

$$\hat{\boldsymbol{\mu}}_l = -\frac{e}{2m} \hat{\mathbf{l}} \tag{18.8}$$

as the orbital magnetic moment operator of the electron.

The magnetic moment operator is expressed most often as

$$\hat{\boldsymbol{\mu}}_l = -\frac{e\hbar}{2m} \left(\frac{\hat{\mathbf{l}}}{\hbar} \right) = -\beta \left(\frac{\hat{\mathbf{l}}}{\hbar} \right) \tag{18.9}$$

where $\beta = e\hbar / 2m$, the Bohr magneton, is equal numerically to 9.272×10^{-24} (joule-m^2)/weber.

The total Hamiltonian of an electron in a spherically symmetric potential field and a uniform field $\mathbf{B}$ can then be taken as the sum of the zero field Hamiltonian (7.5) and the magnetic interaction energy (18.7):

$$\hat{\mathcal{H}} = \left[-\frac{\hbar^2}{2m} \left(\frac{1}{r^2} \right) \frac{\partial}{\partial r} \left(r^2 \frac{\partial}{\partial r} \right) + \frac{\hat{\mathbf{l}}^2}{2mr^2} + V(r) \right] + \frac{\beta}{\hbar} \hat{\mathbf{l}} \cdot \mathbf{B}$$

or

$$\begin{aligned} \hat{\mathcal{H}} &= \hat{\mathcal{H}}_0 + \frac{\beta}{\hbar} \hat{\mathbf{l}} \cdot \mathbf{B} \\ &= \hat{\mathcal{H}}_0 + \frac{\beta}{\hbar} B \hat{l}_z \end{aligned} \tag{18.10}$$

where the z direction is chosen, without loss of generality, to be that of $\mathbf{B}$.

The formal quantum mechanical derivation of the Hamiltonian (18.10) starts with expression for the Hamiltonian of an electron in a vector potential $\mathbf{A}$ and a static scalar potential $V(\mathbf{r})$:

$$\begin{aligned} \hat{\mathcal{H}} &= \frac{1}{2m} (\hat{\mathbf{p}} - e\mathbf{A})^2 + V(\mathbf{r}) \\ &= \hat{\mathcal{H}}_0 + \frac{ie\hbar}{2m} (\nabla \cdot \mathbf{A} + \mathbf{A} \cdot \nabla) + \frac{e^2}{2m} \mathbf{A}^2 \end{aligned}$$

where

$$\hat{\mathcal{H}}_0 = \hat{\mathbf{p}}^2/2m + V(\mathbf{r})$$
$$\mathbf{p} = -i\hbar\nabla$$
$$\mathbf{B} = \nabla \times \mathbf{A}$$

and although not needed in this derivation

$$\mathbf{E} = -\frac{\partial \mathbf{A}}{\partial t} - \nabla V$$

Consider the case of a uniform magnetic field $\mathbf{B} = \mathbf{a}_z B$. The corresponding vector potential $\mathbf{A}$ can be taken as

$$\mathbf{A} = \mathbf{a}_x\left(-\tfrac{1}{2}yB\right) + \mathbf{a}_y\left(\tfrac{1}{2}xB\right)$$

With this choice of $\mathbf{A}$, the Hamiltonian operator becomes

$$\hat{\mathcal{H}} = \hat{\mathcal{H}}_0 + \frac{ie\hbar B}{2m}\left(x\frac{\partial}{\partial y} - y\frac{\partial}{\partial x}\right) + \frac{e^2B^2}{8m}(x^2+y^2)$$

Using the definition of the angular momentum operator $\mathbf{l} = \mathbf{r}\times\mathbf{p} = -i\hbar\mathbf{r}\times\nabla$, we can express $\hat{\mathcal{H}}$ as

$$\hat{\mathcal{H}} = \hat{\mathcal{H}}_0 + \left(\frac{\beta}{\hbar}\right)\hat{l}_z B + \frac{e^2\mathbf{A}^2}{2m}$$
$$= \hat{\mathcal{H}}_0 + \left(\frac{\beta}{\hbar}\right)\hat{\mathbf{l}}\cdot\mathbf{B} + \frac{e^2B^2}{8m}(x^2+y^2) \tag{18.11}$$

where $\beta = |e|\hbar/2m$. This result agrees with (18.10) provided we leave out the term involving (x^2+y^2). This is referred to as the diamagnetic contribution. This term is usually neglected in cases where the angular momentum quantum number l or the spin s are not zero. In such cases it is easy to show that the ratio of the diamagnetic term to $(\beta/\hbar)\mathbf{l}\cdot\mathbf{B}$ is of the order of magnitude of $eB\langle r^2\rangle/6\hbar$ and is negligible for the range of B values that are likely to be encountered in the laboratory, say up to 10^2 weber/m^2.

The operators $\hat{\mathcal{H}}_0$ and $\hat{l}_z$ were shown [see (7.1)] to commute with each other. They possess, consequently, common eigenfunctions $u_{nlm}(r,\theta,\phi)$ (note that here m is an integer and not the electron mass). The (energy) eigenvalues E_{nlm} of the Hamiltonian operator are obtained by solving

$$\hat{\mathcal{H}}u_{nlm} = E_{nlm}u_{nlm}$$

Using (18.10) and the relations $\hat{l}_z u_{nlm} = m\hbar u_{nlm}$, $\hat{\mathrm{H}}_0 u_{nlm} = E_{nl}u_{nlm}$, the last eigenvalue equation becomes

$$\hat{\mathcal{H}}_0 u_{nlm} + \frac{\beta}{\hbar}B\hat{l}_z u_{nlm} = (E_{nl} + m\beta B)u_{nlm}$$

so that

$$\left(\hat{\mathcal{H}}_0 + \frac{\beta}{\hbar}B\hat{l}_z\right)u_{nlm} = (E_{nl} + m\beta B)u_{nlm} = E_{nlm}u_{nlm}$$
$$E_{nlm} = E_{nl} + m\beta B \quad (m = -l, -l+1, \ldots, l) \tag{18.12}$$

According to (18.12),[1] a given zero field quantum level (n, l) with energy E_{nl} will split in a magnetic field B into $(2l+1)$ levels labeled by a different value of m. The $(2l+1)$ orbital degeneracy of a level with energy E_{nl} in a spherically symmetric field is *removed* completely in the presence of a magnetic field.

Numerical Example

A typical commercial magnet can generate easily fields up to, say, 4 weber/m^2 (1 weber/m$^2 = 10^4$ gauss). At $B=1$ weber/m^2 the splitting between two adjacent m levels ($|\delta m|=1$) is

$$\begin{aligned}\delta E &= \beta \\ &= 9.272\times 10^{-24} \text{ joules}\end{aligned}$$

This so-called "Zeeman" splitting is studied most often by applying microwave magnetic fields of frequencies $h\nu = \delta E$, which cause transitions between the magnetically split levels. The requisite frequency in the above example is

$$\begin{aligned}\nu &= \frac{\delta E}{h} \\ &= \frac{9.272\times 10^{-24}}{6.626\times 10^{-34}} \\ &= 1.399\times 10^{10} \text{ Hz}\end{aligned}$$

which falls in a convenient region of the microwave spectrum.

The study of Zeeman splittings of atoms and molecules through their absorption or dispersion of oscillating magnetic fields, an area of research called magnetic resonance, is considered in Section 18.3

18.2 SPIN ANGULAR MOMENTUM

In Section 18.1 we have shown that associated with the angular momentum of a charged particle there is always a magnetic moment and that, according to (18.9), the two corresponding operators are proportional to each other. A direct consequence of this theory is that of the splitting of electronic energy levels in a magnetic field as described by (18.12). This theory predicts, as an example, that the ground state of the hydrogen atom $n=0$ is not split in a magnetic field, since in this case the quantum number l and hence m are zero. Experiments, however, show the $n=0$ level to be split in a magnetic field into *two* sublevels, with the amount of splitting being proportional to the strength of the magnetic field. This observation led Uhlenbeck and Goudsmit[2] in 1926 to assign to the electron an *intrinsic spin angular momentum*. The corresponding

[1]Equation (18.12) applies equally well in cgs units where we take $\beta = 9.272\times 10^{-21}$ and express B in gauss.

operator is designated as $\hat{\mathbf{s}}$ and possesses only two eigenstates: the "spin up" state $|\frac{1}{2}\rangle$ and the "spin down" state $|-\frac{1}{2}\rangle$. The commutation relations observed by $\hat{\mathbf{s}}$ are the same as those of $\hat{\mathbf{l}}$. Using (6.49) we can write

$$\hat{\mathbf{s}}\times\hat{\mathbf{s}} = i\hbar\hat{\mathbf{s}} \tag{18.13}$$

Experiments show that the eigenvalues of s_z—that is, the allowed projections of the intrinsic angular momentum—corresponding to $|\frac{1}{2}\rangle$ and $|-\frac{1}{2}\rangle$ are $+\hbar/2$ and $-\hbar/2$ respectively, so that

$$\hat{s}_z|\pm\tfrac{1}{2}\rangle = \pm\frac{\hbar}{2}|\pm\tfrac{1}{2}\rangle \tag{18.14}$$

$$\hat{\mathbf{s}}^2|\pm\tfrac{1}{2}\rangle = \hbar^2\tfrac{1}{2}(\tfrac{1}{2}+1)|\pm\tfrac{1}{2}\rangle \tag{18.15}$$

These relations are identical to those of (6.37) except that the integer l is replaced here by $s=\frac{1}{2}$ and m by $m_s=\pm\frac{1}{2}$.

The magnetic moment $\hat{\boldsymbol{\mu}}_s$ associated with the intrinsic spin is found, by experiment, to be related to the spin angular momentum $\hat{\mathbf{s}}$ by

$$\hat{\boldsymbol{\mu}}_s = -\frac{e}{m_e}\hat{\mathbf{s}} = -2\beta\left(\frac{\hat{\mathbf{s}}}{\hbar}\right) \tag{18.16}$$

We note, by comparing (18.16) to (18.8) that the constant relating $\hat{\boldsymbol{\mu}}_s$ to $\hat{\mathbf{s}}$ is twice as large as that relating $\hat{\boldsymbol{\mu}}_l$ to $\hat{\mathbf{l}}$. This fact is referred to as the *magnetic spin anomaly*. The complete specification of the eigenstate of an electron in a hydrogenic orbit thus involves the triplet of quantum numbers (n, l, m) that uniquely specifies the orbital wavefunction u_{nlm} as well as the spin quantum number m_s, which can assume the values $\frac{1}{2}$ or $-\frac{1}{2}$. The total eigenfunction is written as

$$\psi_{nlmm_s} = u_{nlm}(\mathbf{r})|m_s\rangle, \qquad |m_s\rangle = |\pm\tfrac{1}{2}\rangle$$

The energies of the hydrogenic ground state $(n=1, l=0, m=0)$ in a magnetic field are given by the eigenvalues of $\hat{\mathcal{H}} = \hat{\mathcal{H}}_0 + (2\beta/\hbar)B\hat{s}_z$:

$$\begin{aligned}\left(\hat{\mathcal{H}}_0 + \frac{2\beta}{\hbar}B\hat{s}_z\right)u_{100}|m_s\rangle &= E_{100m_s}u_{100}|m_s\rangle \\ &= (E_0 + 2\beta m_s B)u_{100}|m_s\rangle \end{aligned} \tag{18.17}$$

Since $m_s=\pm\frac{1}{2}$, the state with energy E_0 (the ground state) is split by the magnetic induction B into two levels:

$$E_{m_s=+1/2} = E_0 + \beta B, \qquad E_{m_s=-1/2} = E_0 - \beta B \tag{18.18}$$

which are separated in energy by

$$E_{1/2} - E_{-1/2} = 2\beta B \tag{18.19}$$

in agreement with experiment.

The intrinsic spin angular momentum of the electron was first postulated following experimental observation (see footnote 2). It was later deduced

[2] G. E. Uhlenbeck and S. Goudsmit, *Naturwiss* **13**, 953 (1925). Also, *Nature* **117**, 264 (1926).

theoretically by Dirac[3] in his relativistic quantum mechanical treatment of the electron. Neutrons and protons also possess an intrinsic spin of $s=\frac{1}{2}$, and are discussed in the next section.

18.3 NUCLEAR SPINS AND NUCLEAR MAGNETIC RESONANCE

Experiments show that, as in the case of electrons, nuclei may possess intrinsic spin angular momentum and associated magnetic moments. In direct analogy with (18.14) and (18.15), we introduce a nuclear spin angular momentum operator $\hat{\mathbf{I}}$ and nuclear spin eigenfunctions $|I, m_I\rangle$ such that as in (6.49)

$$\hat{\mathbf{I}}\times\hat{\mathbf{I}}=i\hbar\hat{\mathbf{I}} \tag{18.20}$$

and

$$\hat{I}_z|I, m_I\rangle = m_I\hbar|I, m_I\rangle \tag{18.21}$$

and

$$\hat{\mathbf{I}}^2|I, m_I\rangle = I(I+1)\hbar^2|I, m_I\rangle \tag{18.22}$$

The projection of the nuclear angular momentum along an arbitrary direction, say z, is $m_I\hbar$, where m_I can assume any one of the values $-I, -I+1, \ldots, I$. The nuclear spin state is thus characterized by the fixed dimensionless numbers I and by m_I. In nuclei with an odd mass number (A), I is a half-odd integer (i.e., $\frac{1}{2}$, $\frac{3}{2}$, etc.), while in nuclei with an even mass number, I is an integer. O^{16} and O^{18}, as an example, have $I=0$, while O^{17} has $I=\frac{5}{2}$. Protons and neutrons have $I=\frac{1}{2}$.

In direct analogy with (18.16), the magnetic moment vector of a nucleus is found to be proportional to its spin. This fact is expressed by writing

$$\hat{\boldsymbol{\mu}}_N = g_N\left(\frac{e\hbar}{2m_p}\right)\frac{\hat{\mathbf{I}}}{\hbar} = g_N\beta_N\frac{\hat{\mathbf{I}}}{\hbar} \tag{18.23}$$

m_p is the mass of the proton. The positive constant $\beta_N = e\hbar/2m_p$ is called the *nuclear magneton.* Its numerical value is 5.049×10^{-27} in MKS units (joule-m^2/weber) (in cgs units $\beta_N=5.049\times10^{-24}$ erg·gauss^{-1}). The constant g_N is determined experimentally and is listed in Table 18.1 along with values of I for some common nuclei. A positive g_N signifies that the magnetic moment and the angular momentum vectors are parallel to each other.

The interaction Hamiltonian of a nucleus and an external magnetic field **B** is

$$\mathcal{H}_N = -\hat{\boldsymbol{\mu}}_N\cdot\mathbf{B} = -\frac{g_N\beta_N}{\hbar}\hat{\mathbf{I}}\cdot\mathbf{B} \tag{18.24}$$

---

[3] P. A. M. Dirac, *Proc. Roy. Soc. A* **117**, 610 (1928).

Table 18.1 The magnetic moments and nuclear magnetic resonance (nmr) frequencies of some nuclei.

Z	*Atom*	*A*	*I*	$g_N I$	*Nuclear Magnetic Resonance Frequency at 10^4 gauss ($1\ weber/m^2$)*
0	neutron	1	$\frac{1}{2}$	-1.913	29.167×10^6
1	H (protons)	1	$\frac{1}{2}$	$+2.79268$	42.576×10^6
		2	1	$+0.85738$	6.536×10^6
6	C	12	0		
		13	$\frac{1}{2}$	$+0.7022$	10.705×10^6
8	O	16	0		
		17	$\frac{5}{2}$	-1.893	5.772×10^6
		18	0		
14	Si	28	0		
		29	$\frac{1}{2}$	-0.5548	8.458×10^6
		30	0		

Source: Varian Associates, NMR Table, 4th ed. (Palo Alto, California).

The corresponding eigenvalues are

$$\begin{aligned} E_{I,m_I} &= \langle I, m_I | \hat{\mathcal{H}}_N | I, m_I \rangle \\ &= -\frac{g_N \beta_N}{\hbar} B \langle I, m_I | I_z | I, m_I \rangle \\ &= -g_N \beta_N B m_I \end{aligned} \tag{18.25}$$

where the direction of **B** was taken without loss of generality as the z axis. The effect of the nuclear magnetic dipole interaction is thus to split the electronic energy levels into $2I+1$ sublevels. The spacing energy between two adjacent levels is

$$h\nu = |g_N| \beta_N B \tag{18.26}$$

The study of transitions between magnetically split nuclear sublevels is the domain of the field of nuclear magnetic resonance.[4] In a typical nuclear magnetic resonance experiment the sample containing the nuclei is placed between the pole faces of an electromagnet and is subjected simultaneously to a radio frequency magnetic field of frequency ν directed at right angles to the constant field **B**. If the applied frequency ν is near the resonance value as

[4] Nuclear magnetic resonance was first demonstrated by Bloch and Purcell. Their original papers, which are very readable are: Bloch, Hansen, and Packard, *Phys. Rev.* **69**, 127 (1946); Purcell, Torrey, and Pound, *Phys. Rev.* **69**, 37 (1946).

given by (18.26), radio frequency power is absorbed in the process of induced transitions between the Zeeman split levels.

An experimental setup for displaying this effect is shown in Fig. 18.1. To get an idea of the range of frequencies involved in nuclear magnetic resonance we consider the case of protons. (A sample of H_2O or glycerine provides a good source of protons.) Taking a magnetic field of $B=1$ weber/m^2 (10^4 gauss) in (18.26) and using the data of Table 18.1, we obtain

$$\begin{aligned}\nu &= \frac{|g_N|\beta_N B}{h} \\ &= \frac{2\times 2.79268\times 5.049\times 10^{-27}\times 1}{6.6257\times 10^{-34}} \\ &= 42.576\times 10^6 \text{ Hz}\end{aligned}$$

Nuclear magnetic resonance has become a very important tool in chemistry and solid state physics. This is due to the extreme sensitivity of the resonant absorption frequencies—that is, the Zeeman splittings—to the nature and symmetry of the environs of the nuclei. This follows from the fact that the total magnetic field "seen" by a nucleus is the sum of the external field as well as that due to other nuclei (with $I\neq 0$) and electrons surrounding it. Small changes in the surroundings such as might occur in chemical coordination in a molecule cause easily observed shifts of the resonance lines. One example of such an application is shown in Fig. 18.2.

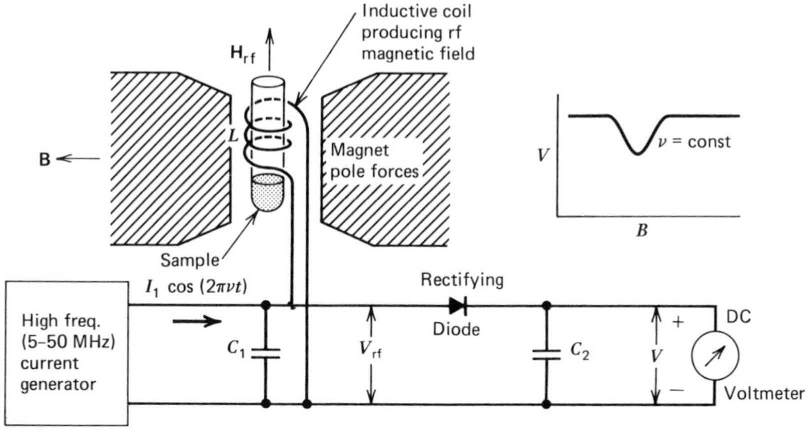

Figure 18.1 A laboratory setup for observing nuclear magnetic resonance. A radio frequency source (rf) supplies the current that, when flowing through the coil L, generates an oscillating magnetic field. As the dc field is swept through resonance, the absorption in the sample causes the "Q" of the LC_1 resonant circuit to drop with a resultant drop of V_{rf}. A rectifying diode transforms V_{rf} to a dc voltage. Absorption of rf power by the nuclear magnetic moments is thus registered as changes in the dc voltage V and is easily observed.

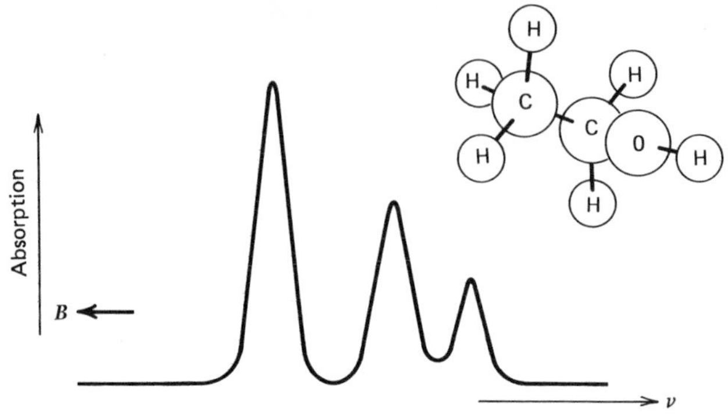

Figure 18.2 Nuclear paramagnetic resonance curve for hydrogen in ethyl alcohol (CH_3CH_2OH). The largest resonance peak is due to the three protons in CH_3, the next to those in CH_2, and the smallest from OH. [After J. T. Arnold, S. S. Dharmati, and M. E. Packard, *J. Chem. Phys.* **19**, 507 (1951).] Figure reproduced from N. F. Ramsey, *Nuclear Moments* (J. Wiley & Sons, New York, 1953), p. 115.

18.4 HYPERFINE INTERACTION

The magnetic fields caused by (orbital and spin) magnetic moments of electrons are sensed by the nuclear magnetic moments and vice versa. The resulting interaction Hamiltonian has the form

$$\hat{\mathcal{H}}_{\text{electron-nuclei}} = A\left(\frac{\hat{\mathbf{I}}}{\hbar}\right)\cdot\left(\frac{\hat{\mathbf{J}}}{\hbar}\right) \tag{18.27}$$

where

$$\hat{\mathbf{J}} = \hat{\mathbf{L}} + \hat{\mathbf{S}} \tag{18.28}$$

is the total (spin + orbital) electron angular momentum operator. The Hamiltonian (18.27) is often called the "hyperfine interaction" term.

The analytic form of (18.27) can be justified if we take the interaction energy as $\hat{\mathcal{H}} = -\hat{\boldsymbol{\mu}}_N \cdot \mathbf{B}_e$ where $\mathbf{B}_e$ is the magnetic induction at the nucleus due to an orbiting (and spinning) electron. A simple application of the Biot–Savart law yields

$$\mathbf{B}_e = \mu_0 \frac{e}{4\pi m_e R^3} \mathbf{J} \tag{18.29}$$

for the magnetic field due to an electron with an angular momentum $\mathbf{J}$ at the center of an orbit of radius R. μ_0 is the magnetic permeability of free space. Although the exact numerical magnitude of A can only be approximated using (18.29), the suggested analytic form of (18.27) is justified.

The magnetic interaction Hamiltonian of an atom or a molecule in a magnetic induction field **B** is taken as the sum of the Zeeman terms and the mutual interaction term

$$\hat{\mathcal{H}}_I = g_J\beta\left(\frac{\hat{\mathbf{J}}}{\hbar}\right)\cdot\mathbf{B} - g_N\beta_N\left(\frac{\hat{\mathbf{I}}}{\hbar}\right)\cdot\mathbf{B} + A\left(\frac{\hat{\mathbf{I}}}{\hbar}\cdot\frac{\hat{\mathbf{J}}}{\hbar}\right) \tag{18.30}$$

The opposite signs of the first two terms in (18.30) reflect the difference in the definition (18.9) and (18.23). The magnetic moment of the electron is taken as

$$\hat{\boldsymbol{\mu}}_e = -g_J\beta\left(\frac{\hat{\mathbf{J}}}{\hbar}\right) \tag{18.31}$$

g_J, the Landè g-factor, is equal to 1 for pure orbital motion ($S=0$) and to 2 for pure spin ($L=0$); in general it is given by[5]

$$g_J = 1 + \frac{J(J+1)+S(S+1)-L(L+1)}{2J(J+1)} \tag{18.32}$$

To find the energy levels of the electron-nucleus system, we need to find the eigenvalues of $\hat{\mathcal{H}}_I$. These eigenvalues can be obtained by a straightforward diagonalization of any matrix representation of $\hat{\mathcal{H}}_I$ as discussed in Section 9.6. An especially convenient set of basis functions for this matrix would be the product set $\psi_J^{M_J}\psi_I^{M_I}$. Two limiting cases, (a) a very weak magnetic field, and (b) a very strong field, can be discussed in simple analytic terms. In the zero field ($B=0$) limit it is convenient to work with the operator

$$\hat{\mathbf{F}} = \hat{\mathbf{I}} + \hat{\mathbf{J}} \tag{18.33}$$

representing the sum of the nuclear and electronic angular momenta. In analogy with Eqs. (18.14) and (18.15), this operator satisfies

$$\hat{\mathbf{F}}^2|F, m_F\rangle = F(F+1)\hbar^2|F, m_F\rangle \tag{18.34}$$

$$\hat{F}_z|F, m_F\rangle = m_F\hbar|F, m_F\rangle \tag{18.35}$$

The functions $|F, m_F\rangle$ can be expressed in terms of linear superposition of product functions

$$|F, m_F\rangle = \sum_{M_I=-I}^{I} a_{F, m_F, m_I, m_J}\psi_J^{M_J=M_F-M_I}\psi_I^{M_I}$$

where a_{F, M_F, m_I, m_J} are the expansion coefficients and only terms with $|M_J| \le J$ are included. $|F, m_F\rangle$ is thus an eigenfunction of $\hat{\mathbf{J}}^2, \hat{\mathbf{I}}^2$ as well as of $\hat{\mathbf{F}}^2$ and $\hat{F}_z$ but not of $\hat{J}_z$ and $\hat{I}_z$. We can consequently use the relation $\hat{\mathbf{F}}^2 = (\hat{\mathbf{I}}+\hat{\mathbf{J}})^2 = \hat{\mathbf{I}}^2 + \hat{\mathbf{J}}^2 + 2\hat{\mathbf{I}}\cdot\hat{\mathbf{J}}$ from which

$$\hat{\mathbf{I}}\cdot\hat{\mathbf{J}} = \tfrac{1}{2}(\hat{\mathbf{F}}^2 - \hat{\mathbf{I}}^2 - \hat{\mathbf{J}}^2) \tag{18.36}$$

[5]This result is derived in most basic texts on atomic physics. See, for example, A. Yariv, *Quantum Electronics*, 1st ed. (John Wiley & Sons, New York, 1967), p. 108.

and

$$\hat{\mathbf{J}}^2|F,m_F\rangle = J(J+1)\hbar^2|F,m_F\rangle$$
$$\hat{\mathbf{I}}^2|F,m_F\rangle = I(I+1)\hbar^2|F,m_F\rangle$$

to obtain

$$\langle F,m_F|\hat{\mathbf{I}}\cdot\hat{\mathbf{J}}|F,m_F\rangle = \frac{\hbar^2}{2}[F(F+1)-I(I+1)-J(J+1)] \quad (18.37)$$

so that at zero magnetic field the nuclear-electronic hyperfine interaction term $(A/\hbar^2)(\hat{\mathbf{I}}\cdot\hat{\mathbf{J}})$ in (18.30) causes the electronic energies to split according to

$$E(F,m_F)\Big|_{B=0} = \frac{A}{2}[F(F+1)-I(I+1)-J(J+1)] \quad (18.38)$$

In the strong field limit we may consider the term $(A/\hbar^2)\hat{\mathbf{I}}\cdot\hat{\mathbf{J}}$ in (18.30) as a perturbation. The zero-order angular momentum eigenfunctions can then be taken as

$$\psi_{JI}^{m_J m_I} = \psi_J^{M_J}\psi_I^{M_I} \quad (18.39)$$

This limit is referred to as the Paschen–Back case. Here **J** and **I** are decoupled from each other so that m_I and m_J but not F or m_F are very nearly good quantum numbers.

Using (18.39) as the basis wave functions, we apply Eq. (11.11*b*) of first-order perturbation theory to calculate the eigenvalues of the complete Hamiltonian (18.30). Taking the perturbation Hamiltonian as

$$\mathcal{H}' = \frac{A}{\hbar^2}(\hat{\mathbf{I}}\cdot\hat{\mathbf{J}})$$

the result is

$$E(I,m_I,J,m_J) = <Im_IJM_J|\frac{g_J\beta B}{\hbar}\hat{J}_z - \frac{g_N\beta_N B}{\hbar}\hat{I}_z + \frac{A}{\hbar^2}\hat{I}_z\hat{J}_z + \frac{A}{\hbar^2}(\hat{I}_x\hat{J}_x + \hat{I}_y\hat{J}_y)|Im_IJM_J\rangle$$

Since the matrix elements involving $\hat{I}_x, \hat{I}_y, \hat{J}_x, \hat{J}_y$ are all zero, we obtain

$$E(I,m_I,J,m_J) = g_J\beta Bm_J - g_N\beta_N Bm_I + Am_Im_J \quad (18.40)$$

As an example of the ideas discussed above, let us consider the effect of an external magnetic field and the hyperfine interaction on the energy of a hydrogen atom in its ground orbital state $^2S_{1/2}$ $(n=0, l=0)$. Here $J=S=\frac{1}{2}$ and from Table 18.1, $I=\frac{1}{2}$. In the zero field $B=0$ case, F is a good quantum number. Since $\mathbf{F}=\mathbf{I}+\mathbf{J}$, the two allowed values of F are 1 and 0. From Eq. (18.38) we obtain

$$E(F=1,m_F) = A/4$$
$$E(F=0,0) = -3A/4 \quad (18.41)$$

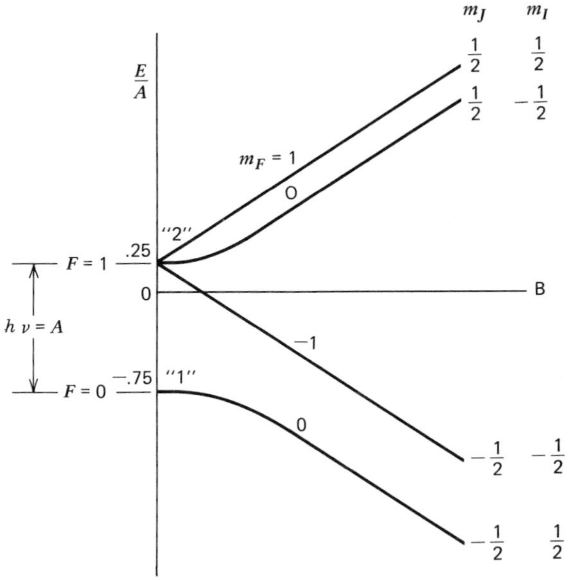

Figure 18.3 The splitting of the ground state $^2S_{1/2}$ ($n=0$, $l=0$) of the hydrogen atom by the hyperfine interaction and by an external magnetic field. The zero field splitting is $h\nu=A$. $\nu=1.420405\times10^2$ Hz ($\lambda=21.121$ cm). The hydrogen maser stimulated transition is "2"→"1."

These energies are shown in Fig. 18.3 at $B=0$. The separation A was determined by a resonance transition technique[6] as

$$\nu = A/h = 1.420405\times10^9 \text{ Hz}$$

The corresponding wavelength is

$$\lambda = c/\nu = 21.121 \text{ cm}$$

This is the well known "21-cm line" emitted by interstellar hydrogen. It is used by radio astronomers to study various aspects of interstellar physics.

Figure 18.3 also shows the high field region described by (18.40). Notice the change in the labeling of the levels from those of F and m_F at the low fields to m_J, m_I at high fields. The energy levels in this region are described by (18.40) (with $J=S=\frac{1}{2}$, $I=\frac{1}{2}$, $A/h=1.420405\times10^9$).

The Hydrogen Maser

The energy level splitting at zero magnetic field (Fig. 18.3) is used in the hydrogen maser oscillator,[7] an example of which is sketched in Fig. 18.4.

---

[6] J. Nafe, E. Nelson, and I Rabi, *Phys. Rev.* **71**, 914 (1947).
[7] D. Kleppner, M. Goldenberg, and N. Ramsey, *Phys. Rev.* **126**, 603 (1962).

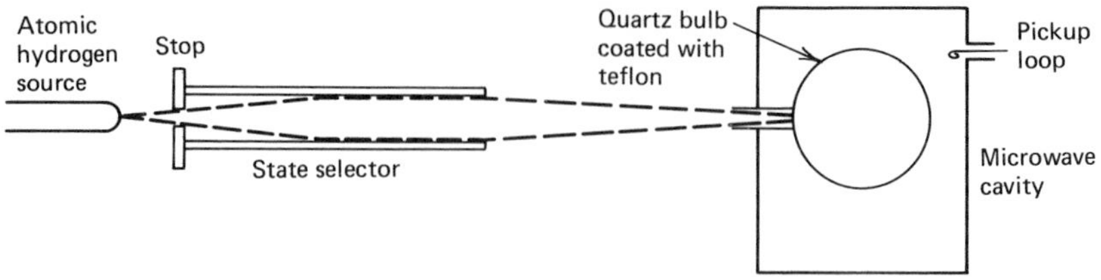

Figure 18.4 A schematic diagram of a hydrogen maser. The desired atoms with $m_J = \frac{1}{2}$ are directed by the inhomogeneous magnet (state selector) into an opening in the microwave cavity. The stimulated emission due to $|1,0\rangle \rightarrow |0,0\rangle$ (here we use the low field designation since in the microwave cavity $B=0$) is coupled to the detection apparatus by a pickup loop.

Atomic hydrogen is extracted from a radiofrequency discharge source and is made to pass through a region of strong inhomogeneous magnetic field (state-selector). Since the force on a magnetic dipole moment in an inhomogeneous field is proportional to the projection of the magnetic moment along the field gradient, hydrogen atoms in the two upper states $|F=1, m_F=1\rangle$, $|F=1, m_F=0\rangle$, which in high fields have $m_J = \frac{1}{2}$, are separated from those in the remaining states that have $m_J = -\frac{1}{2}$. The separated atoms are directed through an opening into a microwave cavity that is resonant at $\nu = 1.42405 \times 10^9$ Hz. The oscillating magnetic field in the cavity induces transition of atoms in the $F=1, m_F=0$ state, labeled as "2" in Fig. 18.3, to $F=0, m_F=0$ (level "1"). The radiation emitted in the process adds coherently to that of the oscillating field. If the radiation losses of the cavity are sufficiently low (high "Q" cavity) and if the flux of excited state hydrogen atoms is sufficiently large, the power emitted by stimulated emission exceeds that which is lost, and the "maser" (microwave amplification by stimulated emission of radiation) oscillation is self-sustaining.

The hydrogen maser oscillation is endowed with a remarkable frequency stability. This is due to a number of reasons including: (a) the relative insensitivity of the upper (2) and lower (1) laser levels to residual magnetic fields, since $dE_{2,1}/dB = 0$ (see Fig. 18.3); (b) the long spontaneous lifetime of atoms in "2"; and (c) the high quality factor (Q) of the microwave cavity. Hydrogen masers are consequently used as atomic clock time standards that have already demonstrated frequency stability of better than one part in 10^{14}.

Emission of the 1420.405-MHz ("21-cm") radiation from interstellar neutral hydrogen in our galaxy was first reported by Ewen and Purcell in 1951.[8] The radiation is produced in the process of spontaneous transitions ($t_{\text{spont}} \sim 11$ million years) of atoms from the $|F=1, m_F=0\rangle$ state to $|F=0, m_F=0\rangle$. The excitation to the upper state is accomplished by thermalization in the background blackbody radiation ($T \sim 3.5°$ K) of interstellar space.

[8]H. I. Ewen and E. M. Purcell, Radiation from galactic hydrogen at 1420 Mc/s, *Nature* **168**, 356 (1951).

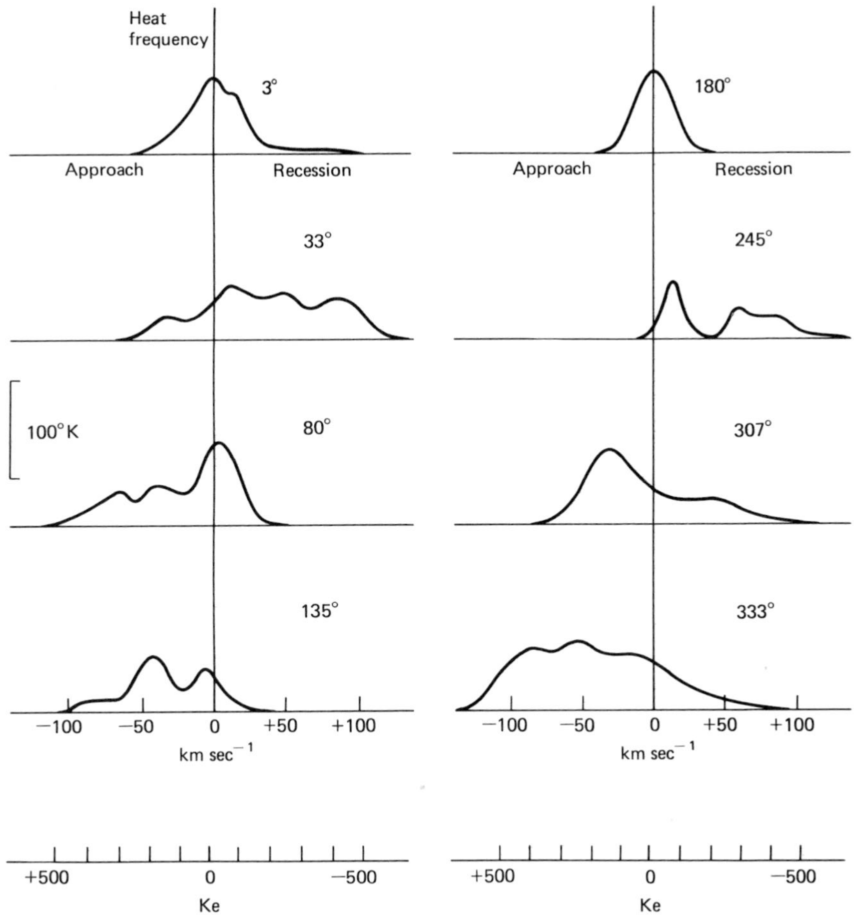

Figure 18.5 Hydrogen line profiles at different longitudes in the plane of our galaxy. [After F. Kerr and G. Westerhout, *Galactic Structure*, Vol. 5 (Univ. of Chicago Press, 1964), Ch. 8.]

Observations of the 21-cm line have enabled radio astronomers to study, for the first time, the spiral structure of our galaxy. The frequency ($\nu =$ 1420.405 MHz) of a stationary hydrogen atom is Doppler shifted upward or downward, depending on whether the emitting atom is moving toward the observer or away from him. The shift is

$$\Delta\nu = \pm \frac{v}{c}\nu$$

where v is the component of the velocity of the atom relative to the observer. Examples of actual spectra are shown in Fig. 18.5.

18.5 ELECTRON PARAMAGNETIC RESONANCE

The field of electron paramagnetic resonance involves the study of transitions between the magnetically split energy levels of electrons. These transitions are induced by high-frequency, typically microwave, magnetic fields in the presence of large steady magnetic fields. The latter field is responsible for the energy level splitting.

As the simplest illustration of the principles involved, we consider the case of an electron in its orbital ground state $n=l=0$. In the presence of a magnetic field B along the z direction, the Hamiltonian of the electron is given by

$$\hat{\mathcal{H}} = \hat{\mathcal{H}}_0 + \frac{2\beta}{\hbar} B\hat{s}_z \tag{18.42}$$

where $\hat{\mathcal{H}}_0$ is the Hamiltonian when $B=0$. The eigenfunctions of the electron consist of the product of the orbital and spin functions:

$$\psi_{\pm 1/2} = u_{100}|\pm\tfrac{1}{2}\rangle \tag{18.43}$$

where u_{100} denotes the orbital ground state ($n=1, l=m=0$). The energy levels of the $|\pm\frac{1}{2}\rangle$ states are given according to (18.18) as

$$\left\langle \psi_{\pm 1/2} \middle| \hat{\mathcal{H}}_0 + \frac{2\beta}{\hbar} B\hat{s}_z \middle| \psi_{\pm 1/2} \right\rangle = E_0 \pm \beta B \tag{18.44}$$

where $E_0 = \langle u_{100}|\hat{\mathrm{H}}_0|u_{100}\rangle$ is the electron energy when $B=0$. The energy splitting is shown in Fig. 18.6.

If, in an addition to a steady magnetic field B, the electron is subjected to a time varying magnetic field,

$$\mathbf{B}(t) = \mathbf{B}_1 \cos \omega t \tag{18.45}$$

then according to (11.33) we may expect the field to induce transitions between the states $|\pm\frac{1}{2}\rangle$ provided the frequency ω satisfies the resonance condition $\hbar\omega \simeq E_{-1/2} - E_{1/2} = 2\beta B$ and the perturbation Hamiltonian has a nonvanishing matrix element connecting the states $|\frac{1}{2}\rangle$ and $|-\frac{1}{2}\rangle$.

To ensure the latter condition we take $\mathbf{B}(t)$ to be parallel to the x direction so that it is perpendicular to the dc magnetic field. The total

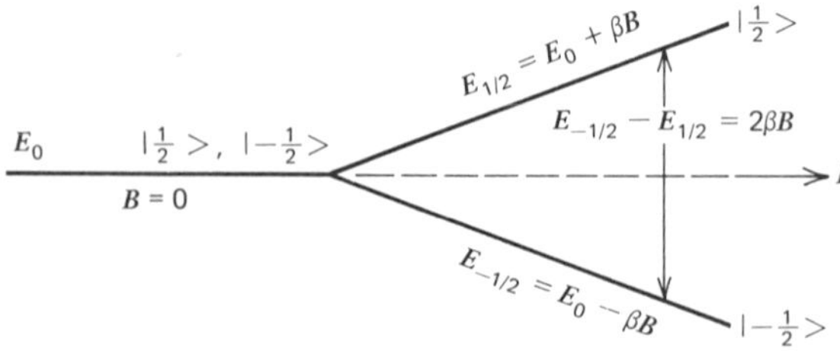

Figure 18.6 The energy levels of an electron spin in an external magnetic field.

Hamiltonian is then given by adding a term

$$\hat{\mathcal{H}}'(t) = [2\beta \mathbf{B}_1(t) \cdot \mathbf{s}]/\hbar \tag{18.46}$$

to (18.42). Using (18.45) we rewrite the perturbation of the Hamiltonian due to the radio frequency field $\mathbf{B}_1(t)$ as

$$\hat{\mathcal{H}}'(t) = \frac{\beta}{\hbar} B_1 \hat{s}_x (e^{i\omega t} + e^{-i\omega t}) \tag{18.47}$$

The transition rate per atom from $|-\frac{1}{2}\rangle$ to $|+\frac{1}{2}\rangle$, and from $|\frac{1}{2}\rangle$ to $|-\frac{1}{2}\rangle$ is given according to the Fermi golden rule (11.33) as

$$W \equiv W_{-1/2 \leftrightarrow 1/2} = \frac{2\pi}{\hbar^3} \beta^2 B_1^2 |\langle \tfrac{1}{2}|\hat{s}_x|-\tfrac{1}{2}\rangle|^2 \delta(2\beta B - \hbar\omega) \tag{18.48}$$

We note that had we chosen $\mathbf{B}(t)$ in the z direction (i.e., the direction of the steady magnetic field), then no transitions would take place, since $\langle \frac{1}{2}|\hat{s}_z|-\frac{1}{2}\rangle = 0$. The matrix element of (18.48) can be evaluated using (9.24),

$$\hat{s}_x = \tfrac{1}{2}(\hat{s}^+ + \hat{s}^-) \tag{18.49}$$

and (9.32),

$$\begin{aligned} \hat{s}^- |-\tfrac{1}{2}\rangle &= 0 \\ \hat{s}^+ |-\tfrac{1}{2}\rangle &= \hbar|\tfrac{1}{2}\rangle \end{aligned} \tag{18.50}$$

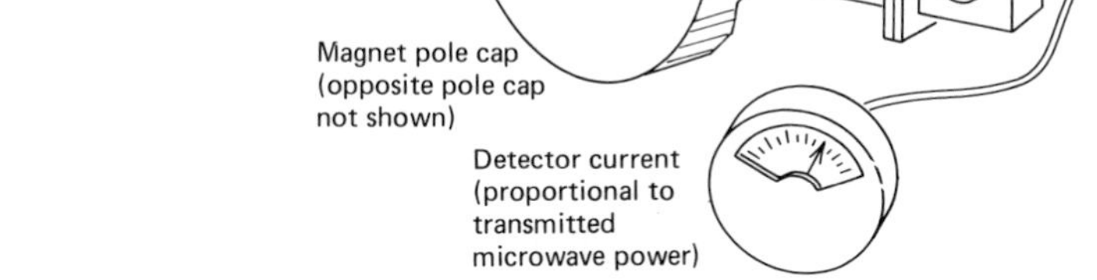

(*a*)

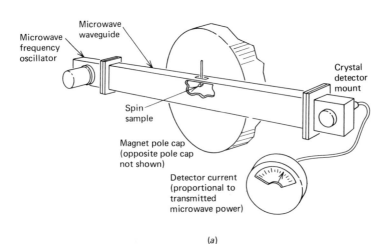

Figure 18.7 (*a*) A typical arrangement of a paramagnetic resonance setup. The electron spin sample is placed in a microwave guide where it is subjected to the high-frequency transverse magnetic field. The "z" directed constant magnetic field is provided by the magnet (one pole face is shown). The transmitted power is measured by power meters consisting of a crystal detector and a dc current meter.

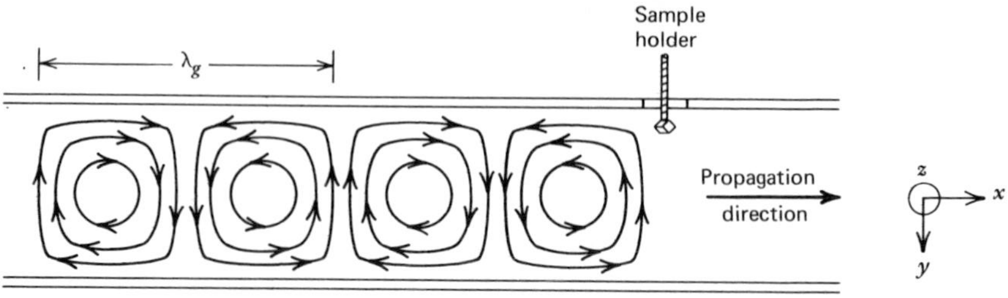

(b)

Figure 18.7 (b) The microwave field pattern propagating through the waveguide. A paramagnetic crystal places along an interior edge of the waveguide experiences a harmonic magnetic field an the wave passes by. Near the walls this field is directed as shown and is thus perpendicular to the z directed dc field generated by the electromagnet (see Fig. 18.7a).

so that (18.48) becomes

$$W=\frac{\pi\beta^2}{2\hbar}B_1^2|\langle\tfrac{1}{2}|\tfrac{1}{2}\rangle|^2\delta(2\beta B-\hbar\omega)$$

$$=\frac{\pi\beta^2}{2\hbar}B_1^2\delta(2\beta B-\hbar\omega) \tag{18.51}$$

In the experimental observation of electron paramagnetic resonance one places the sample containing the electrons inside a microwave waveguide (or resonator) where it is subjected to a microwave magnetic field $B_1 \cos \omega t$. The steady magnetic field B is provided by an electromagnet as shown in Fig. 18.7. The usual practice is to hold the frequency *constant* as the magnetic field B is swept through the resonance region.

As the resonance condition $2\beta B=\hbar\omega$ is approached, the induced transitions cause an absorption of energy from the microwave field at a rate of

$$P_{(\text{watts})}=(N_2-N_1)W\hbar\omega V \tag{18.52}$$

where W is the induced transition rate per atom given by (18.51), N_2, N_1 are the numbers (per unit volume) of electrons in the upper $|\tfrac{1}{2}\rangle$ and lower $|-\tfrac{1}{2}\rangle$ levels, respectively, and V is the sample volume. In practice one observes absorption over a finite region of the dc magnetic field B. This reflects mainly the spatial inhomogeneity of the applied field, causing atoms in different parts of the sample to satisfy the resonance condition at slightly different times. Such inhomogeneities may be due to magnet imperfections or to variations in the atomic surroundings of atoms such as may be caused by neighboring atoms with magnetic moments.

We can account for such a state of affairs by introducing the lineshape function $g[2\beta(B-B_0)]$. The probability that any given atom will actually see a field between B and $B+dB$ when the nominal applied field is B_0, is given by $g[2\beta(B-B_0)]\, d(2\beta B)$. It follows immediately that

$$\int_{-\infty}^{\infty} g[2\beta(B-B_0)]\, d(2\beta B)=1$$

since each atom must "see" some value of the field between $-\infty$ and ∞. The function $g(2\beta(B-B_0)]$ is typically a bell-shaped function peaked at $B=B_0$ with a characteristic width equal to the root square deviation of B averaged over the volume of the atomic sample (and multiplied by 2β).

The average transition rate W_{ave} per atom is obtained by multiplying (18.51) by $g[2\beta(B-B_0)]$ and integrating:

$$W_{\text{ave}}=\frac{\pi\beta^2 B_1^2}{2\hbar}\int_{-\infty}^{\infty} g[2\beta(B-B_0)]\,\delta(2\beta B-\hbar\omega)\, d(2\beta B)$$

$$=\frac{\pi\beta^2 B_1^2}{2\hbar} g(\hbar\omega-2\beta B_0) \tag{18.53}$$

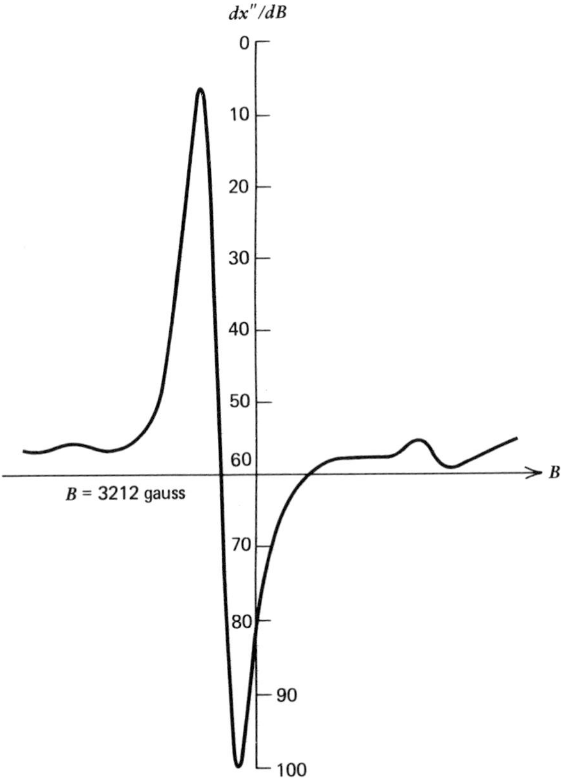

Figure 18.8 The derivative of the absorption trace (absorbed power vs. applied magnetic field) in a typical microwave EPR setup.

so that the plot of absorption against the applied magnetic field B_0 at a fixed ω will trace the broadening function g. An experimental plot of the derivative of the absorbed power obtained in a setup such as in Fig. 18.7 is shown in Fig. 18.8.

PROBLEMS

1. Calculate the spontaneous lifetime for the hyperfine transition $|F=1, m_F=0\rangle \to |F=0, m_F=0\rangle$ of atomic hydrogen in the $n=1$, $l=0$ $(^2S_{1/2})$ at zero external magnetic field.

   Clue: Use the approach of Section 12.6, replacing the perturbation Hamiltonian (12.63) with the magnetic dipolar term

$$\hat{\mathcal{H}}' = -\hat{\boldsymbol{\mu}}\cdot\mathbf{B} = 2\beta\sqrt{\frac{\hbar\omega}{V\mu}}\left(\frac{s_z}{\hbar}\right)(a_l^{+} + a_l)\cos k_l z$$

   as appropriate to magnetic dipole transition. Take

$$|F=1, m_F=0\rangle = \frac{1}{\sqrt{2}}\left(\alpha_{1/2}^{-1/2}\beta_{1/2}^{1/2} + \alpha_{1/2}^{1/2}\beta_{1/2}^{-1/2}\right)$$

$$|F=0, m_F=0\rangle = \frac{1}{\sqrt{2}}\left(\alpha_{1/2}^{-1/2}\beta_{1/2}^{1/2} - \alpha_{1/2}^{1/2}\beta_{1/2}^{-1/2}\right)$$

   where α_j^m and β_j^m are the electronic and nuclear spin eigenfunctions.

   Answer: $t_{\text{spont}} \sim 10$ million years!

2. Solve for the hyperfine levels of the ground-state hydrogen atom $(^2S_{1/2})$ as shown in Fig. 18.3. This can be done by diagonalizing the Hamiltonian (18.30) in any convenient representation. Check your result against the result quoted in Ramsey's *Nuclear Moments* (J. Wiley and Sons, New York, 1953) and references B35 and M17 therein.

3. Derive Eq. (18.29).

4. Obtain an approximate value for the ratio of the diamagnetic term to that of the paramagnetic term $[(\beta/\hbar)\mathbf{l}\cdot\mathbf{B}]$ in the magnetic Hamiltonian (18.11), replacing (x^2+y^2) with typical expectation values $\langle x^2+y^2\rangle$ for hydrogenic atoms. For what values of B is the ratio smaller than 10^{-3}?

CHAPTER NINETEEN

Charge Transport in Semiconductors

It is a matter of debate whether the biggest impact on society in the second half of the twentieth century is from advances in nuclear physics or from the applications of semiconductor physics in the area of computers and information manipulation. My own vote would go to the semiconductors.[1]

It is thus appropriate to conclude this book with three chapters dealing with some of the basic concepts involved in semiconductor devices. This will also afford us an opportunity to bring together a number of basic concepts that were considered separately in the early portions of this book. These include the subjects of electron energy eigenstates, energy bands, group velocity of electrons in crystals, and Fermi–Dirac statistics.

19.1 CARRIERS IN INTRINSIC SEMICONDUCTORS

A problem of central importance in semiconductors is the calculation of the density of charge carriers in a crystal. Let us consider, for example, a semiconductor crystal with energy bands as shown in Fig. 19.1. The chemical potential is μ and the energy gap is E_g. The zero energy reference is taken arbitrarily to correspond to the top of the valence band.

The probability that an electron state at ε is occupied by an electron is given by the Fermi–Dirac distribution law (16.6)[2]

$$f(\varepsilon)=\frac{1}{e^{(\varepsilon-\mu)/kT}+1} \tag{19.1}$$

If we assume that the energy of the electrons in the conduction band is related

---

[1]A nuclear conflagration may tip the scale in favor of nuclear physics.

[2]In this chapter, to conform with popular usage, we use E to describe the electric field, E_g the energy gap, and ε the electron energy.

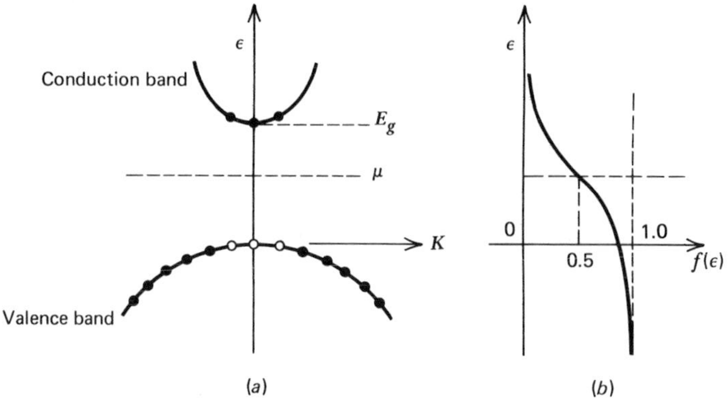

Figure 19.1 (a) A semiconductor with a small density of electrons (solid circles) in the conduction band and holes (empty circles) in the valence band. (b) The Fermi–Dirac distribution function $f(\varepsilon)$ plotted on the same vertical energy scale as (a).

to K, as in free space, by a parabolic relation

$$(\varepsilon - E_g) = \frac{\hbar^2 K^2}{2m_e}$$

where m_e is the effective mass defined in Eq. (17.47), we can use (16.7) to write

$$D_c(\varepsilon) = \frac{1}{2\pi^2}\left(\frac{2m_e}{\hbar^2}\right)^{3/2}(\varepsilon - E_g)^{1/2} \tag{19.2}$$

for the density of electron states (electrons/unit volume-unit energy) in the conduction band.

In a similar fashion we have

$$D_v(\varepsilon) = \frac{1}{2\pi^2}\left(\frac{2m_h}{\hbar^2}\right)^{3/2}(-\varepsilon)^{1/2} \tag{19.3}$$

for the density of electron states in the valence band, m_h is the effective mass of a valence band electron. Since in our choice of the energy reference (Fig. 19.1) electrons in the valence band have negative energies, the term $(-\varepsilon)$ in (19.3) is positive.

Our next task consists of obtaining expressions for the density of electrons in the conduction band and for the density p of holes in the valence band. The determination of n and p is important, since the conduction band electrons and valence band holes are responsible for electric current conduction when an electric field is applied to the semiconductor.

Electron Density

The number dn of electrons in the conduction band (per m^3) whose energies lie between ε and $\varepsilon + d\varepsilon$ is the product of the density of states and the occupation probability, $dn = D_c(\varepsilon)f(\varepsilon)\,d\varepsilon$, so that the total density (electrons/m^3) is

$$n = \int_{E_g}^{\infty} D_c(\varepsilon)f(\varepsilon)\,d\varepsilon$$

As will be shown in what follows, the chemical potential μ falls in most cases within the forbidden gap so that the condition $(\varepsilon - \mu) \gg kT$ is satisfied for $\varepsilon > E_g$. We may thus approximate the Fermi–Dirac distribution law (19.1) by

$$f(\varepsilon) \simeq e^{(\mu-\varepsilon)\beta} \tag{19.4}$$

where $\beta \equiv (kT)^{-1}$. The last integral becomes

$$n \simeq \int_{E_g}^{\infty} \frac{1}{2\pi^2}\left(\frac{2m_e}{\hbar^2}\right)^{3/2}(\varepsilon - E_g)^{1/2} e^{(\mu-\varepsilon)\beta}\,d\varepsilon \tag{19.5}$$

Defining $u = (\varepsilon - E_g)\beta$ we can transform (19.5) to

$$n = \left(\frac{1}{2\pi^2}\right)\left(\frac{2m_e}{\hbar^2}\right)^{3/2} \frac{e^{(\mu-E_g)\beta}}{\beta^{3/2}} \int_0^{\infty} u^{1/2}e^{-u}\,du$$

The definite integral is equal to $\sqrt{\pi}/2$, so that

$$n = 2\left(\frac{2\pi m_e kT}{h^2}\right)^{3/2} e^{-(E_g-\mu)/kT}$$

$$= N_c e^{-(E_g-\mu)/kT} \tag{19.6}$$

where N_c, the "effective density of states" in the conduction band, is given by

$$N_c = 2\left(\frac{2\pi m_e kT}{h^2}\right)^{3/2} \tag{19.7}$$

EXAMPLE:

In silicon at $T = 300$ K and with $m_e = 0.2$ m (see Table 19.1) we obtain $N_c = 2.2 \times 10^{18}$ cm^{-3}.

The Density of Holes

The number of holes (per m^3) in an energy interval $d\varepsilon$ of the valence band is equal, by definition, to the number of unoccupied electron states in $d\varepsilon$. The probability that a state ε is not occupied is equal to unity minus the probability that the state is occupied. We denote this probability as $f_h(\varepsilon)$:

$$f_h(\varepsilon) = 1 - f(\varepsilon)$$

$$= 1 - \frac{1}{e^{(\varepsilon-\mu)/kT} + 1}$$

$$= \frac{e^{\varepsilon-\mu}}{e^{(\varepsilon-\mu)\beta} + 1} \simeq e^{(\varepsilon-\mu)\beta} \tag{19.8}$$

where the approximation is valid when $(\mu-\varepsilon)\gg kT$. Using a procedure identical to that leading to (19.6), we write

$$p=\int_{-\infty}^{0} D_v(\varepsilon)f_h(\varepsilon)\,d\varepsilon$$

$$=\int_{-\infty}^{0}\frac{1}{2\pi^2}\left(\frac{2m_h}{\hbar^2}\right)^{3/2}(-\varepsilon)^{1/2}e^{(\varepsilon-\mu)\beta}\,d\varepsilon$$

and finally

$$p=2\left(\frac{2\pi m_h kT}{h^2}\right)^{3/2}e^{-\mu/kT}$$

$$=N_v e^{-\mu/kT}$$

$$N_v=2\left(\frac{2\pi m_h kT}{h^2}\right)^{3/2} \tag{19.9}$$

is the "effective density of states" in the valence band.

From (19.6) and (19.9) we obtain

$$np=4\left(\frac{2\pi kT}{h^2}\right)^{3}(m_e m_h)^{3/2}e^{-E_g/kT} \tag{19.10}$$

Since the last expression does not depend on the chemical potential μ, it follows that the product of the electron and hole densities in a bulk semicon-

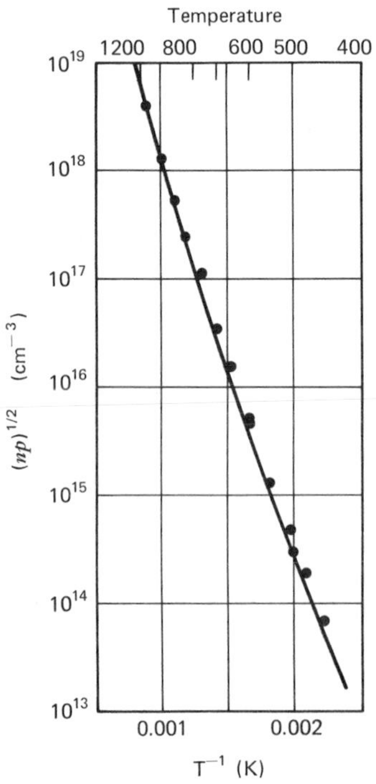

Figure 19.2 A plot of $\log(np)^{1/2}$ vs. T^{-1} in the intrinsic range of the carrier concentration in silicon. The points are experimental. The solid curve is a plot of the relation

$$(np)^{1/2}=const\;T^{3/2}\times\exp\left(-\frac{1.21}{2kT/e}\right)$$

From the curve and Eq. (19.10) we deduce $E_g=1.21$ eV. [After F. J. Morin and J. P. Maita, *Phys. Rev.* **96**, 28 (1954).]

ductor *pure* or *doped* is a constant that in a given material is a function of the temperature alone.

An experimental confirmation of (19.10) is shown in Fig. 19.2, which shows a plot of the measured values of the log of $(np)^{1/2}$ vs. T^{-1}.

Intrinsic Semiconductors

In an intrinsic (undoped) semiconductor $n=p$, since each electron in the conduction band must cause a vacancy (hole) in the valence band. This is due to the fact that, as discussed in Section 17.2, the number of valence electrons in the constituent atoms is exactly the number needed to fill the valence band of the crystal. Electrons in the conduction band are thus due to thermal excitation [described by the Fermi–Dirac law (16.6)] from the valence band. Using (19.10) we obtain

$$\begin{aligned} n_i &= p_i \\ &= \sqrt{np} \\ &= 2\left(\frac{2\pi kT}{h^2}\right)^{3/2} (m_e m_h)^{3/4} e^{-E_g/2kT} \end{aligned} \tag{19.11}$$

The value μ of the chemical potential in an intrinsic semiconductor can be obtained by equating n (19.6) to p (19.9), resulting in

$$\mu = \frac{E_g}{2} + \tfrac{3}{4}kT \ln \frac{m_h}{m_e} \tag{19.12}$$

so that the chemical potential is very nearly at midgap.

To obtain an idea for the number of intrinsic carriers in semiconductors consider, as an example, the crystal GaAs at $T=300$ K. Using the data from Table 19.1, $m_e=0.07$ m, $m_h=0.6$ m, $E_g=1.43$ eV, we obtain

$$\begin{aligned} n_i &= 2\left(\frac{2\pi k(300)}{h^2}\right)^{3/2} (0.07\times 0.6\ \mathrm{m}^2)^{3/4} \exp\left(-\frac{1.43e}{2k(300)}\right) \\ &= 2.246\times 10^6\,\mathrm{cm}^{-3} \end{aligned}$$

The intrinsic density of carriers (electrons and holes) in semiconductors is thus many orders of magnitude smaller than it is in metals ($\gtrsim 10^{22}$ electrons/cm^3). We may conclude that the intrinsic conductivity of semiconductors with $E_g \sim 1$ eV or higher is very small except for very elevated temperatures. As a matter of fact the actual conductivity of even the purest available semiconductors in most cases far exceeds the intrinsic values and is due to the residual traces of impurity atoms. In practical semiconductor devices, doping with impurities is used to control the conductivity. It is thus essential to understand, next, how the presence of impurity atoms affects the conductivity.

Table 19.1 Some Basic Properties of Semiconductor Crystals

	Energy Gap (eV)				
Crystal	$T=0$	*300 K*	m_e/m	m_h/m	$\varepsilon/\varepsilon_0$
Diamond	5.4				
Ge	0.75	0.67	0.1	0.04	16.0
Si	1.20	1.11	0.2	0.16	12.0
InAs	0.36	0.34	0.024	0.41	12.5
InSb	0.26	0.16	0.012	0.25	15.9
InP	1.29	1.35	0.073	~0.2	12.1
GaAs	1.52	1.43	0.07	0.6	11.5

19.2 THE IONIZATION ENERGY OF IMPURITY ATOMS

Let us contemplate what happens when an impurity atom is substituted for an atom of an otherwise perfectly regular crystal. To be specific, we consider the case of a pentavalent atom such as arsenic inside a germanium crystal. The germanium atom has four ($4s^2 4p^2$) valence electrons. These can be used to form tetrahedral $s-p$ orbtials in the manner discussed in Section 7.4 and illustrated schematically in Fig. 19.3. In the germanium crystal the tetrahedral bonding results in the basic diamond crystal structure as shown in Fig. 19.4.

Consider next what happens when an impurity atom such as arsenic substitutes for one germanium atom. The arsenic atom has five ($4s^2 4p^3$) valence electrons. Four of these electrons are "used up" in the tetrahedral bonding to the nearest four germanium atoms. The fifth electron is then loosely bound to the As^+ ion core in a manner depicted in Fig. 19.5. The arsenic atom that possesses an extra electron (compared to the germanium

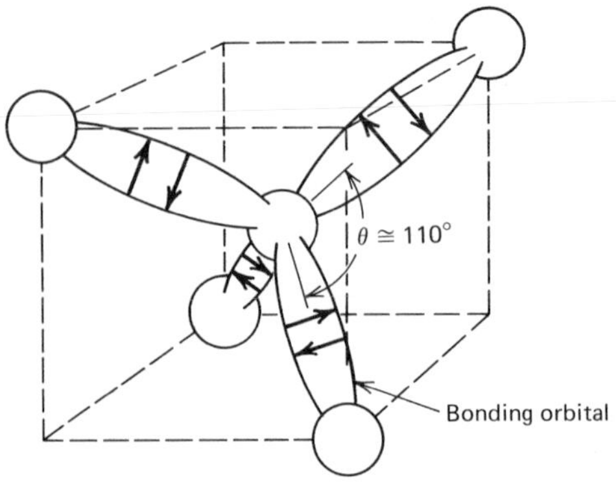

Figure 19.3 The bonding orbitals and the basic structure of a germanium crystal.

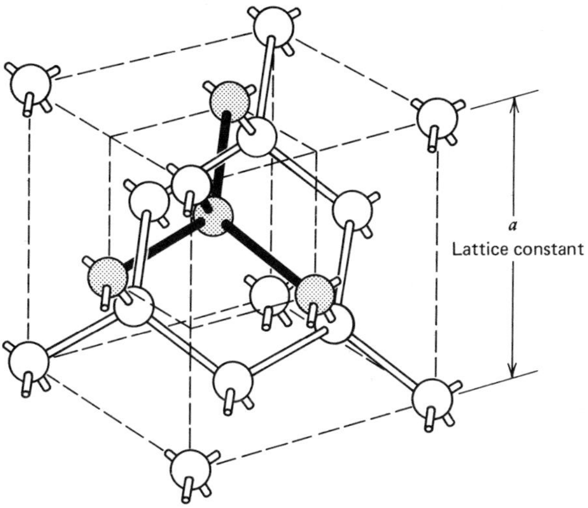

Figure 19.4 Fundamental (primitive) unit cell of the diamond-type crystal structure. The basic tetrahedral bonding of Fig. 19.3 is shown as dark bars.

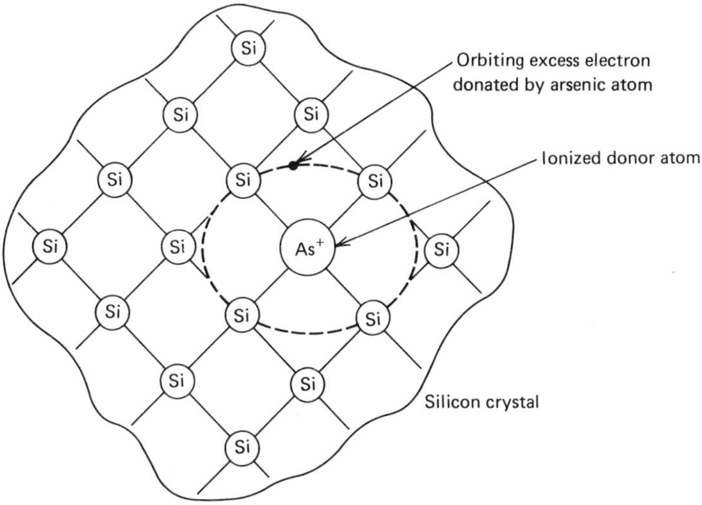

Figure 19.5 An N-type impurity (arsenic in this case) atom uses four of its five valence electrons in tetrahedral bonding to the nearest silicon atoms while the fifth electron is loosely bound to the ionized As^+ core. The energy needed to remove the fifth electron from its binding orbit about its parent atom to infinity—the ionization energy of the impurity atom—is E_d. (See Fig. 19.6.)

atom it replaced) is called a donor atom and the doped semiconductor is called N (for negative) type semiconductor.

The problem of solving for the wavefunction and energies of the fifth electron is thus similar to that of the hydrogen atom. The As^+ core plays the role of the proton, while the fifth electron is similar to that of the single electron in the hydrogen atom.[3]

There are two basic differences between this case and the case of the hydrogen atom, which was considered in Chapter 7: (1) the electron in a crystal has an effective mass m_e rather than a free electron mass m; (2) the electron in a crystal moves in a medium that possesses a static dielectric constant ε, while the hydrogenic electron "sees" the vacuum constant ε_0. The electron in a crystal is thus subject to a weaker Coulomb attraction. We can consequently apply directly the results of Section 7.2, replacing everywhere ε_0 by ε and m by m_e. The ionization energy of the donor atom electron is given by (the modified) Eq. (7.28) as

$$\begin{aligned} E_d &= \frac{e^4 m_e}{2(4\pi\varepsilon)^2 \hbar^2} \\ &= \frac{e^4 m}{2(4\pi\varepsilon_0)^2 \hbar^2} \frac{m_e/m}{(\varepsilon/\varepsilon_0)^2} \\ &= 13.64 \frac{m_e/m}{(\varepsilon/\varepsilon_0)^2}\ \text{eV} \end{aligned} \tag{19.13}$$

Typical values of m_e/m are ~ 0.1, while $\varepsilon/\varepsilon_0$ in semiconductor is ~ 10 (see Table 19.1). Ionization energies of donor impurity atoms in semiconductors are thus $\sim (m_e/m)(\varepsilon_0/\varepsilon)^2 \sim 10^{-3}$ times the hydrogen ionization energy. Typical values run between 10 and 20 meV. The measured ionization energy of phosphorus atoms in germanium, as an example, is 12 meV.

We also note that the ground-state orbit radius of the electron is given by

$$\begin{aligned} a_d &= \frac{\hbar^2 4\pi\varepsilon}{m_e e^2} \\ &= \frac{\hbar^2 4\pi\varepsilon_0}{m e^2} \frac{\varepsilon/\varepsilon_0}{m_e/m} \\ &= a_0 \frac{\varepsilon/\varepsilon_0}{m_e/m} \end{aligned}$$

where a_0 (the Bohr radius) $= 5.2917 \times 10^{-11}$ m.

---

[3]This model is not strictly valid since, due to the finite spatial extent of the four bonding electrons, the positively charged core cannot be approximated by a point charge as in the hydrogen atom. The approximation, however, is useful for the higher lying levels and for the calculation of the ionization energy of the donor atom.

In the above example we obtain $a_d \sim 100a_0$, so that the donor electrons "roam" over a large number of unit cells.

In the energy band diagram of a semiconductor we show the ground state of the extra electron that is donated by the impurity atom a distance E_d below the conduction band—that is, inside the forbidden gap. The conduction band is thus equivalent to the region of continuum states above the ionization level in the case of the hydrogen atom. By increasing the energy of an electron in the ground hydrogenic state $n=1$ of an impurity atom by E_d, it is placed at the bottom of the conduction band, where its wavefunction is no longer localized about the donor atom.

If the impurity atom has fewer valence electrons than the crystal atoms, we refer to it as an acceptor. An example of doping by an acceptor would be the incorporation of trivalent indium in a silicon or germanium crystal. The bonding of the impurity atom is completed by removing an electron from the valence band, leaving behind an electron vacancy. This positively charged "hole" is attracted to the negative In^- core, and the resulting "hydrogenic" ionization energy is given by

$$E_a = \frac{e^4 m_h}{2(4\pi\varepsilon)^2 \hbar^2}$$

$$= \frac{e^4 m}{2(4\pi\varepsilon_0)^2 \hbar^2} \frac{m_h/m}{(\varepsilon/\varepsilon_0)^2} \qquad (19.13a)$$

where m_h is the effective hole mass. Physically, the process of ionizing an acceptor atom consists of removing an electron from the top of the valence band and placing it in a bound state of the acceptor atom. Equivalently, using the exact formal analogy between electrons and holes, we may think of this process as one of removing a hole from its localized ground-state orbit about the negatively charged acceptor ion and placing it in the valence band. (Recall that in our energy diagrams the hole energy increases downwards.) The energy expended in the process is E_a. The ground state of the acceptor atom is shown a distance E_a above the valence band maximum. The ground states of the donor and acceptor atoms thus fall inside the forbidden gap, as shown in Fig. 19.6.

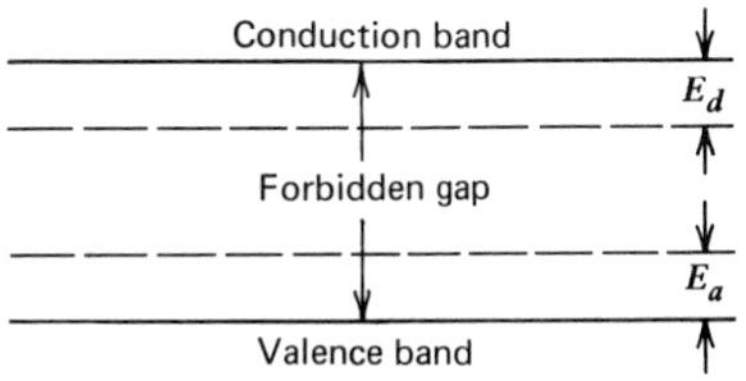

Figure 19.6 The energy levels of impurity atoms in a semiconductor. The ground state of a donor impurity is E_d below the conduction band minimum while that of an acceptor atom is E_a above the valence band edge.

19.3 CARRIER CONCENTRATION IN DOPED SEMICONDUCTORS

The control of the density of mobile charge carriers (electrons and holes) in semiconductors is of key importance in achieving the necessary electric properties in semiconductor device performance. Let us consider first the case of an N-type semiconductor doped with a density of N_d donor atoms per unit volume. Let the ionization energy of the donor atoms be E_d as shown in Fig. 19.7. The probability that a donor atom is neutral—that is, that the extra electron is bound in its ground hydrogenic state—is given by the Fermi–Dirac function

$$f(E_2)=\frac{1}{1+\exp[\beta(E_2-\mu)]}\equiv\frac{1}{1+\exp x} \tag{19.14}$$

where $x\equiv\beta(E_2-\mu)$, $\beta\equiv(kT)^{-1}$, and $E_2=E_g-E_d$ is the energy of the ground-state electron measured from the top of the valence band. The density of ionized donors is given by the total density N_d minus the density $N_d f(E_2)$ of neutral donors:

$$N_d^+ = N_d[1-f(E_2)] \tag{19.15}$$

$$= N_d\frac{1}{1+e^{-x}}$$

$$= N_d\frac{1}{1+e^{\beta(\mu-E_g+E_d)}} \tag{19.16}$$

If $n\gg p$ (an assumption that must be justified at the end of the derivation), then since $np=n_i^2=\text{const}$ [see (19.10)], $n\gg n_i$ and the great majority of the conduction band electrons must be due to the ionization of donor atoms. It follows that

$$n\simeq N_d^+ \tag{19.17}$$

but n is also given by (19.6) as

$$n=N_c e^{\beta(\mu-E_g)} \tag{19.18}$$

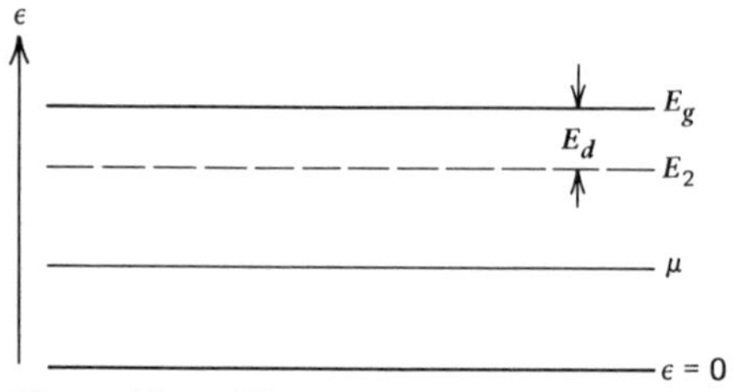

Figure 19.7 The energy levels used in the derivations in Section 19.3.

so that using (19.15),

$$N_c e^{\beta(\mu-E_g)} = \frac{N_d}{1+e^{\beta(\mu-E_g+E_d)}} \tag{19.19}$$

A simple manipulation leads to

$$\left(e^{\beta\mu}\right)^2 e^{-\beta(E_g+E_2)} + e^{\beta\mu}e^{-\beta E_g} - \frac{N_d}{N_c} = 0$$

and

$$e^{\beta\mu} = \frac{-1+\sqrt{1+4(N_d/N_c)e^{\beta(E_g-E_2)}}}{2e^{-\beta E_2}} \tag{19.20}$$

The general result (19.20) assumes a simple form in the extremes of high and low temperatures.

"High" Temperature

In this limit, which is the one most often encountered in practice,

$$4\frac{N_d}{N_c}e^{(E_g-E_2)/kT} \ll 1 \tag{19.21}$$

Using the first two terms in the Taylor expansion of the radical in (19.20) gives

$$e^{\beta\mu} \simeq \frac{N_d}{N_c}e^{\beta E_g} \tag{19.22}$$

so that

$$\mu \simeq E_g - kT\ln\frac{N_c}{N_d} \tag{19.23}$$

Using (19.22) in (19.18) leads to

$$n \simeq N_c e^{-\beta E_g}\frac{N_d}{N_c}e^{\beta E_g} = N_d \tag{19.24}$$

It follows that in the "high" temperature limit essentially *all the donor atoms are ionized and the density of conduction band electrons is nearly equal to the density* N_d *of donor atoms.*

We now need to check our initial assumption $n \gg n_i$. The numerical example at the end of the last section showed that $n_i \sim 10^6$ cm^{-3}. It follows that since $n \simeq N_d$, our assumption is valid if one employs donor densities $N_d > 10^8$ cm^{-3}. This is always true in practice.

EXAMPLE:

It may be interesting to investigate the effect of doping on the position of the chemical potential of silicon. Using the value $N_c = 2.2\times10^{18}$ cm^{-3} (see the

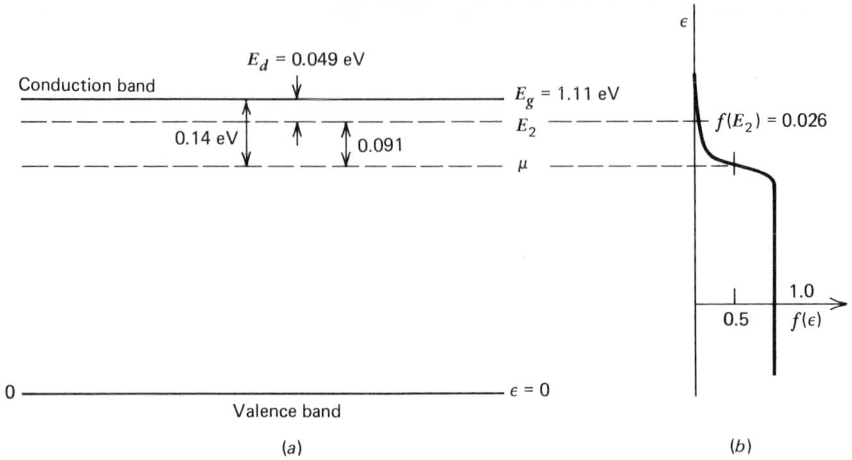

Figure 19.8 (*a*) The position of the Fermi level (chemical potential) μ, and the ionization energy for silicon doped with $N_d \sim 10^{16}$ of arsenic atoms. (*b*) The Fermi–Dirac distribution junction.

example in Section 19.1) and a typical value of $N_d = 10^{16}$ cm^{-3} and $T =$ 300 K, we obtain

$$kT\ln\frac{N_c}{N_d} = 0.14\,\text{eV}$$

so that, according to (19.23), the chemical potential μ is 0.14 eV below the edge of the conduction band. The chemical potential in an undoped semiconductor is, according to (19.12), a distance $E_g/2 \sim 0.55$ eV below the conduction band edge, so that the doping in this case resulted in an upward shift of μ from its midgap position in an undoped crystal by 0.41 eV. The calculated position of μ and other relevant energies are shown in Fig. 19.8. The probability $f(E_2)$ of finding the donor atoms in their un-ionized (ground) state is

$$f(E_2) = \frac{1}{e^{(E_2-\mu)/kT}+1} = \frac{1}{e^{0.091/kT}+1} = 0.026$$

It follows that 97.4% of the donor atoms are ionized. Since each ionized donor atom contributes one electron to the conduction band, it follows that $n \simeq N_d^+$, as assumed above.

"Low" Temperature

Here the temperature is low enough so that

$$4\frac{N_d}{N_c}e^{(E_g-E_2)/kT} \gg 1$$

In this limit we obtain from (19.20),

$$e^{\beta\mu} = \left(\frac{N_d}{N_c}\right)^{1/2} \exp\left[\beta\left(\frac{E_g - E_2}{2}\right)\right] e^{\beta E_2}$$

$$= \left(\frac{N_d}{N_c}\right)^{1/2} \exp\left[\beta\left(\frac{E_g + E_2}{2}\right)\right] \quad (19.25)$$

and from (19.18)

$$n = N_c e^{-\beta(\mu - E_g)}$$

$$= (N_d N_c)^{1/2} e^{-E_d/2kT} \quad (19.26)$$

where $E_d = E_g - E_2$ is the ionization energy of the donor impurity atom.

Solving (19.25) for μ gives

$$\mu = E_g - \frac{1}{2}\left(E_d + kT \ln \frac{N_c}{N_d}\right) \quad (19.27)$$

19.4 SCATTERING OF ELECTRONS IN SEMICONDUCTOR CRYSTALS

In Chapter 17 we found that under the influence of an external electric field **E** the electron motion is described by

$$m_e \frac{d\mathbf{v}_g}{dt} = -e\mathbf{E} \quad (19.28)$$

where m_e is the effective mass, $-e$ is the electron charge, and $\mathbf{v}_g$ is the group velocity of the electron. In an idealized perfect and rigid lattice the motion of the carriers is unimpeded and the resulting electrical conductivity will approach infinity.

In a real crystal the motion of the electron is interrupted continuously by collisions due to a variety of physical mechanisms. Any deviation (perturbation) $\Delta V(\mathbf{r})$ of the potential field from that of a perfect crystal can cause it to scatter electrons. Quantum mechanically we view the process of scattering as a transition of an electron from an initial Bloch state

$$\psi_{\mathbf{K}}(\mathbf{r}) = u_{\mathbf{K}}(\mathbf{r}) e^{i\mathbf{K}\cdot\mathbf{r}} \quad (19.29)$$

to a final state

$$\psi_{\mathbf{K}'}(\mathbf{r}) = u_{\mathbf{K}'}(\mathbf{r}) e^{i\mathbf{K}'\cdot\mathbf{r}} \quad (19.30)$$

which is due to $\Delta V(\mathbf{r})$. The scattering rate is proportional, according to Fermi's golden rule (11.32), to

$$W_{\text{scat}} \propto |\langle \psi_{\mathbf{K}'} | \Delta V | \psi_{\mathbf{K}} \rangle|^2 \quad (19.31)$$

The scattering perturbation $\Delta V(\mathbf{r})$ may be due to a variety of mechanisms, which include: (1) lattice vibrations; in this case the scattering is accompanied by emission or absorption of phonons as demonstrated in Figs.

19.9a and 19.9b, the phonon energy is supplied or taken up by the scattered electron; (2) scattering by an ionized (or neutral) impurity atom or from a lattice defect such as a vacancy as shown in Fig. 19.10. Most of these scattering events are nearly elastic, since the objects of collisions (lattice defects, impurity atoms) are many times more massive than the electron (or hole). The effect of the collision is thus mainly one of randomizing the direction of travel of the charge carrier, while leaving its energy essentially constant. We may thus view the motion of an electron (or hole) in a crystal as a series of connected straight lines with a mean displacement of zero. These segments have an average length, "mean free path," of

$$l = v_T \tau \tag{19.32}$$

where $\tau \equiv W_{\text{scat}}^{-1}$ is the mean time between collisions, and v_T is the thermal velocity

$$v_T = \sqrt{\frac{2kT}{m_e}}$$

The resulting motion is sketched graphically in Fig. 19.11a.

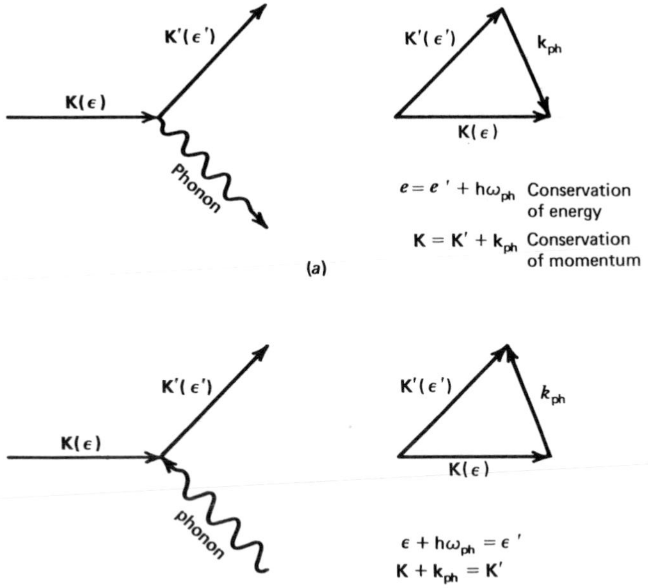

Figure 19.9 (a) Electron scattering accompanied by phonon emission. An electron with "momentum" **K** and energy ε scatters to a state $(\mathbf{K}', \varepsilon)$ emitting in the process a phonon with a wave vector $\mathbf{k}_{\text{ph}}$. (b) Electron scattering by phonon absorption. Same as (a) except the phonon is absorbed in the process of electron scattering. This is shown in a reversal in the sign of the $\mathbf{k}_{\text{ph}}$. The momenta of the colliding particles (electrons, phonons) are proportional to their **K** or $(\mathbf{k}_{\text{ph}})$ vectors so that conservation of momentum requires the closure of the triangles.

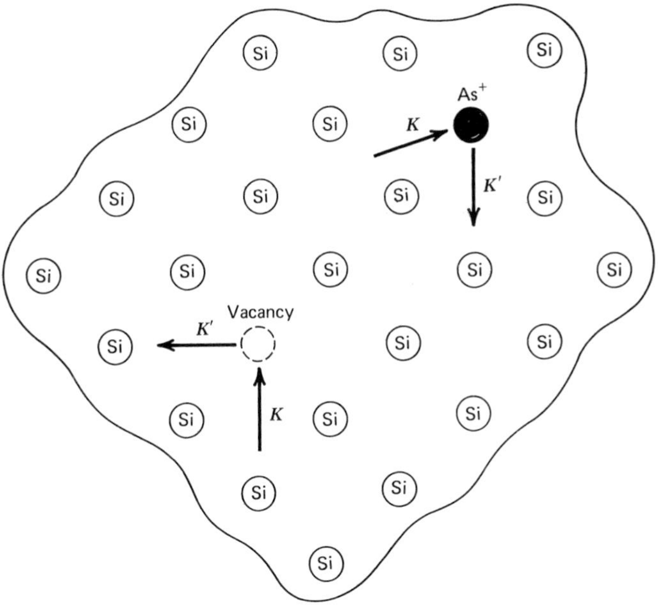

Figure 19.10 Scattering of electrons in a crystal lattice (silicon in this case) from an ionized substitutional impurity (As^+) or from a localized lattice defect (shown as a vacancy).

If an external electric field is applied to the crystal, the electron (or hole) is accelerated between scattering events and acquires a mean displacement in the direction of the force, as shown in Fig. 19.11*b*. If the mean time between collisions is τ, then the drift velocity of the electron is

$$v_e \simeq -\frac{|e|E}{m_e}\tau_n = -\mu_n E \tag{19.33}$$

$$\mu_n = \frac{|e|\tau_n}{m_e} \tag{19.34}$$

where the subscripts n and e refer to electrons. (Here we follow conventional usage.) The constant μ_n is called the electron mobility.

A similar relation applied to the motion of holes yields

$$v_h = \mu_p E$$

$$\mu_p = \frac{|e|\tau_p}{m_h} \tag{19.35}$$

The total current flowing per unit area of semiconductor is then

$$i = -n|e|v_e + p|e|v_h = (n|e|\mu_n + p|e|\mu_p)E$$

where n and p are the electron and hole densities, respectively. Using $i = \sigma E$,

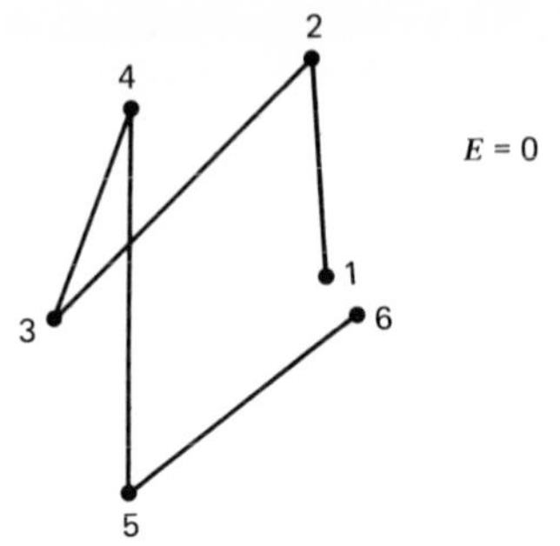

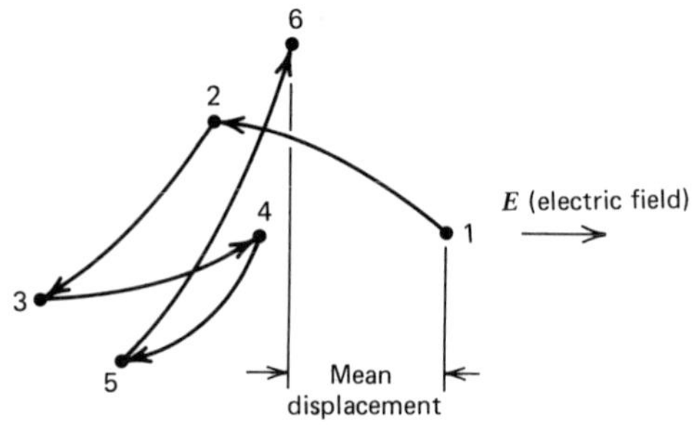

Figure 19.11 Random thermal motion of an electron interrupted by repeated scattering: (*a*) No applied field—the average displacement is zero. (*b*) With an applied electric field the electron is displaced along the direction of the force.

where σ is the conductivity, we obtain

$$\sigma = \frac{ne^2\tau_n}{m_e} + \frac{pe^2\tau_p}{m_h}$$

$$= \left(n|e|\mu_n + p|e|\mu_p\right) \tag{19.36}$$

The experimental values of carrier mobilities for a few semiconductor crystals are listed in Table 19.2. Using these values as well as those of the effective masses m_e given in Table 19.1, we find that in silicon

$$(\tau_n)_{T=300} = \frac{\mu_n m_e}{e} \simeq 1.8\times 10^{-13}\,\text{s}$$

while in InSb

$$(\tau_n)_{T=300} \simeq 5.3\times 10^{-13}\,\text{s}$$

Table 19.2 Measured Values of Room Temperature Carrier (Electron and Hole) Mobilities in a Number of Crystals

	Mobility (cm^2/volt-s)	
Crystal	*Electrons*	*Holes*
Diamond	1800	1200
Si	1600	600
Ge	4500	2000
GaAs	7000	300
Cu	35	—
AgCl	50	—
InAs	33,000	—
InSb	78000	—
InP	4600	—

The mobility in MKS units (m^2/volts-s) is obtained by dividing the practical unit (cm^2/volt-s) in the table by 10^4.

The mean distance traversed by an electron between collisions is thus

$$l_{si} \sim v_T \tau_n = \sqrt{\frac{2kT}{m_e}}\, \tau_n \simeq 2 \times 10^5 \times 1.8 \times 10^{-13}$$

$$= 3.6 \times 10^{-8}\,\text{m} = 360\,\text{Å}$$

19.5 DIFFUSION AND RECOMBINATION

One consequence of the very high frequency of collisions of charge carriers in semiconductors is that when viewed on a scale large compared to a mean free path (which we found was only a few hundred angstroms), the transport of carriers is subject to the law of diffusion. To derive the law of diffusion, we refer to Fig. 19.12. Consider the balance of carriers that cross some arbitrary

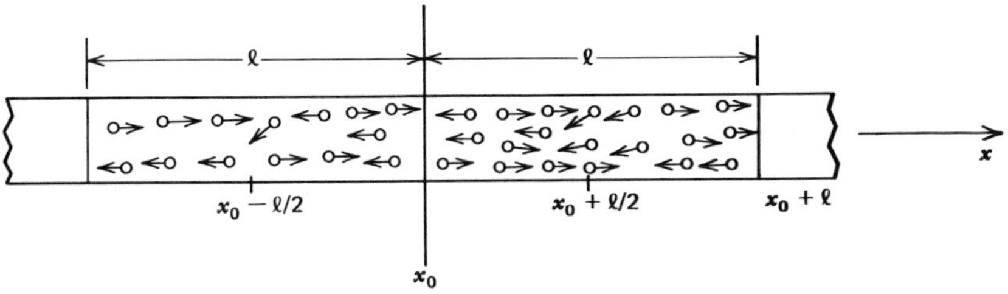

Figure 19.12 Carrier diffusion in a collision dominated motion. The mean free path is l.

plane, say the one at x_0, of area A during one mean collision time τ. The number of particles crossing from left to right in a time τ is

$$N_{L\to R}=\frac{lA}{2}n\left(x_0-\frac{l}{2}\right) \tag{19.37}$$

while that crossing in the opposite direction is

$$N_{R\to L}=\frac{lA}{2}n\left(x_0+\frac{l}{2}\right)$$

l is the mean free path, $n(x)$ is the carrier density, and $(l/2)nA$ is thus the total number of particles transported in a time τ. The factor of $\frac{1}{2}$ accounts for the fact that on the average only half of the particles are moving toward the plane. The mean density of the particles which cross x_0 from left to right is taken as that at $x_0-l/2$, while for the reverse direction we use the density at $x_0+l/2$. The net number of particles crossing from left to right in a time τ is thus

$$\begin{aligned}N_{L\to R}-N_{R\to L}&=\left[n\left(x_0-\frac{l}{2}\right)-n\left(x_0+\frac{l}{2}\right)\right]lA\\&\simeq-\frac{1}{2}\frac{\partial n}{\partial x}l^2A\end{aligned}$$

Let j indicate the particle *flux* defined as the *net* number of particles flowing across a unit area per second. We divide the last result by τA, obtaining

$$j=-\frac{l^2}{2\tau}\frac{\partial n}{\partial x}=-D\frac{\partial n}{\partial x} \tag{19.38}$$

where D, the diffusion coefficient, is given by

$$D=\frac{l^2}{2\tau} \tag{19.39}$$

The diffusion coefficients for electrons and holes will be designated by D_n and D_h, respectively.

The Einstein Relation

The diffusion coefficient

$$D_n=\frac{l_n^2}{2\tau_n}$$

and the mobility

$$\mu_n=\frac{e\tau_n}{m_e}$$

of some charge carrier, say the electron, are both functions of the mean scattering time τ_n. Using the above definition as well as the relations

$$l_n^2=v_T^2\tau_n^2$$

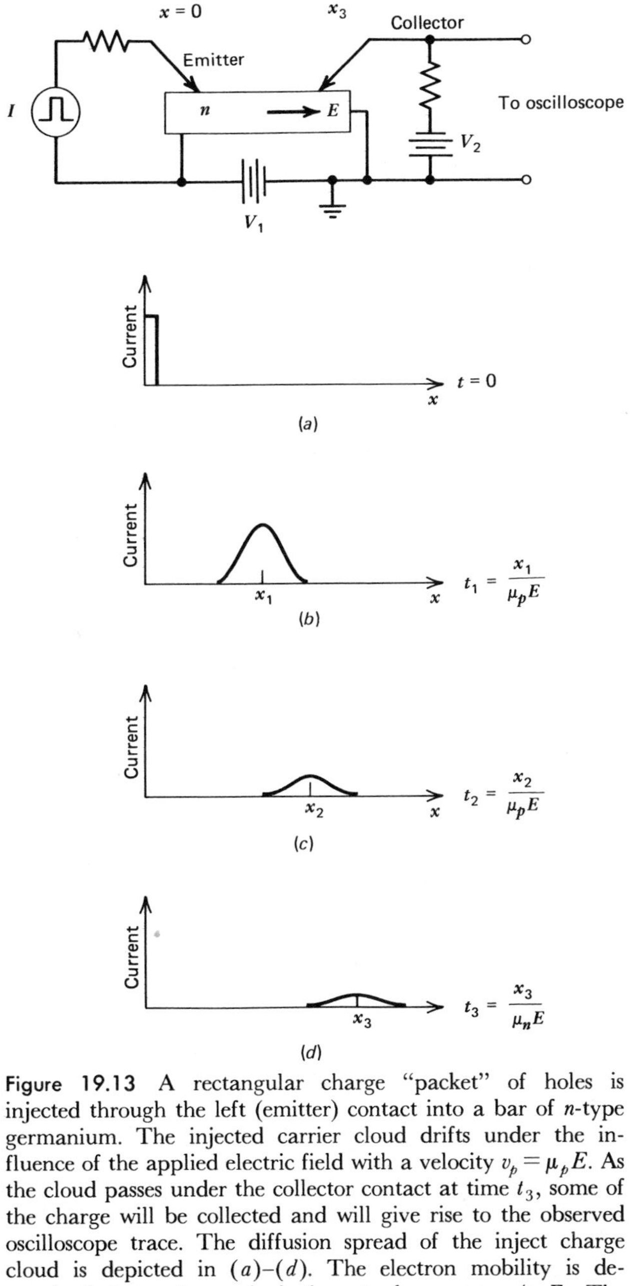

Figure 19.13 A rectangular charge "packet" of holes is injected through the left (emitter) contact into a bar of n-type germanium. The injected carrier cloud drifts under the influence of the applied electric field with a velocity $v_p = \mu_p E$. As the cloud passes under the collector contact at time t_3, some of the charge will be collected and will give rise to the observed oscilloscope trace. The diffusion spread of the inject charge cloud is depicted in (a)–(d). The electron mobility is determined from the arrival time t_3 by $\mu_p = x_3/t_3E$. The diffusion causes the injected packet to spread as it travels. This spread is noted as a change in the current pulse as a function of the x position of the collector. [The original experiment is described by D. R. Haynes and W. Shockley in *Phys. Rev.* **75**, 631 (1949).]

and

$$v_T^2 = \frac{2kT}{m_e}$$

we obtain

$$\frac{D_n}{\mu_n} = \frac{kT}{e} \tag{19.40}$$

which is known as the Einstein relation. It relates the diffusion coefficient and mobility of any charge carrier that is near thermal equilibrium. This relation will play an important role in our treatment of current flow in *p-n* junctions in the next chapter.

It may be interesting to check the Einstein relation in the case of silicon, as an example. Using Table 19.2 we obtain $\mu_n = 0.16$ m^2/volt-s. The corresponding diffusion coefficient is $D_n = k_T\mu_n/|e| = 41.2 \times 10^{-4}$ m^2/s, while the experimentally measured values are in the range of 4×10^{-3} to 5×10^{-3} m^2/s.

The drift of minority carriers in a semiconductor as well as their diffusion were first demonstrated in a classical experiment by Haynes and Shockley, which is sketched in Fig. 19.13.

Recombination of Electrons and Holes

Early in this chapter we obtained expressions for the thermal equilibrium densities of electrons in the conduction band (n) and of holes (p) in a semiconductor. The equilibrium, however, is a dynamic one since electrons and holes are continuously annihilated and generated. The basic process of annihilation is shown in Fig. 19.14*a*. It involves a transition of an electron from the conduction band to an empty state in the valence band. This process

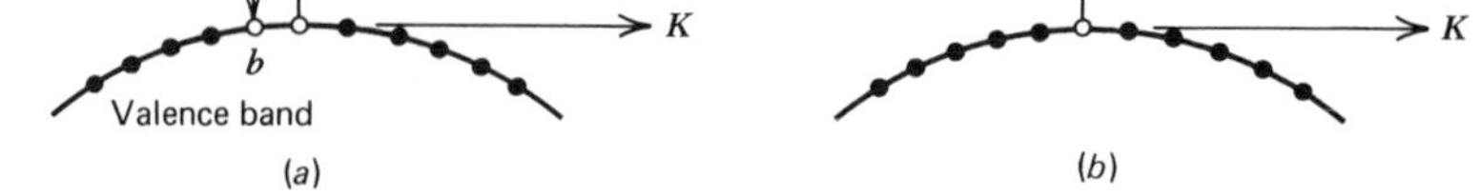

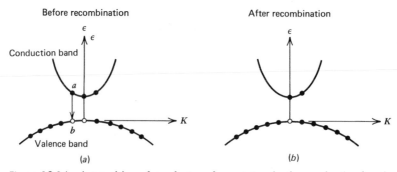

Figure 19.14 A transition of an electron from state a in the conduction band to an empty state b in the valence band results in the simultaneous annihilation of a conduction electron and a hole. The process is called electron-hole recombination. In the example we have (a) three electrons and two holes before recombination and (b) two electrons and one hole after recombination.

results in a reduction of the electron and hole populations, each by one, and is called electron-hole recombination. This process is the same as the spontaneous transitions considered in Chapter 12, and the lifetime of an electron due to this transition is called the recombination lifetime τ_r and is calculated using the same formalism as that leading to (12.71). If the density $n(x)$ of electrons at point x in a P-type semiconductor, as an example, is allowed to deviate from its equilibrium value n_p, then in the absence of any other agencies we expect it to return to its equilibrium value in a manner described by

$$\frac{\partial n}{\partial t} = -\frac{n - n_p}{\tau_r} \tag{19.41}$$

or, defining the excess electron density $n_E \equiv n - n_p$,

$$\frac{\partial n_E}{\partial t} = -\frac{n_E}{\tau_r}$$

so that

$$n_E(t) = n_E(0)e^{-t/\tau_r} \tag{19.42}$$

The Carrier Transport Equation

We are now in a position to consider what happens to the density $n(x, t)$ of electrons and of holes $p(x, t)$ under some very general conditions. We start by considering the differential volume dV shown in Fig. 19.15. Let the x-directed flux of some particle species (electrons or holes) be $j(x, t)$ (particles per unit time per unit area). A simple inventory of the changes occurring within dV during one unit time can be expressed in words as: "The increase in the number of particles within dV per unit time is equal to the excess of incoming particles (of the same species) over those leaving plus the rate at which particles are generated within dV minus the rate at which they are annihilated." This statement takes the mathematical form of

$$\frac{\partial}{\partial t} n(x, t)\, dx = j(x, t) - j(x + dx, t) + G\, dx - R\, dx \tag{19.43}$$

where G and R are, respectively, the particle generation and recombination rates (particles/s$-$m^3), and j is the particle flux (particles/s$-$m^2).

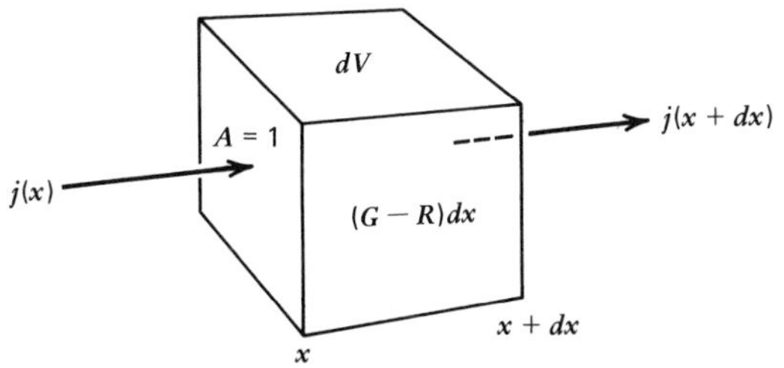

Figure 19.15 The elementary volume used to derive the carrier transport equation (19.45). $G - R$ is the net rate of particle generation per unit volume.

Since

$$j(x,t)-j(x+dx,t)=-\frac{\partial j}{\partial x}dx$$

we rewrite (19.43) as[4]

$$\frac{\partial n}{\partial t}=-\frac{\partial j}{\partial x}+G-R \tag{19.44}$$

If we apply (19.44) to the case of electrons, then the particle flux j is the sum of a drift and diffusion terms

$$j=-n\mu_n E-D_n\frac{\partial n}{\partial x}$$

and (19.44) becomes

$$\frac{\partial n}{\partial t}=\frac{\partial}{\partial x}\left(n\mu_n E+D_n\frac{\partial n}{\partial x}\right)+G-R \tag{19.45}$$

The corresponding equation for holes is

$$\frac{\partial p}{\partial t}=\frac{\partial}{\partial x}\left(-p\mu_p E+D_p\frac{\partial p}{\partial x}\right)+G-R \tag{19.46}$$

(The G and R rates for electrons and holes are of course not necessarily the same.) In the case of minority electrons in a bulk field-free P-doped material, we put $E=0$ and, using (19.41),

$$G-R=-\frac{n(x)-n_p}{\tau_r}$$

so that in steady state $(\partial/\partial t=0)$

$$D_n\frac{\partial^2 n}{\partial x^2}=\frac{n(x)-n_p}{\tau_r} \tag{19.47}$$

In terms of $n_E\equiv n(x)-n_p$, the excess electron density (19.47) becomes

$$\frac{\partial^2 n_E}{\partial x^2}=\frac{n_E}{D_n\tau_r} \tag{19.48}$$

whose solution is

$$n_E(x)=Ae^{x/L_n}+Be^{-x/L_n} \tag{19.49a}$$

where L_n, the diffusion length of excess minority electrons, is

$$L_n=\sqrt{D_n\tau_r} \tag{19.49b}$$

Diffusion in a Semi-Infinite Slab

In this case, which we will encounter in our study of p-n junctions, the carrier density at some plane, say $x=0$, is clamped at some value $n_E(0)$. We must take the constant A in (19.49) as zero, since a finite value of A would lead to a

---

[4] In three dimensions (19.45) becomes $\partial n/\partial t=-\nabla\cdot\mathbf{j}+G-R$.

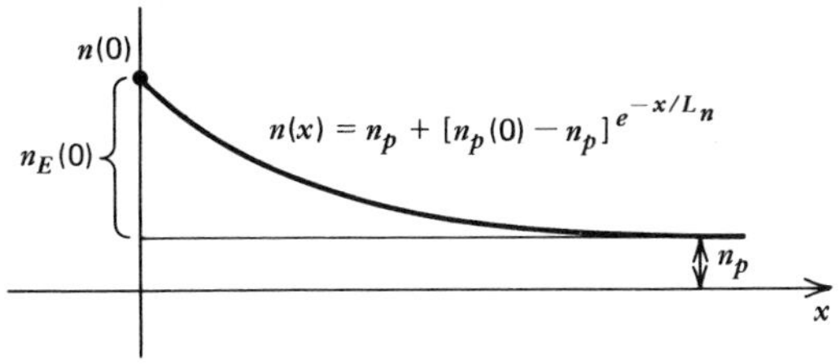

Figure 19.16 The solution of the diffusion equation (19.48) for the density of electrons in a semi-infinite *P*-type semiconductor crystal.

physically unacceptable solution that is exponentially increasing. The solution for the density profile is then

$$n_E(x>0) = n_E(0)e^{-x/L_n} \tag{19.50}$$

so that $n(x)$ decays exponentially with distance to its bulk value n_p, as illustrated in Fig. 19.16.

PROBLEMS

1. Using the measured room temperature conductivity of copper, estimate the carrier collision lifetime.
2. Assume an arbitrary small number of donor atoms present in a semiconductor. What is the position of the Fermi energy in the limit of $T \to 0$ K?
3. An ingot of GaAs is grown from a melt containing 100 gram of GaAs and 10^{-5} gram of zinc.
   (a) Calculate the density of Zn atoms.
   (b) Calculate the ionization energy for the Zn acceptor atoms (assume $\varepsilon = 12\varepsilon_0$).
   (c) Calculate the position of the Fermi level at low and high temperatures.
   (d) Calcualte n and p.
4. Consider the geometry of Fig. 19.17. A sample of *N*-type silicon is subjected to an electronic field $E_z = V_0/l$ and a magnetic field $\mathbf{B}_0$ as shown.
   (a) Show that the voltmeter will indicate a voltage

   $$V = \mu\left(\frac{V_0}{l}\right)B_0 d$$

   where μ is the electron mobility.
   (b) Show that

   $$\frac{E_y}{IB_0} = \frac{A}{ne}$$

   where I is the total current, $E_y = v/d$ is the transverse field, and A is the cross sectional area normal to z.

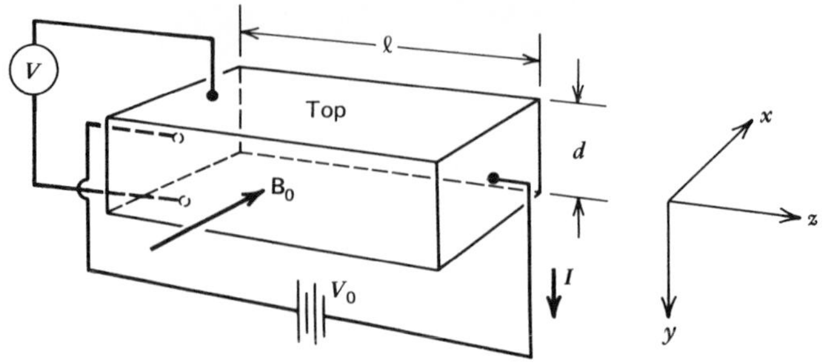

Figure 19.17

Clue: The field E_y developed in the crystal is that for which the transverse (y) electric and magnetic forces on the drifting electrons are equal (in magnitude) and opposite.

(c) What is the transverse field developed in a sample of GaAs, with $l = 1$ cm, $B_0 = 1$ weber/m^2, $V_0 = 10^3$ volts, and $\mu = 0.7$ m^2/volt-s.

5. Obtain an expression for the surface charge density (Coulomb/m^2) that is induced in the top (and bottom) surface of the sample in the experiment of Fig. 19.17.

6. A 1-cm bar of germanium with a mobility of $\mu_n = 4500$ cm^2/volt-s is connected to a power supply with $V_0 = 100$ volt.
   (a) How long will it take an electron to drift the full distance?
   (b) How many collisions will the average electron suffer during one transit?
   (c) What is the total energy received by the electron from the field and lost to the lattice?

7. Consider an intrinsic bar of GaAs at 77 K with $\mu_n = 10^5$ cm^2/volt-s and $\mu_p = 10^4$ cm^2/volt-s.
   (a) Using an absorption coefficient for incident light of $\alpha = 10$ cm^{-1}, calculate the conductivity of the sample immediately following an impulse of light with an energy fluence of 1 joule/cm^2. (Assume that each absorbed photon generates an electron-hole pair and that the product α(typical crystal dimension)$\ll 1$.
   (b) Assuming a continuous irradiation with an intensity of 10 watts/cm^2, what is the steady state sample conductivity if the electron-hole recombination time is $\tau_r = 10^{-7}$ s.
   (c) Can you use the experimental situation described in part (a) to measure the recombination time τ_r?

8. (a) Show that an electron with an effective mass m_e will rotate about a uniform magnetic field **B** at an angular velocity of $\omega_c = eB/m_e$.
   (b) Show how a measurement of ω_c—the cyclotron resonance frequency—coupled with that of the transverse voltage in Problem 4, can be used to determine the scattering time τ.

CHAPTER TWENTY

The *P*-N Semiconductor Junction. The *P*-N-*P* Junction Transfer

The reference to applications in the title of this book reflects an attempt to carry the development of basic quantum mechanics from fundamental postulates and derivations all the way to basic applications. We have done this already in the description of laser oscillation and the hydrogen maser.

In this chapter we make use of some of the background developed in earlier chapters to describe the principles involved in the operation of the semiconductor junction transistor. This device, invented in 1946 by Bardeen, Brattain, and Shockley,[1] must rank among the most important applications of science to modern day technology. It is responsible directly for the development of integrated electronics, and thus for the revolution taking place in the processing and handling of information by computers as well as in numerous applications involving electronic control of machines and processes.

Before taking up the description of the transistor we need to understand the basic mechanisms involved in the current flow across the junction separating a *P*-doped semiconductor from one that is *N* type.

---

[1] J. Bardeen and W. H. Brattain, *Phys. Rev.* **75**, 1208 (1949); W. Shockley, *Bell System Tech. J.* **28** (1949).

20.1 THE CARRIER AND POTENTIAL PROFILES IN A *P*-N JUNCTION

There are a number of ways *P*-N junctions can be prepared. A commonly applied method is to diffuse a large density of acceptor atoms into an N-type semiconductor. The diffused layer where $N_A > N_D$ (concentration of acceptor atoms larger than that of donor atoms) is thus rendered *P* type. The junction is formed at the transition plane, where $N_A = N_D$.

Figure 20.1*a* depicts an idealized semiconductor junction with an abrupt transition at $x=0$ from an excess N_A of acceptor atoms over donors to an excess N_D (at $x>0$) of donors over acceptors.

To understand what happens near the chemical junction ($x=0$), let us imagine that the *P* and N sides of the junction are initially far apart and are subsequently brought together. As soon as contact is established, a situation exists where at $x<0$ the mobile charge carriers are all holes, while at $x>0$ they are electrons. The process of diffusion will drive holes from the *P* to the N region and electrons from the N into the *P* side. The electrons that move across the junction from right to left leave behind a band of width b of positively charged donor ions. The situation is illustrated in Figs. 20.1*a* and 20.1*c*. In a similar manner, the holes that have relocated from $x<0$ to $x>0$ create a band of width a of negatively charged acceptor ions. This transfer of charge ceases when the electrostatic potential barrier set up by the net charge in the space charge layer $-a<x<b$ becomes strong enough to impede any further (net) migration. This electrostatic potential barrier is shown in Fig. 20.1*e*. It is referred to as the "built-in" voltage or the "diffusion" potential. The region $-a<x<b$ is referred to as the depletion layer, since the electric field that exists there sweeps it clean of mobile electrons and holes. Since the sample is in thermal equilibrium, it must be describable in terms of a single-valued chemical potential μ, which is shown as a horizontal line in Fig. 20.1*f*. Far enough from the junction plane $x=0$, the chemical potential must assume the same position relative to the band edges as it does in the bulk material. This requirement can be reconciled with the constancy of μ by a bending of the conduction and valence band edges near the junction, as shown in Fig. 20.1*f*. Since the band edges correspond to electron energies and the energy of an electron (at rest) is the negative of the potential, their shape is a mirror image of the electrostatic potential curve $eV(x)$ of Fig. 20.1*e*.

The Contact Potential ϕ

In the junction region the total hole particle current density (particles/unit time-unit area) is given by the sum of the diffusion and drift terms first described in Section 19.5:

$$j_h(x) = -D_p\frac{\partial p}{\partial x} + pv_h$$

$$= -D_p\frac{\partial p}{\partial x} - p\mu_p\frac{dV}{dx} \tag{20.1}$$

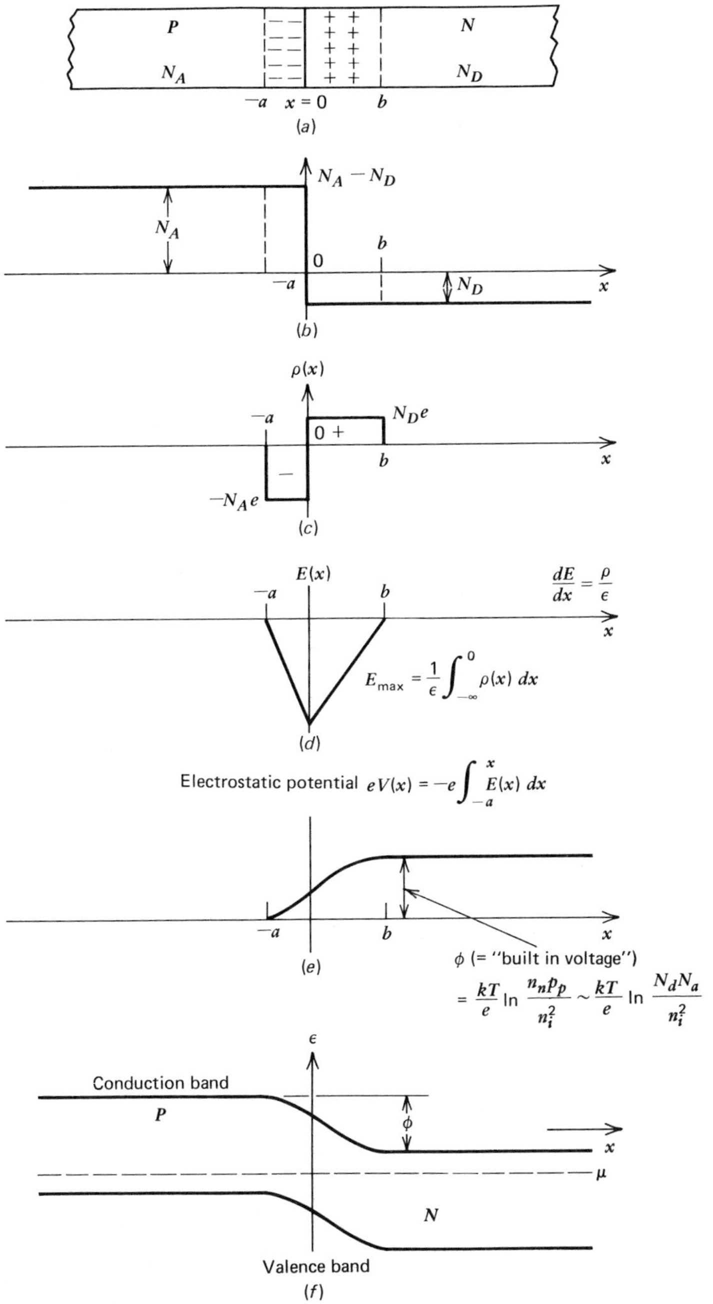

Figure 20.1 The *P*-N semiconductor junction. (*a*) The region ($-a < x < 0$) of uncompensated ionized acceptors and that ($0 < x < b$) of ionized donors. (*b*) The doping profile $N_A - N_D$. (*c*) The uncompensated charge profile due to ionized impurity atoms in the depletion layer. (*d*) The electric field profile. (*e*) The electrostatic potential. (*f*) The energy band diagram.

where we used the relation

$$v_h = \mu_p E = -\mu_p \frac{dV}{dx}$$

The electron (particle) current density is described similarly by

$$j_e(x) = -D_n \frac{\partial n}{\partial x} + n\mu_n \frac{dV}{dx} \tag{20.2a}$$

At thermal equilibrium the net electron and hole current must each be equal to zero at any plane x

From (20.1) we obtain

$$D_p \frac{dp}{\partial x} = -p\mu_p \frac{dV}{dx} \tag{20.2b}$$

or

$$-\frac{e}{kT}\frac{dV}{dx} = \frac{1}{p}\frac{dp}{dx} = \frac{d}{dx}\ln p$$

where we used the Einstein relation (19.40) $D_p/\mu_p = kT/e$. Integrating the last expression between any two points x_1 and x_2 leads to

$$V(x_2) - V(x_1) = \frac{kT}{e}\ln\frac{p(x_1)}{p(x_2)} \tag{20.3}$$

so that

$$\frac{p(x_2)}{p(x_1)} = e^{-e[V(x_2)-V(x_1)]/kT} \tag{20.4}$$

If the point x_1 is taken at $x = -a$, where the hole density p_p is at its equilibrium bulk value in the p material, and if x_2 is taken at b, where $p = p_n$ (the equilibrium hole distribution in the bulk n material), then from (20.3) the potential difference across the junction is

$$\phi \equiv V(b) - V(-a) = \frac{kT}{e}\ln\frac{p_p}{p_n} \tag{20.5}$$

An entirely similar derivation starting with (20.2a) leads to

$$\phi = \frac{kT}{e}\ln\frac{n_n}{n_p} \tag{20.6}$$

Multiplying (20.6) and (20.5) by n_n/n_n, using the relation (19.10) $n_p n_n = n_i^2$, and the "high" temperature result (19.24) $n_n \simeq N_d$ and $p_p \simeq N_a$, we obtain a simple result:

$$\phi = \frac{kT}{e}\ln\frac{N_D N_A}{n_i^2} \tag{20.7}$$

The density of electrons as a function of x is given by a relation similar to (20.4):

$$\frac{n(x_2)}{n(x_1)} = e^{e[V(x_2)-V(x_1)]/kT} \tag{20.8}$$

where the different sign of the exponent compared to (20.4) reflects the fact that the electrons are attracted to regions of large electrostatic potential $V(x)$.

By taking $x_1 = -a$ and $x_2 = b$, we obtain from (20.4) and (20.8)

$$p_n = p(b) = p(-a)e^{-e\phi} = p_p e^{-e\phi}$$
$$n_p = n(-a) = n(b)e^{-e\phi} = n_n e^{-e\phi} \tag{20.9}$$

When an external voltage source V_a is applied across a *P*-N junction, the potential barrier across it becomes $\phi - V_a$. The sign of V_a is taken as positive when the *P* side is connected to the positive terminal of the power supply and the N side to the negative terminal. Another important consequence is that a net current flows across the junction.

20.2 THE *P*-N JUNCTION WITH AN APPLIED VOLTAGE

The condition of thermal equilibrium is disturbed when an external voltage, which we denote by V_a, is applied across the junction. The resulting band profiles are shown in Figs. 20.2*a*–20.2*c*. In (*a*) $V_a = 0$, in (*b*) $V_a > 0$, and in (*c*) $V_a < 0$. The sense of V_a is as defined above.

The application of a voltage V_a across the semiconductor sample causes the potential barrier across the junction to change from ϕ to $\phi - V_a$. This results in a reduced electrostatic barrier $\phi - |V_a|$ in the case of a positive bias ($V_a > 0$) and in a large barrier $\phi + |V_a|$ for $V_a < 0$, as shown in Figs. 20.2*b* and 20.2*c*, respectively. We note from these figures that the chemical potential difference between the *P* and N sides must be equal to eV_a. This is due to the fact that the chemical potential corresponds to the energy needed to add one electron to the medium and this energy must change exactly by eV_a in going from the *P* to the N regions.

The change in the height of the potential barrier from ϕ to $\phi - V_a$ must be accompanied, as will be shown later, by a change in the width $(a+b)$ of the depletion layer as shown in figs. 20.2*b* and 20.2*c*.

In obtaining the carrier distribution and currents across the junction we start, again, with Eq. (20.1). The total hole current $j_h(x)$ is given as in (20.1) by

$$j_h(x) = -D_p \frac{dp}{dx} - p\mu_p \frac{dV}{dx}$$

but since $V_a \neq 0$, we can no longer take $j_h(x)$ as zero. Some very simple considerations (see Problem 3) show, however, that

$$\left| D\frac{dp}{dx} \right| \simeq |pv_h| \gg j_h(x) \tag{20.10}$$

that is, the drift and diffusion components of the hole current are individually many orders of magnitude bigger than the net hole current. We can thus, to a very high degree of accuracy, write as in (20.2*b*)

$$D_p \frac{dp}{dx} = -p\mu_p \frac{dV}{dx} \tag{20.11}$$

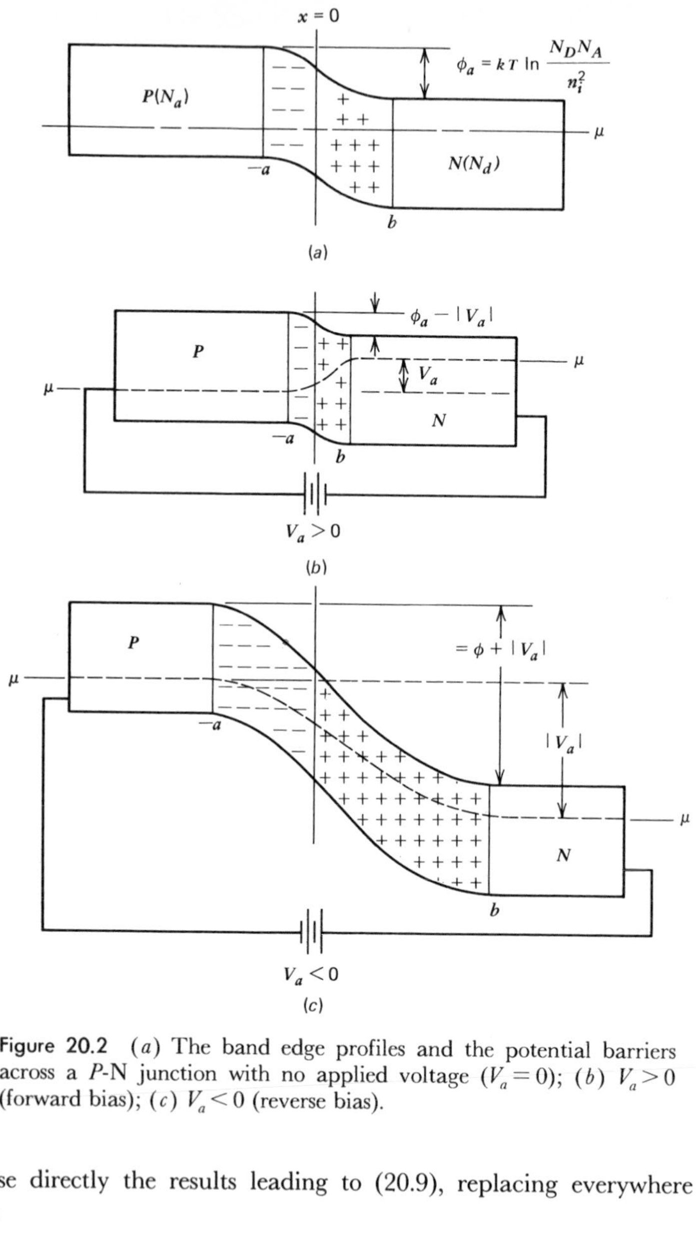

Figure 20.2 (*a*) The band edge profiles and the potential barriers across a *P*-N junction with no applied voltage ($V_a = 0$); (*b*) $V_a > 0$ (forward bias); (*c*) $V_a < 0$ (reverse bias).

and use directly the results leading to (20.9), replacing everywhere ϕ by $\phi - V_a$:

$$p(b) = p_p e^{-e(\phi - V_a)/kT} = p_n e^{eV_a/kT}$$

$$n(-a) = n_n e^{-e(\phi - V_a)/kT} = n_p e^{eV_a/kT} \qquad (20.12)$$

so that the minority carrier (electrons in the *p* region and holes in the *n* region) density at the *edges* of the depletion layer is modified with respect to its bulk value by the factor $\exp(V_a/kT)$. The excess minority carrier density at the depletion layer edges is thus

$$p_E(b) \equiv p(b) - p_n = p_n(e^{eV_a/kT} - 1) \tag{20.13a}$$

$$n_E(-a) \equiv n(-a) - n_p = n_p(e^{eV_a/kT} - 1) \tag{20.13b}$$

Note that for $V_a = 0$, $p_E(b) = 0$; while for $V_a \gtrless 0$, $p_E(b) \gtrless 0$. Similar results obtain for $n_E(-a)$. It should also be recalled that $x = -a$ and $x = b$, the depletion layer boundaries, are not fixed but depend on V_a.

Equation (20.13) is referred to as the *law of the junction* and is the practical starting point for the description of most junction devices.

The Current Flow in *P*-N Junctions

We are now in a position to derive an expression for the total current flowing across a *P*-N junction. At $x \geq b$ the electric field is zero $[V(x) = \text{const}]$ so that the distribution of holes is governed by the law of diffusion alone. We apply directly the result of (19.49) to obtain

$$\begin{aligned} p_E(x \geqslant b) &= p_E(b) e^{-(x-b)/L_p} \\ &= p_n(e^{eV_a/kT} - 1) e^{-(x-b)/L_p} \end{aligned} \tag{20.14}$$

and similarly,

$$\begin{aligned} n_E(x < -a) &= n_E(-a) e^{(x+a)/L_n} \\ &= n_p(e^{eV_a/kT} - 1) e^{(x+a)/L_n} \end{aligned} \tag{20.15}$$

The distribution of holes and electrons in a *P*-N junction is shown in Fig. 20.3.

In steady state the total current flowing through the junction is independent of x. We may thus choose to evaluate it at the edge $x = b$ of the depletion layer. This current is made up of a component due to holes and one due to electrons. The first of these is given by

$$\begin{aligned} I_h(x = b) &= -eD_p A \left.\frac{\partial p}{\partial x}\right|_{x=b} \\ &= \frac{eD_p A}{L_p} p_E(b) \end{aligned} \tag{20.16}$$

where use was made of the fact that $dp_E/dx = dp/dx$ and of (20.14). The electron component of the current must also be evaluated at $x = b$. If no recombination of electron and holes takes place within the depletion layer, $-a \leq x \leq b$, then the number of electrons crossing the plane $x = b$ in a given period of time must be equal to that crossing $x = -a$, so that, according to (20.15),

$$\begin{aligned} I_e(x = b) &= I_e(x = -a) \\ &= eD_n A \left.\frac{\partial n}{\partial x}\right|_{x=-a} \\ &= \frac{eD_n A}{L_n} n_E(-a) \end{aligned} \tag{20.17}$$

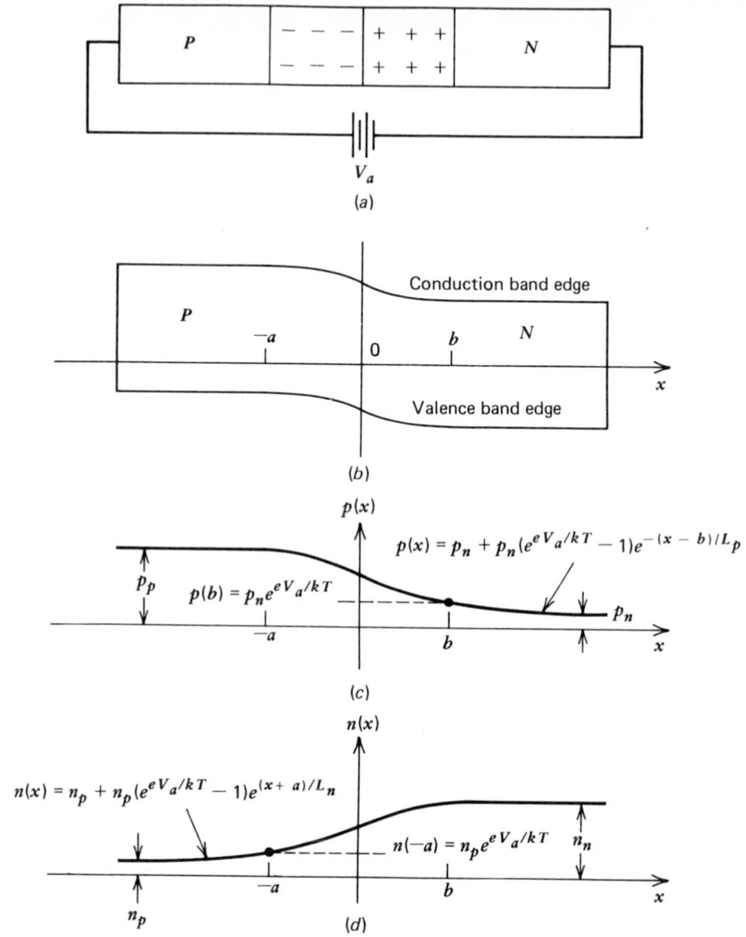

Figure 20.3 Carrier profiles in a forward biased *P*-N junction. (*a*) The junction and the depletion layer with its net space charge. (*b*) The energy band diagram. (*c*) The hole density profile. Note that according to the law of the junction $p(b) = p_n \exp(|e|V_a k/T)$. At $x > b$, $p(x)$ decays exponentially by diffusion to a value P_n. (*d*) The profile of electron density.

The total junction current density is thus

$$\begin{aligned} I &= I_e(x=b) + I_h(x=b) \\ &= \frac{eD_n A}{L_n} n_p\left(e^{eV_a/kT} - 1\right) + \frac{eD_p A}{L_p} p_n\left(e^{eV_a/kT} - 1\right) \\ &= eA\left(\frac{D_n n_p}{L_n} + \frac{D_p p_n}{L_p}\right)\left(e^{eV_a/kT} - 1\right) \end{aligned}$$

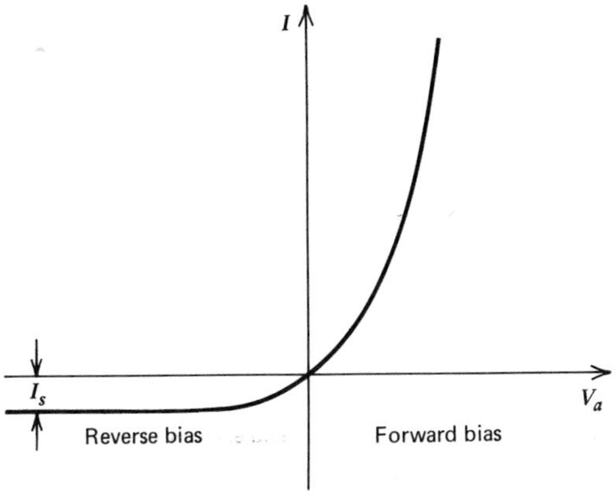

Figure 20.4 An experimental plot of the voltage current characteristics of a typical *P*-N junction.

or

$$I = I_S\left(e^{eV_a/kT} - 1\right) \tag{20.18}$$

where the *saturation* current I_S is

$$I_S = eA\left(\frac{D_n n_p}{L_n} + \frac{D_p p_n}{L_p}\right) \tag{20.19}$$

Equation (20.18) describes the basic rectifying characteristics of *P*-N junctions. At a forward bias ($V_a > 0$) the junction current is large and the resistance V_a/I becomes small. When the applied bias is negative ($V_a < 0$), the current approaches the saturation value $-I_s$, which is very small in comparison with the forward current. As a result, the diode impedance under reverse bias is very large. These nonlinear voltage-current characteristics are the basis for the application of these diodes in numerous applications such as digital logic circuits and microwave mixers, to name two examples.

An experimental curve of the voltage current characteristics of a typical *P*-N junction is shown in Fig. 20.4.

Junction Fields and Capacitance

In many applications that include the use of *P*-N junctions as solar cells, optical detectors, impatt oscillators, and as microwave parametric amplifiers, it is necessary to obtain the electric field distribution in the depletion layer of the junction.

From Gauss law $\nabla \cdot \mathbf{E} = \rho/\varepsilon$ and the relation $E = -dV/dx$, we obtain for the geometry of Fig. 20.3

$$\frac{d^2V}{dx^2} = \frac{eN_a}{\varepsilon} \quad \text{for } -a < x < 0 \tag{20.20}$$

and

$$\frac{d^2V}{dx^2} = -\frac{eN_d}{\varepsilon} \quad \text{for } 0<x<b \tag{20.21}$$

The constant ε is the low frequency dielectric constant of the semiconductor. The depletion layer is fully devoid of carriers so that the net charge is due to the ionized acceptor atoms and is $\rho = -eN_a$ for $-a<x<0$ and $\rho = eN_d$ for $0<x<b$. (We assumed that all the impurity atoms in the depletion layer are ionized.) The net charge outside the depletion layer is zero.

The boundary conditions are

$$E = -\frac{dV}{dx} = 0 \quad \text{at } x=-a \text{ and } x=b \tag{20.22}$$

that is, the field vanishes at the edges of the depletion layer and is continuous at $x=0$.

In addition, the total potential drop across the depletion layer is

$$V(b) - V(-a) = \phi - V_a \tag{20.23}$$

The solutions of (20.20) become

$$V(x) = \frac{eN_a}{2\varepsilon}(x^2 + 2ax) \quad (-a<x<0) \tag{20.24}$$

$$V(x) = -\frac{eN_d}{2\varepsilon}(x^2 - 2bx) \quad (0<x<b) \tag{20.25}$$

which, using the continuity condition, yields

$$N_a a = N_d b \tag{20.26}$$

that is, the depletion layer contains a positive charge $N_d ae$ on the N side $(-a<x<0)$ due to ionized donor atoms, which is equal in magnitude to that of the negative charges $-N_a be$ on the P side $(0<x<b)$. Condition (20.23) gives

$$\phi - V_a = \frac{e}{2\varepsilon}(N_d b^2 + N_a a^2)$$

which together with (20.26) leads to

$$a = (\phi - V_a)^{1/2}\left(\frac{2\varepsilon}{e}\right)^{1/2}\left(\frac{N_d}{N_a(N_a+N_d)}\right)^{1/2} \tag{20.27}$$

$$b = (\phi - V_a)^{1/2}\left(\frac{2\varepsilon}{e}\right)^{1/2}\left(\frac{N_a}{N_d(N_a+N_d)}\right)^{1/2} \tag{20.28}$$

Differentiation of (20.24) and (20.25) yields

$$E = -\frac{e}{\varepsilon}N_a(x+a) \quad (-a<x<0)$$

$$E = -\frac{e}{\varepsilon}N_d(b-x) \quad (0<x<b) \tag{20.29}$$

The maximum field occurs at the chemical junction $x=0$,

$$E_{\max} = -\frac{2(\phi - V_a)}{a+b}$$

so that for $V_a < 0$ (reverse bias) extremely large fields ($\simeq 10^5$ V/cm) can exist near the chemical junction.

The total charge per unit area on one side of the depletion layer is

$$Q = eN_a a = eN_d b$$

and is, according to (20.27) or (20.28), a function of the applied voltage V_a. The junction thus possesses a capacitance per unit area:

$$\begin{aligned} C &= \frac{dQ}{dV_a} \\ &= -eN_a \frac{da}{dV_a} \\ &= \left(\frac{e\varepsilon}{2}\right)^{1/2} \left(\frac{N_a N_d}{N_a + N_d}\right)^{1/2} \left(\frac{1}{\phi - V_a}\right)^{1/2} \\ &= \frac{\varepsilon}{a+b} \end{aligned} \tag{20.30}$$

This nonlinear dependence of C on V_a is the basis for the application of the P-N junction transistor as a microwave parametric amplifier.[2]

20.3 THE *P*-N-*P* JUNCTION TRANSISTOR

The basic structure of a P-N-P transistor is illustrated in Fig. 20.5. A single high purity crystal, usually silicon, is doped with P- and N-type impurities so as to result in the P-N-P sandwich shown in the figure. The device can be viewed as two P-N junctions connected back to back by means of the common B region, which is called the base. The left P-N junction is forward biased by means of an applied positive voltage V_{BE}. The second junction is reverse biased by means of an applied voltage V_{BC}. The side of the device connected to the positive side of V_{BE} is called the emitter (E), while the P side of the reverse biased junction is termed the collector (C). The reason for this terminology will be made clear by the description that follows.

Viewed as a three-terminal device with currents I_E, I_B, and I_C, it follows that in steady state

$$I_E + I_B = I_C \tag{20.31}$$

that is, the net inflowing (or outgoing) current is zero.

Before embarking on a detailed mathematical description of the transistor, it will be useful to understand its behavior on a qualitative basis. To be

---

[2] M. Uenohara, "Low Noise Amplification," in *Handbuch der Physik* **23** (Springer-Verlag, Berlin, 1969), pp. 1–81.

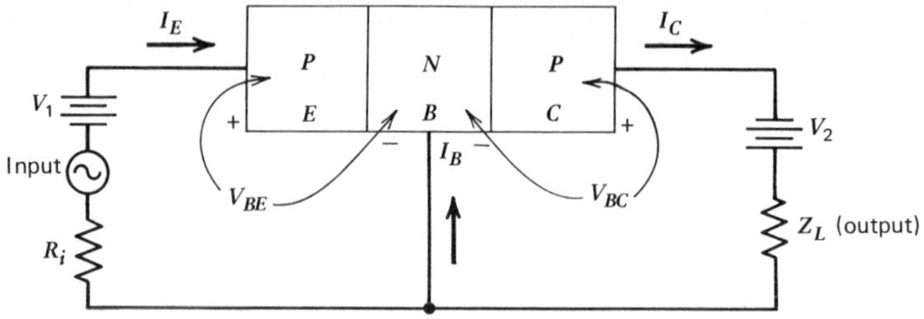

Figure 20.5 The basic configuration and biasing circuit of a *P*-N-*P* junction transistor.

specific, we will describe how a transistor can be used to amplify weak currents.

The forward bias V_{BE} of the emitter junction results in a positive forward current I_E flowing through the junction in accordance with the description in Section 20.2. Since the emitter is usually doped much more heavily than the base, this current is due almost exclusively to holes crossing from E to B. Once in the base B, the holes become minority carriers. If the base is sufficiently thin, then most, but not all of the holes injected from the emitter diffuse to the base collector junction, which is strongly reverse biased. The effect of the large potential drop in depletion layer B due to the reverse bias is that the arriving holes are swept into the collector region (C) and constitute the current I_C.

The reader new to electronics may appreciate a more detailed physical explanation. The holes swept by the potential drop in depletion layer B into the collector C region give rise to an excess positive charge in that region. This charge attracts electrons through the external wire contact from the battery V_2 until the net charge is again zero. The flow of these electrons constitutes the output current I_C from the collector.

A small fraction of the holes injected across the emitter junction do not survive the trip across the base. These holes recombine with base majority electrons. It follows immediately that $I_E > I_C$. Each recombination event results in the annihilation of one electron and one hole. Since in steady state the number of electrons in the base is constant, the annihilated electrons must be replenished. This gives rise to the base current I_B. The polarity of I_B in Fig. 20.5 agrees with common convention. In the example shown, electrons must flow into the base (to replace those annihilated by recombination) so that $I_B < 0$.

In a typical transistor the base is sufficiently narrow so that most of the holes injected across the emitter junction are collected by the reverse biased collector junction. Let us assume, as an example, that for each 21 injected holes, 20 are collected while one recombines in the base. We thus have

$$I_E = 21, \quad I_C = 20, \quad I_B = -1$$

in some current units.

The connection shown in Fig. 20.5 is referred to as the common base configuration, since the base is common, electrically, to both the input and output sides. Although the input and output currents are nearly equal, there is a large power amplification, since the output impedance Z_L can exceed that of the input R_i by a large factor. This is merely a reflection of the fact that the emitter junction is forward biased so that, according to Fig. 20.4, small voltage increments ΔV_E cause large current increments ΔI_E. In the reverse biased collector junction the opposite is true. It follows that

$$\frac{\Delta V_C}{\Delta I_C} \gg \frac{\Delta V_E}{\Delta I_E}$$

Since the input power to the transistor is $\sim I_E^2 R_i$ while the output power is $\sim I_C^2 Z_L$ and $I_E \simeq I_C$, the transistor acts as a power amplifier.

The description of the transistor operation given above views the emitter current as a cause and I_C and I_B as the response. It is perfectly legitimate, however, to view I_B as the *input* (cause) and I_C as the *resulting output*. In the case in the above example an increment ΔI_B of the base current causes the collector current to change by $\Delta I_C = -20\Delta I_B$. This is the basis for using the transistor as a current amplifier. In this case, a weak input current is fed into the base, while the amplified current emerges from the collector terminal.

The Transistor Currents

The qualitative description of the operation of the transistor given above makes it clear that an important characteristic of the device is the relation between the emitter, base, and collector currents. To derive this relation we refer to Fig. 20.5. We assume that due to the heavy doping of the emitter and collector regions relative to that of the base ($p^+ \gg n$), the currents flowing across the two junctions are made up predominantly of holes. As a first step we need to solve for the hole density in the base region. Since this region ($0 < x < w$) has no field, the general solution for the excess hole density must satisfy the diffusion equation, and consequently is given according to (19.49) by

$$P_E(x) = Ae^{-x/L_p} + Be^{x/L_p}$$

The constants A and B are determined by the boundary values of $p_E(0)$ and $p_E(w)$. These in turn are given by the law of the junction (20.13).

The application of (20.13) in conjunction with the last equation gives

$$\begin{aligned} A + B &= p_E(0) \\ &= p_n(e^{eV_1/kT} - 1) \end{aligned}$$

and

$$\begin{aligned} Ae^{-w/L_p} + Be^{w/L_p} &= p_E(w) \\ &= p_n(e^{-eV_2/kT} - 1) \simeq -p_n \end{aligned}$$

The last approximation is valid, since $eV_2 \gg kT$. Solving the last two equations for A and B and using the result in the expression for $p_E(x)$ leads, after some algebra, to

$$\underset{0<x<w}{p_E(x)} = \frac{p_n}{\sinh(w/L_p)}\left((e^{eV_1/kT}-1)\sinh\frac{w-x}{L_p} - \sinh\frac{x}{L_p}\right)$$

The hole current in the base is thus

$$I_h(x) = -D_p eA\frac{dp_E}{dx}$$

$$= \frac{D_p eAp_n}{L_p\sinh(w/L_p)}\left((e^{eV_1/kT}-1)\cosh\frac{w-x}{L_p} + \cosh\frac{x}{L_p}\right) \quad (20.32)$$

Here and in what follows A is the area of the junction normal to the current flow direction. The emitter current is given by

$$I_E \simeq I_h(0) = \frac{D_p eAp_n}{L_p}\left((e^{eV_1/kT}-1)\coth\frac{w}{L_p} + \frac{1}{\sinh(w/L_p)}\right) \quad (20.33a)$$

These currents are shown in Fig. 20.6.

It is instructive to consider at this point two limiting cases. In the first instance the base width w is large compared with the hole diffusion length L_p

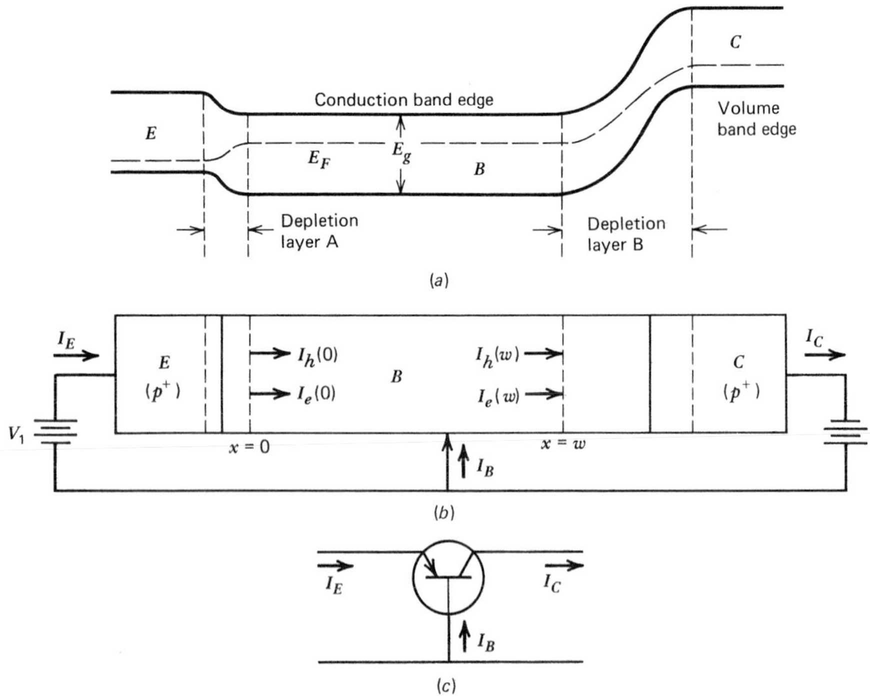

Figure 20.6 The valence and conduction bands' edges of a typical P-N-P transistor are shown in (a). The physical arrangement and the applied voltages are shown in (b). The conventional engineering symbol for the transistor is shown in (c).

($w \gg L_p$). This case is illustrated in Fig. 20.7*a*. The emitter and collector junctions are largely independent of each other. The distribution of holes $p(x)$ and the hole current $I_p = -eD_p\, dp/dx$ obey the single junction laws (20.14) and (20.16). In particular, the collector current is independent of the emitter current. In this mode the device cannot serve as a transistor.

The situation when $w \ll L_p$ is illustrated in Fig. 20.7*b*. In this the fraction of the injected holes that recombine in the base region is negligible so that the distribution of $p(x)$ approaches a straight line.

If $p(x)$ were to be described by a straight line, then the emitter current $-D_p eA\, dp/dx]_{x=0}$ and the collector current $-D_p eA\, dp/dx]_{x=w}$ would be equal. The slight deviation of $p(x)$ from exact linearity causes the slope at $x=w$ to be somewhat smaller than at $x=0$; that is, the collector current is somewhat smaller than the injected emitter current. The difference is equal to the base current.

Under typical bias condition $eV_1 \gg kT$ and in narrow base transistors $w \ll L_p$, we obtain from (20.33*a*) to second order in (w/L_p):

$$I_E \simeq I_h(0) \frac{D_p eAp_n}{w} e^{eV_1/kT}\left(1+\frac{w^2}{2L_p^2}\right) \tag{20.33b}$$

The collector current is given by

$$\begin{aligned} I_c &\simeq I_h(w) \\ &\simeq \frac{D_p eAp_n}{w} e^{eV_1/kT} \end{aligned} \tag{20.34}$$

where the approximation involves the same assumptions used in (20.33*b*).

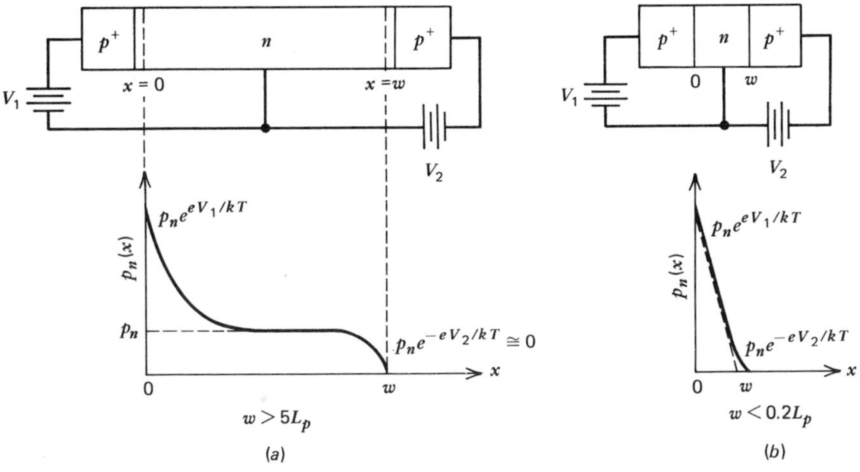

Figure 20.7 (*a*) A wide base transistor $w \gg L_p$, the minority carrier distributions $p_n(x)$ near the emitter, and the collector junction are independent of each other. (*b*) A narrow base transistor. Here the recombination of subjected holes in the base is insignificant resulting in a nearly linear $p_n(x)$. The collector current is thus nearly equal to the emitter current.

The base current is obtained as

$$I_B = I_C - I_E \simeq = -\frac{D_p e A p_n w}{2L_p^2} e^{eV_1/kT}$$

$$= -\frac{e p_n A w}{2\tau_r} e^{eV_1/kT} \tag{20.35}$$

The second equality used the relation (19.49*b*) $L_{n,p}^2 = D_{n,p}\tau_r$.

It is interesting as well as instructive to obtain expression (20.35) for the base current using a more direct physical reasoning. The base current is due physically to electrons entering the base region in order to replace those lost by recombination with the holes in transit. The number of such recombinations per unit time is equal to

$$N_{\text{recomb}} = \frac{\text{total no. of excess holes in base}}{\tau_r}$$

$$= A \int_{x=0}^{w} \frac{p_E(x)\,dx}{\tau_r} \tag{20.36}$$

In the narrow base limit $w \ll L_p$ the expression (20.32) for $p_E(x)$ is that of a straight line with a value $p_n(e^{eV_1/kT} - 1)$ at $x = 0$ and $-p_n$ at $x = w$. The integral of (20.35) yields

$$N_{\text{recomb}} = \frac{p_n A w}{2\tau_r}\left(e^{eV_1/kT} - 2\right) \approx \frac{p_n A w}{2\tau_r} e^{eV_1/kT} \quad (eV_1 \gg kT)$$

and after using $I_B = -eN_{\text{recomb}}$ we obtain (20.35).

An important device parameter of a transistor is its common base current gain α:

$$\alpha = \frac{\partial I_c}{\partial I_E} = \frac{I_c}{I_E}$$

which relates the change in the output (collector) current due to a change in the input (emitter) current. Using (20.33*a*) and (20.34) yields

$$\alpha = \frac{1}{1 + w^2/2L_p^2} \tag{20.37}$$

If, as an example, the ratio $w/L_p = 0.2$, then $\alpha = 0.98$. In this case 98% of the holes injected into the emitter exit from the collector terminal.

If the same transistor were to be used as a current amplifier, then the input current will be fed into the base and the output will be the collector current. The relevant figure of merit in this case is

$$\beta = \frac{I_c}{I_B}$$

$$= \frac{I_c}{I_c - I_E}$$

$$= -\frac{\alpha}{\alpha - 1} \simeq -2\left(\frac{L_p}{w}\right)^2 \tag{20.38}$$

In the above example, $\beta \simeq -50$.

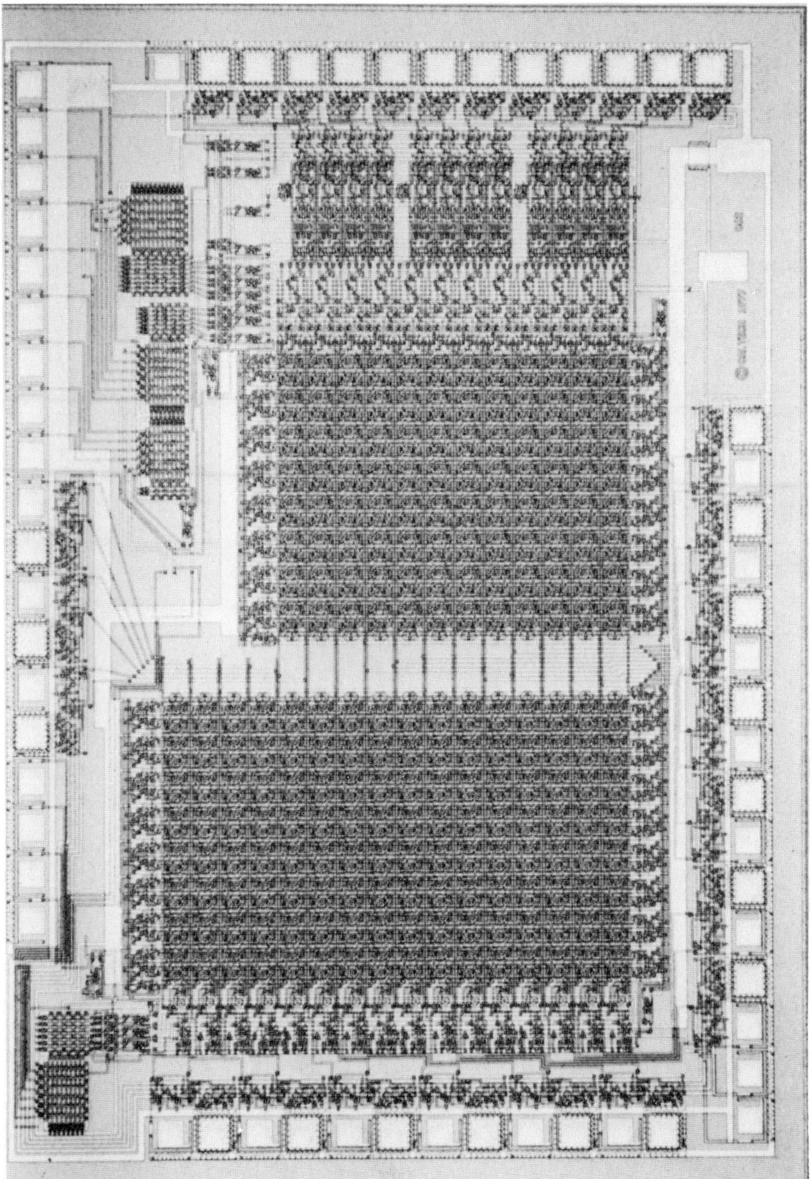

Figure 20.8 Photograph of a modern large-scale integrated circuit fabricated on a single crystal silicon chip. (Courtesy C. A. Mead, Caltech.)

Again the author recalls his own difficulty as a student in understanding the actual physical mechanisms involved in the current amplification β. The following may help: Consider a single electron injected into the base region from a current source. This extra electron will lower slightly the potential barrier in the depletion region A (Fig. 20.6a) for hole injection from region E to B. This will result, in our example, in an *additional* 49 holes diffusing across the base to the collector, thus increasing the current I_C. The 50th hole will recombine with an electron and annihilate it thus restoring the potential distribution to its initial state. The net result is that *fifty* holes flowed into the collector region due to only *one* electron injected into the base—that is, $\beta = -50$. A modern integrated electronic circuit on a single silicon chip is shown in Fig. 20.8. Another important class of transistors, the field effect transistors, are discussed in connection with problem 8.

PROBLEMS

1. Solve for the field and potential distribution in a P-N junction in which the excess impurity density changes linearly in a distance d from a value of N_D on the n side to a value N_A on the p side.
2. Show that in the depletion region of a semiconductor with zero applied bias $p(x)n(x) = n_i^2$. Use thermodynamic reasoning to show that with no applied bias ($V_a = 0$) the net current flow across a P-N junction is zero. Justify, using order of magnitude estimates, the inequality (20.10).
3. Show that in the presence of an applied voltage V_a the carrier concentrations $n(x)$, $p(x)$ in the depletion layer are given by

$$n(x)p(x) = n_i^2 e^{eV_a/kT}$$

4. Derive the equation for the depletion layer thickness, capacitance, and barrier heights as a function of voltage in a metal semiconductor junction (N type) in which the Fermi level at the interface is exactly at midgap.
5. Show that for $N_d \gg N_a$ the concentration $N_a(x)$, where x is the edge of the depletion layer in the p side, is given by

$$N_a(x) = \frac{2}{e\varepsilon}\left(\frac{d(1/C^2)}{dV}\right)^{-1}\Bigg|_{V=V_a}$$

   where C is the junction capacitance per unit area and V_a is the applied external voltage. Since x is a function of V_a, the above relation indicates how $N_a(x)$ may be obtained experimentally from a measurement of C vs. V_a.
6. Draw a figure similar to Fig. 20.6 but for a N-P-N transistor. Give a qualitative description of its operation.

7. Derive the α parameter of a *P*-N-*P* transistor in which the electron component of the emitter junction current I_E cannot be neglected—that is, in which the fraction of electrons moving from right to left across depletion layer *A* in Fig. 20.6 is not negligible compared to that of holes moving from left to right.

8. Consider the device (Metal-Semiconductor Field Effect Transistor) shown in the figure. A doped (*N* type) layer of thickness $t(\sim 2\ \mu\text{m})$ is grown expitaxially (single crystal) on top of an insulating substrate of the same crystal structure. Heavily doped (metal-like) *N* type regions, "source" and "drain", serve as electrodes. A depletion layer of thickness w constricts the electron flow to a channel of height d. Since W depends on V_{GD}, the voltage across the metal-semiconductor junction, the total current I_{SD} is controlled by the gate voltage V_{GD}.

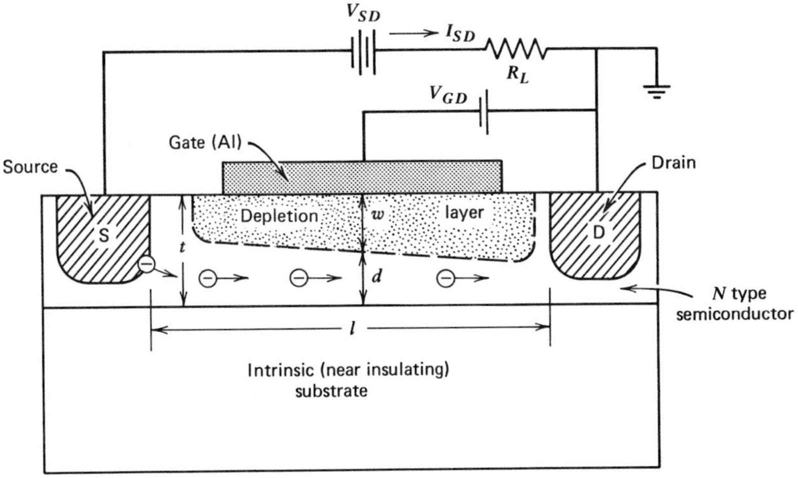

Given the doping density N_D and the conductivity σ of the *N* region, use the discussion of problem 4 to derive an expression for the transconductance.

$$g_m = \left.\frac{\partial I_{SD}}{\partial V_{GD}}\right|_{V_{SD}=\text{const.}}$$

of the field effect transistor.

CHAPTER TWENTY-ONE

The Semiconductor Injection Laser

There is now a major revolution afoot in the communications industry. The transmission of information (telephone conversations, television programming, computer and peripheral equipment hookup) by means of electric signals propagating in metallic conductors is giving way to pulses of light propagating in hair-thin silica fibers. The overwhelming advantages of this new mode of communication are (1) the extreme low loss (<1 db/km) of the signal; (2) the potentially immense information bandwidth that makes it possible to use extreme short pulses ($<10^{-10}$s), each carrying one bit turned on and off at rates exceeding 10^{10} bits/s; and (3) the small diameter (10–100 μm) of the silica fibers.

One of the most important factors that helped make optical fiber communication a reality is the invention in 1962 of the semiconductor injection laser. These are the only lasers pumped directly by an electric current. They are consequently easily incorporated into the electronic circuitry used in the communication technology. In addition, their output wavelengths (0.8 $\mu m < \lambda < 1.6$ μm) fall in the optimal (low loss, low group velocity dispersion) transmission region of the silica fibers.

Our basic understanding of semiconductor physics at this point should be sufficient for a qualitative description of the fundamental principles of operation of semiconductor lasers. For a more detailed and quantitative description the reader is referred to more specialized texts, some of which are listed at the end of the chapter.

21.1 OPTICAL ABSORPTION AND STIMULATED EMISSION IN SEMICONDUCTORS

In very highly doped semiconductors the Fermi level is forced into either (a) the conduction band for donor impurity doping or (b) into the valence band

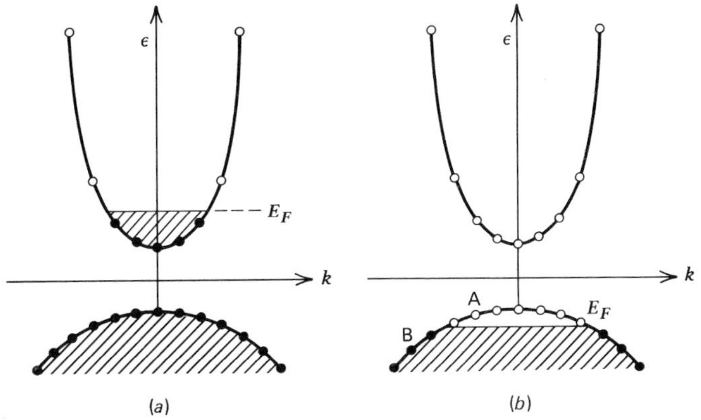

Figure 21.1 (*a*) Energy band of a degenerate *N*-type semiconductor at 0 K. (*b*) A degenerate *P*-type semiconductor at 0 K. The cross-hatching represents regions in which all the electron states are filled. Empty circles indicate unoccupied states.

for acceptor impurity doping. We refer to the semiconductor under these conditions as being "degenerate." This situation is demonstrated by Fig. 21.1. According to the Fermi–Dirac distribution law,

$$f(\epsilon)=\frac{1}{e^{(\epsilon-E_F)/kT}+1} \tag{21.1}$$

at 0 K, all the electron states whose energies lie below E_F are filled, while those above it are unoccupied, as shown in the figure. In this respect the degenerate semiconductor behaves like a metal where the conductivity does not disappear at low temperatures. The unoccupied states in the valence band are referred to as holes, and they are treated like the electrons except that their charge, corresponding to an electron deficiency, is positive and their energy is measured downward.

The reason that the hole energy increases downward can become clear from Fig. 21.1(*b*). When we say that a hole is excited from state *A* to state *B*, which is lower in the diagram, we really mean that an electron is excited from *B* to *A*, which requires an investment of positive energy.

Band-to-Band Transitions and Absorption in Semiconductors

As in ordinary lasers, the amplification of radiation in semiconductors is the exact opposite of absorption. It thus involves a reversal of the position of occupied and unoccupied energy states. In either case we need to understand the nature of the transitions involved.

The interaction Hamiltonian of an electron in a semiconductor and an optical electric field of the form

$$\mathcal{E}(\mathbf{r}, t) = \tfrac{1}{2}\mathcal{E}_0 e^{i(\omega t - \mathbf{k}_{\text{opt}} \cdot \mathbf{r})} + \text{c.c.} \tag{21.2}$$

is given, assuming $\mathcal{E}_0 = \mathbf{i}\mathcal{E}_0$, by

$$\mathcal{H}' = \frac{e\mathcal{E}_0 x}{2}\left(e^{i(\omega t - \mathbf{k}_{\text{opt}} \cdot \mathbf{r})} + \text{c.c.}\right) \tag{21.3}$$

If we apply the result of time-dependent perturbation theory (11.32) to calculate the transition of an electron from the valence band denoted by sub-"vee" to the conduction band (sub-"cee") under the influence of an electric field $\mathcal{E}(r, t)$, we must use the perturbation matrix element

$$\mathcal{H}'_{vc} = \frac{e\mathcal{E}_0}{2} \int u^*_{v\mathbf{k}}(\mathbf{r}) u_{c\mathbf{k}'}(\mathbf{r}) x e^{(i\mathbf{k}' - \mathbf{k} - \mathbf{k}_{\text{opt}}) \cdot \mathbf{r}} \, dv \tag{21.4}$$

where we used (17.35) for the form of the electron wavefunctions. The rapid phase fluctuation of the factor $\exp[i(\mathbf{k}' - \mathbf{k} - \mathbf{k}_{\text{opt}}) \cdot \mathbf{r}]$ will cause $\mathcal{H}'_{vc}$ to be vanishingly small except when

$$\mathbf{k}' - \mathbf{k} \cong \mathbf{k}_{\text{opt}} \tag{21.5}$$

in the optical region $k_{\text{opt}} \sim 2\pi/\lambda \sim 10^5 \text{ cm}^{-1}$, while for the electrons we have —except in the immediate vicinity of the band extrema—$|\mathbf{k}' - \mathbf{k}| \sim k \sim 10^8$ cm^{-1}. We may thus assume $k_{\text{opt}} \simeq 0$ and write the condition for appreciable transition rates as

$$\mathbf{k}' = \mathbf{k} \tag{21.6}$$

so that transitions occur mostly between initial and final electron states with the same **k** vector. This is referred to as conservation of crystal momentum.

Consider next the absorption coefficient $\alpha(\omega)$ of a plane electromagnetic wave of radian frequency ω propagating in a bulk undoped (intrinsic) semiconductor with an energy diagram as shown in Fig. 21.2. As discussed above, the transition conserves "momentum" and is represented, consequently, by a vertical arrow. The sample is assumed to be in thermal equilibrium, and the Fermi level near mid-gap is far enough from the band edges that all levels in the valence band are full and those in the conduction band are empty. We find it most convenient to start with (11.32)

$$W_{ab} = \frac{2\pi}{\hbar} |\mathcal{H}'_{ab}|^2 \delta(E_b - E_a - \hbar\omega) \tag{21.7}$$

for the probability rate for a transition from state a to b. According to Fig.

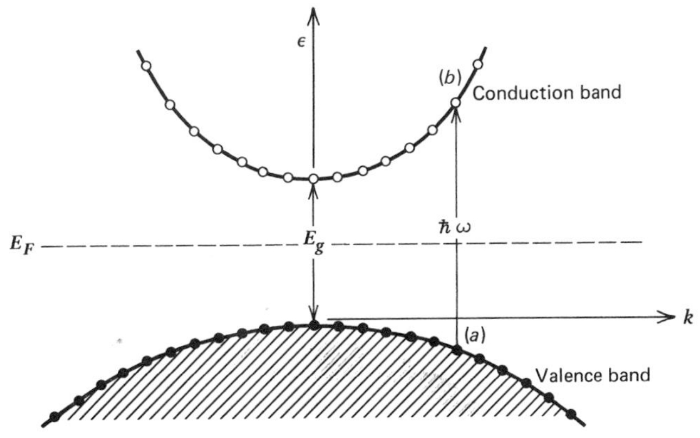

Figure 21.2 The absorption of a photon in a semiconductor due to a transition of an electron from an occupied state (*a*) in the valence band to an empty state (*b*) in the conduction band.

21.2 we have

$$E_b - E_a = \frac{\hbar^2 k^2}{2}\left(\frac{1}{m_c} + \frac{1}{m_v}\right) + E_g$$

for the transition indicated by an arrow in the figure. The probability rate for this *single* transition can thus be written as

$$W_k = \frac{2\pi}{\hbar} |\mathcal{H}'_{vc}(k)|^2 \delta\left(\frac{\hbar^2 k^2}{2m_r} + E_g - \hbar\omega\right) \tag{21.8}$$

where $m_r = m_v m_c/(m_v + m_c)$ is the reduced effective mass. The total number N of transitions per second in a crystal of volume V is given by multiplying (21.8) by $\rho(k)$, the number of states per unit k (in V) and then integrating over all values of k. Using (15.43) we write

$$N = \frac{2V}{\pi\hbar} \int_0^\infty |\mathcal{H}'_{vc}(k)|^2 \delta\left(\frac{\hbar^2 k^2}{2m_r} + E_g - \hbar\omega\right) k^2\, dk \tag{21.9}$$

Introducing the variable X by

$$X = \frac{\hbar^2 k^2}{2m_r} + E_g - \hbar\omega$$

the last integral becomes

$$N = \frac{2V}{\pi\hbar} \int |\mathcal{H}'_{vc}(k)|^2 \frac{m_r}{\hbar^2} \delta(X) \sqrt{\frac{2m_r}{\hbar^2}(X + \hbar\omega - E_g)}\, dX$$

$$= \frac{V}{\pi} |\mathcal{H}'_{vc}(k)|^2 \frac{(2m_r)^{3/2}}{\hbar^4} (\hbar\omega - E_g)^{1/2} \tag{21.10}$$

where $\hbar^2k^2/2m_r + E_g = \hbar\omega$. The absorption coefficient $\alpha(\omega)$ is given by

$$\alpha(\omega) = \frac{\text{power absorbed per unit volume}}{\text{power crossing a unit area}}$$

$$= \frac{N\hbar\omega/V}{\varepsilon_0 n \mathcal{E}_0^2/2}$$

where n is the index of refraction, c the velocity of light in vacuum, and $\mathcal{E}_0$ is the field amplitude. Using (21.10) we obtain

$$\alpha_0(\omega) = \frac{\omega e^2 x_{vc}^2 (2m_r)^{3/2}}{2\pi\varepsilon_0 nc\hbar^3}(\hbar\omega - E_g)^{1/2} \tag{21.11}$$

where we replaced, in conformity with (21.4), $\mathcal{H}'_{vc}(k)$ by $e\mathcal{E}_0 x_{vc}/2$ with $x_{vc} \equiv \langle u_{vk}|x|u_{ck}\rangle$. In practice the numerical coefficients are lumped together and $\alpha_0(\omega)$ is expressed as

$$\alpha_0(\omega) = K(\hbar\omega - E_g)^{1/2} \tag{21.12}$$

where K can be determined from absorption data. In gallium arsenide (GaAs), for example, the following data apply:

$$E_g \simeq 1.5 \text{ eV}$$

$$m_v = 0.1 \text{ m}_{\text{electron}}$$

$$m_c = 0.065 \text{ m}_{\text{electron}}$$

$$K \simeq 6 \times 10^3 \text{ cm}^{-1} (\text{eV})^{-1/2}$$

so that at a frequency whose photon energy, as an example, exceeds the gap energy E_g by 0.01 eV, the absorption coefficient is $\alpha_0(\omega) = 6 \times 10^3 \times 10^{-1} = 600 \text{ cm}^{-1}$.

Now assume that by some means we can prepare a semiconducting crystal in which the states up to some level in the conduction band are all occupied and those above a certain level in the valence band are empty. This situation is illustrated in Fig. 21.3, where at 0 K all the conduction states up to the quasi-Fermi level E_{Fc} are occupied while all the valence band states down to the quasi-Fermi level E_{Fv} are empty. The calculation of the absorption coefficient at $\hbar\omega$ is *identical* to that leading to (21.11) except that, since the upper states are full while the lower ones are empty, the sign of the absorption coefficient is reversed; that is, the radiation is amplified rather than absorbed. We thus have for the semiconductor depicted by Fig. 21.3,

$$\begin{aligned}
&\alpha(\omega) = \alpha_0(\omega) = 0 && \hbar\omega < E_g && \\
&\alpha(\omega) = -\alpha_0(\omega) = -K(\hbar\omega - E_g)^{1/2} && E_g < \hbar\omega < E_{Fc} - E_{Fv} && \text{Amplification} \\
&\alpha(\omega) = \alpha_0(\omega) = K(\hbar\omega - E_g)^{1/2} && \hbar\omega > E_{Fc} - E_{Fv} && \text{Absorption}
\end{aligned} \tag{21.13}$$

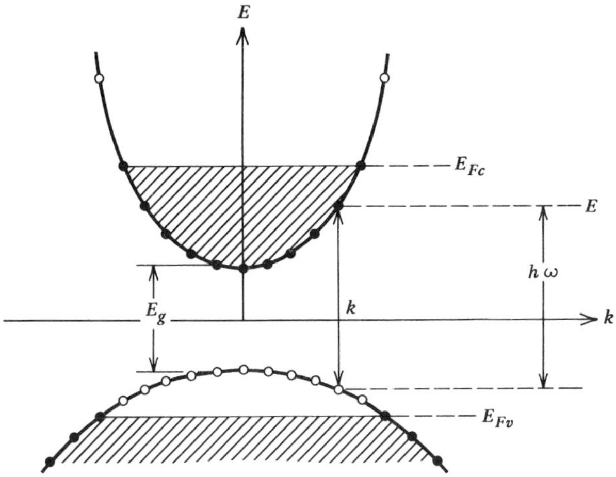

Figure 21.3 A semiconductor with degenerate electron and hole populations. At a frequency $\hbar\omega$, as shown, the sign of the absorption is reversed with respect to that of Fig. 10.25 and amplification results. Full dots denote states occupied by electrons while empty circles denote vacant states.

This is the argument used by Dumke[2] in predicting and estimating the possibility of laser action in semiconductors. This argument shows that in the case of the GaAs example considered above, in which $\hbar\omega - E_g = 0.01$ eV, the exponential loss coefficient—that is, α_0 in $I = I_0 e^{-\alpha_0 z}$—is 600 cm^{-1}. This will result in an exponential gain constant of $g_0 = 600\ \text{cm}^{-1}$ in a semiconductor with inverted populations. This is the reason why lasers utilize direct semiconductors with their intense absorption (hence amplification) characteristics.

We can extend the arguments of the last section to a system at some finite temperature. In calculating the rate of upward transitions we must multiply (21.9) by the probability that the lower state a is occupied, times the probability that the upper level b is empty, since clearly no transitions can originate in an empty state or terminate in a state that is already full. This factor is given by $f_v(E_a)[1 - f_c(E_b)]$. To obtain the *net* rate of upward (absorbing) transition, we must subtract from the $a \to b$ rate the rate for downward transitions. The matrix element and integration involved in this second rate are identical to that leading to (21.9), except that the relevant occupancy factor now becomes $f_c(E_b)[1 - f_v(E_a)]$, which is the probability that the upper level b (in the conduction band) is occupied while that of the lower level a is not.

The excess of upward (absorption) over downward (amplifying) transitions per second is now given by (21.9) modified by inclusion of occupancy

[2]W. P. Dumke, Interband transitions and maser action, *Phys. Rev.* **127**, 1559 (1962).

factors:

$$N_{a \to b} - N_{b \to a} = \frac{2V}{\pi\hbar} \int_0^\infty |\mathcal{H}'_{vc}(k)|^2 [f_v(1-f_c) - f_c(1-f_v)]$$

$$\times \delta\left(\frac{\hbar^2 k^2}{2m_r} + E_g - \hbar\omega\right) k^2\, dk$$

$$= \frac{2V}{\pi\hbar} \int_0^\infty |\mathcal{H}'_{vc}(k)|^2 [f_v(E_a) - f_c(E_b)]\, \delta\left(\frac{\hbar^2 k^2}{2m_r} + E_g - \hbar\omega\right) k^2\, dk \tag{21.14}$$

where

$$E_b - E_a = E_g + \frac{\hbar^2 k^2}{2m_r} = \hbar\omega \tag{21.14a}$$

Before proceeding we pause to draw some important conclusions from (21.14). The condition for amplification is that $N_{a \to b} < N_{b \to a}$, which, according to (21.14), requires that

$$f_c(E_b) - f_v(E_a) > 0 \tag{21.15}$$

Using (21.2) with quasi-Fermi levels E_{Fv} for the valence band and E_{Fc} for the conduction band, the last equation yields

$$E_{Fc} - E_{Fv} > \hbar\omega \tag{21.16}$$

This is the so-called Bernard–Duraffourg inversion condition for semiconductors. We note that in thermal equilibrium $E_{Fc} = E_{Fv}$, and the system is thus either transparent or absorbing at *all* frequencies.

Returning to (21.14), the integration is identical to that leading to (21.11), thus resulting in

$$\alpha(\omega) = \alpha_0(\omega)[f_v(E_a) - f_c(E_b)]$$

so that the gain constant $\gamma(\omega) = -\alpha(\omega)$ is

$$\gamma(\omega) = \alpha_0(\omega)[f_c(E_b) - f_v(E_a)] \tag{21.17}$$

where E_a and E_b satisfy (21.14*a*).

The maximum gain attainable in a given semiconductor at ω is thus numerically equal to $\alpha_0(\omega)$—the zero temperature absorption of the intrinsic material. To obtain this gain we need to excite the material so that $f_c(E_b) = 1$ and $f_v(E_a) = 0$; that is, the upper laser level (conduction band) is fully occupied $[f_c(E_b) = 1]$ while the valence band final state is empty $[f_v(E_a) = 1]$.

Since $\alpha_0(\omega)$ is the (intrinsic) zero temperature limit of the absorption, the temperature and excitation dependence of the gain $\gamma(\omega)$ must be contained in the factor $[f_c(E_b) - f_v(E_a)]$ in (21.17). This point may be better appreciated by referring to Fig. 21.4, which is a graphical representation of (21.17). Two

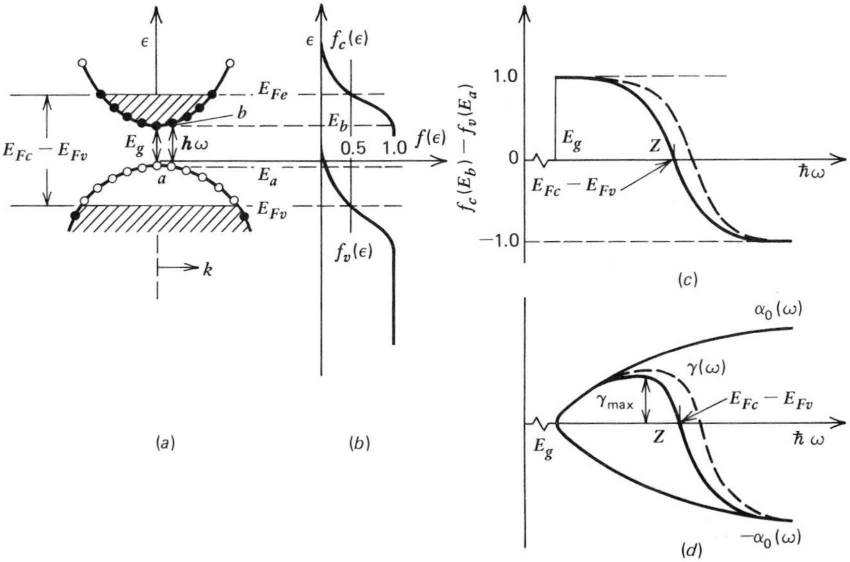

Figure 21.4 A graphical step-by-step construction of the gain profile $\gamma(\omega)$ according to (10.7)–(10.19). The dashed curves in (c) and (d) correspond to a higher excitation level. This causes the point Z to move to the right resulting in an increase of γ_{max} and of the frequency at which it occurs.

important points stand out. The effect of larger excitation is to increase the number of carriers in the two bands and thus to increase the maximum gain γ_{max} as indicated by the dashed curves in Figs. 21.4c and 21.4d. An increase in temperature broadens the transition region of the Fermi functions. This, according to Fig. 21.4d, reduces γ_{max}.

The *P*-N Junction Laser The first laser action in semiconductors was obtained in degenerate gallium arsenide *P*-N junctions. Under conditions of high current injection in such a junction, there exists a region near the depletion layer that contains, simultaneously, a degenerate population of electrons and holes. Figure 21.5 shows a degenerate *P*-N junction. With zero-applied bias as in Figure 21.5a (or for low-bias voltages), the condition $E_{F_c} - E_{F_v} > \hbar\omega$ is not satisfied and no amplification can result. With an applied forward voltage nearly equal to the gap voltage E_g/e, as shown in Fig. 21.5b, there exists an "active" region containing both degenerate electron and hole populations. For a typical frequency ω, such as shown in the figure, the gain condition $\hbar\omega < E_{F_c} - E_{F_v}$ is satisfied, and that portion of the radiation at ω that is confined to the active region is amplified.

The thickness of the active layer in this simple case can be approximated on the basis of the distance that the electrons, injected into the P region, can diffuse before recombining with a hole—that is, before making a transition to the valence band. Using a diffusion coefficient of $D=10$ cm^2/sec, and a

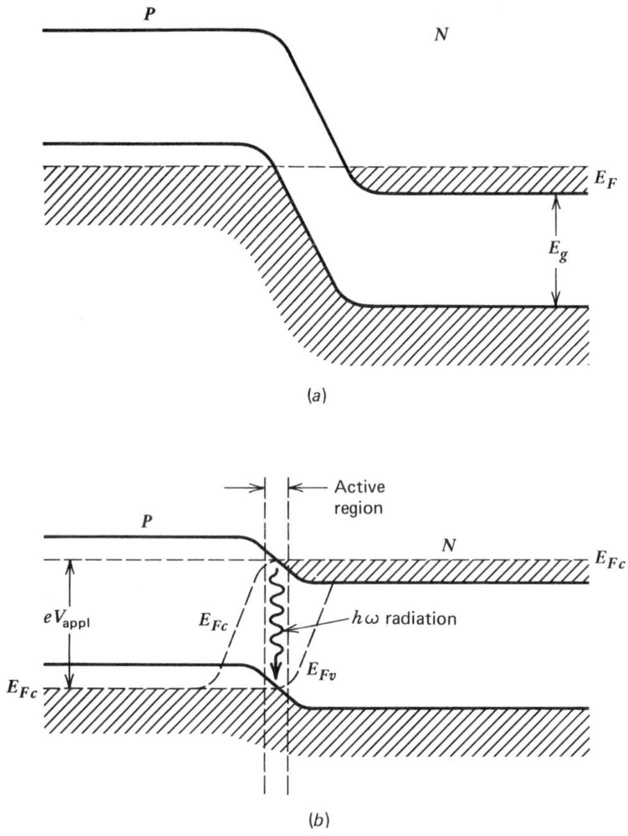

Figure 21.5 (a) Degenerate P-N junction at zero-applied bias. (b) At a forward bias voltage $V_{\text{appl}} \simeq E_g/e$. The region containing both electrons and holes is called the active region. The oscillatory arrow indicates a recombination of an electron with a hole in the active region leading to an emission of a photon with energy $\hbar\omega \sim (E_{F_c} - E_{F_v})$. The active region extends into the plane of the figure.

recombination time of $\sim 10^{-9}$ sec, we obtain $\sqrt{Dt_{\text{recomb}}} \simeq 10^{-4}$ cm for the thickness of the active region.

Early GaAs laser junctions were obtained by diffusing the acceptor atoms, such as Zn, from a very high surface concentration $N_A \sim 10^{20}$ cm^{-3}, into an N-type GaAs crystal doped, typically, by 10^{18} atoms/cm^{-3}. Te atoms were used often as the donor dopant. More recent lasers use epitaxially grown layers with different doping to form GaAlAs-GaAs heterojunctions. This point is discussed below.

In Fig. 21.6 is shown a sketch of a simple P-N junction laser. The optical resonator which is necessary for oscillation is formed by polishing a pair of opposite crystal faces. In GaAs these are often the naturally cleaving (110) faces.

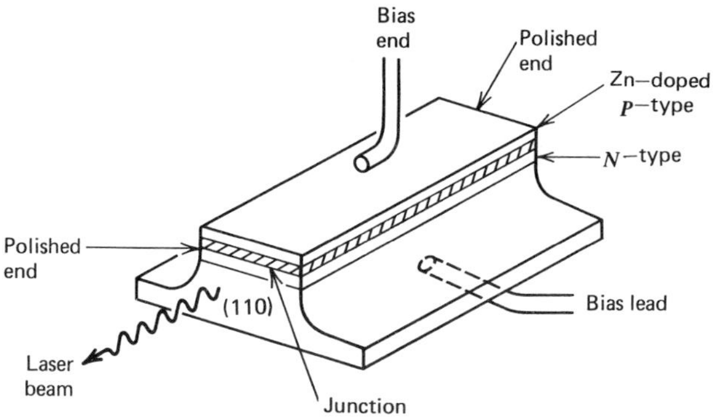

Figure 21.6 A *P*-N junction laser in GaAs. Two parallel cleavage planes serve as the resonator reflectors.

Modern semiconductor injection lasers utilize a multilayer structure grown by liquid or vapor phase epitaxy. In these structures the GaAs active layer in which light is generated by stimulated electron-hole recombination is sandwiched between $Ga_{1-x}Al_xAs$ layers. $Ga_{1-x}Al_xAs$ is a single crystal material in which a fraction x of the gallium atoms is replaced by aluminum atoms. This ternary (three-atom) crystal has a higher energy gap than GaAs and a smaller index of refraction. This situation is sketched in Fig. 21.7. The multilayer structure is grown epitaxially so that it consists of a single uninterrupted crystal.

The main reason for the use of the $Ga_{1-x}Al_xAs$ layers is that due to the increase in the index of refraction it becomes possible for laser light rays propagating inside the central GaAs layer to be totally internally reflected at the interfaces with the two $Ga_{1-x}Al_xAs$ bounding layers. This confines the laser radiation very effectively to the active layer, whose thickness is typically 2000 Å and prevents it from spreading by diffraction. This confinement increases the laser light intensity and consequently the laser gain.

The process of light confinement in a high index slab is described rigorously by using the concept of dielectric waveguide mode. The understanding of these modes is important when dealing with injection lasers. For further reading, see the bibliography.

The recombination efficiency of a semiconductor junction is defined as the probability that injected carriers recombine through the emission of a photon. In some of the good semiconductor materials (GaAs, GaInAsP, GaAlAs) the recombination efficiency exceeds 50%. From Fig. 21.5*b* it follows that the energy of a photon given off in the process of electron-hole recombination is very nearly equal to E_g—the energy gap. Since the applied forward voltage V_a at the high current needed to reach threshold is very nearly the gap voltage E_g/e, it follows that the ratio of the energy of an emitted photon to the energy imparted—by the power supply—to one electron flowing through the circuit is nearly unity. The net result is that injection lasers are efficient converters of electric to light power.

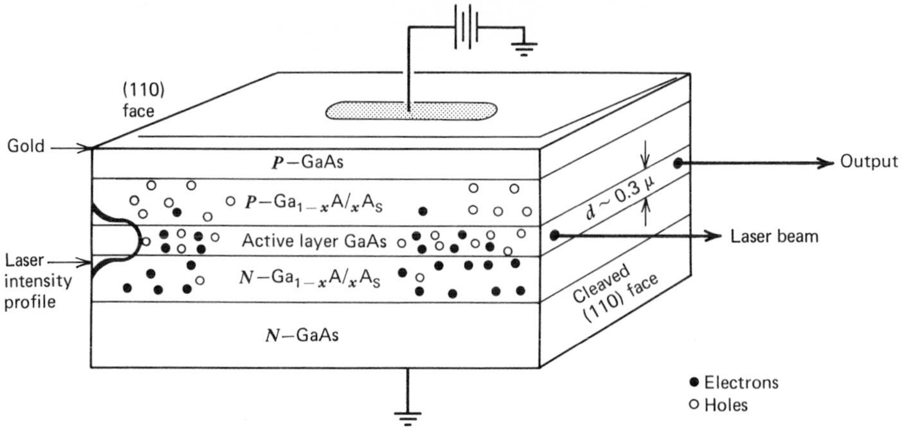

Figure 21.7 A typical double heterostructure GaAs-GaAlAs laser. Electrons and holes are injected into the active GaAs layer from the N and P $Ga_{1-x}Al_xAs$ layers, respectively. Frequencies near $\nu = E_g/h$ are amplified by stimulating electron-hole recombination.

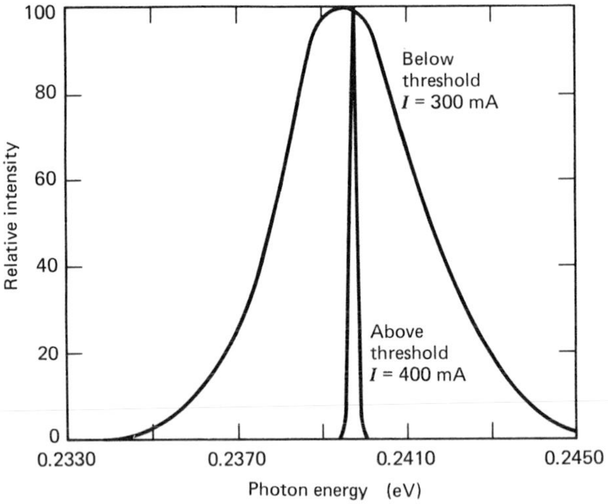

Figure 21.8 Infrared-emission spectra of a forward-biased InSb diode at 1.7°K below and above threshold. The broad spectrum below threshold is due to the spontaneous recombination of electrons and holes. Width of line above threshold is limited by the resolution of spectrometer. The above threshold plot has been reduced in its vertical scale so as to fit in the same plot. The actual area under this plot is much larger than under the "below threshold" curve.

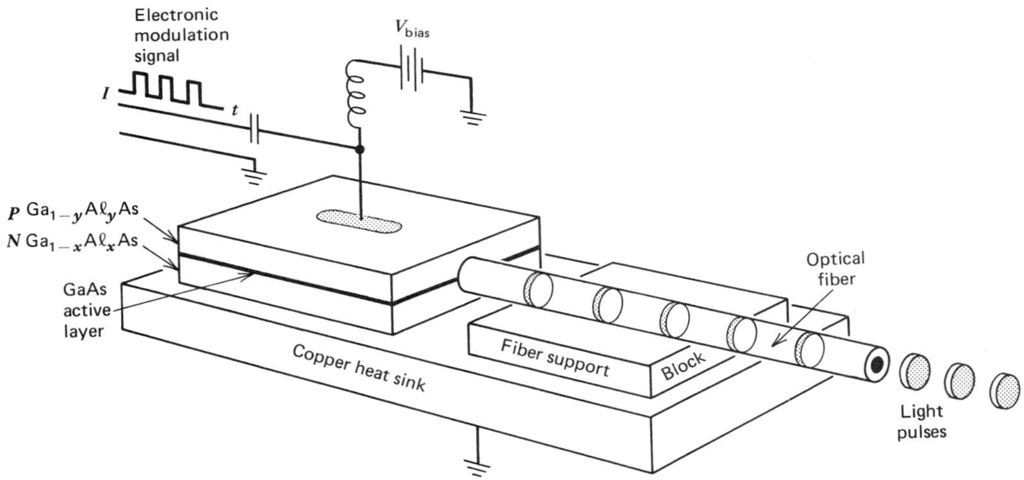

Figure 21.9 A GaAs/GaAlAs double heterostructure semiconductor laser is modulated by a binary electrical current signal. By turning a semiconductor laser on and off, this modulation is replicated as a series of optical pulses that are fed into an optical fiber, which then carries them to the receiving station.

A typical spectrum of the output light above and below threshold is shown in Fig. 21.8.

We conclude with a sketch, Fig. 21.9, showing a diode laser powering an optical fiber communication channel.

PROBLEMS

1. A somewhat idealized model for the dielectric waveguiding taking place in the double heterojunction diode laser is shown in Fig. 21.10. The scalar wave equation obeyed by the electromagnetic modes (of radiation with a frequency $\omega/2\pi$) is

$$\left(\nabla^2+\frac{\omega^2}{c^2}n_i^2\right)E=0$$

Taking the modes of the structure in the form of

$$E_l=A_l e^{i(\omega t-\beta_l z)}f_l(x)$$

Solve for the mode functions $f_l(x)$ and the propagation constants β_l. Show that for sufficiently small values of $(d/\lambda)\sqrt{n_1^2-n_2^2}$, only one confined mode—that is, a mode with an exponential decreasing x dependence in the cladding layers—can exist.

Hint: This problem is formally equivalent to that of the electron in a finite potential well, which is treated in Section 4.2. The boundary conditions are also similar to those of Section 4.2, namely, that E_y and $\partial E_y/\partial x$ are continuous across the interface and must vanish at $x=\pm\infty$.

2. (a) Solve for the mode function $f_0(x)$ and propagation constant β_0 of the lowest order mode of the waveguide shown in Fig. 21.11.
   (b) Plot the transverse mode profile.
   (c) What fraction of the mode power is propagating inside the active region?

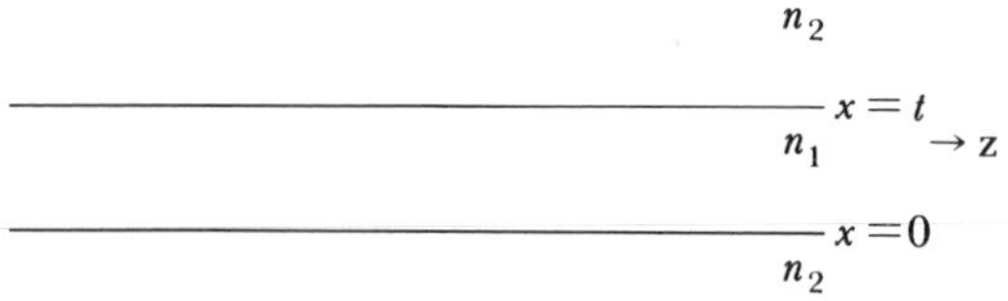

Figure 21-10 A two dimensional slab waveguide.

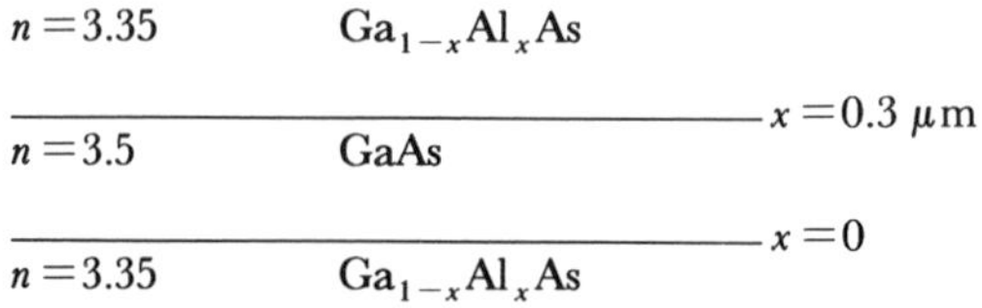

Figure 21-11 A typical GaAs/GaAlAs waveguide used in double heterostructure lasers.

Bibliography

Davisson, C., and L. H. Germer, The scattering of electrons by a single crystal of nickel, *Nature* **119**, 558 (1927); Diffraction of electrons by a crystal of nickel, *Phys. Rev.* **30**, 705 (1927).

de Broglie, L., Investigations on the theory of quanta. *Ann. Phys.* **3**, 22 (1925).

Einstein, A. On a heuristic point of view concerning the generation and transformation of light, *Ann. Phys.* **17**, 132 (1905); especially Section 8.

Einstein, A., and L. Infeld. *The Evolution of Physics.* New York: Simon & Schuster, 1961.

Gamow, G., "Mr. Tompkins in Wonderland" and "Mr. Tompkins Explores the Atom." Both stories are included in *Mr. Tompkins in Paperback.* New York: Cambridge University Press, 1965.

Gray, P. E., L. Campbell, and L. Searle. *Electronic Principles* New York: John Wiley & Sons, 1969.

Hansch, T., A. L. Schawlow, and G. Series. The spectrum of atomic hydrogen. *Scientific American* **240**, No. 3 (1979).

"Lasers and light." readings from *Scientific American* (1969).

Miller, S. E., E. A. J. Marcatili, and T. Li. Research toward optical transmission systems. *Proc. IEEE* **61**, 1703 (1973).

Svelto, O. J. *Principles of Lasers.* New York, Plenum Press, 1976).

Sze, S. M. *Physics of Semiconductor Devices.* New York: Wiley-Interscience, 1969.

Thomson, G. P., and A. Reid, Diffraction of cathode rays by a thin film, *Nature* **119**, 890 (1927); P. Thomson, Experiments on the diffraction of cathode rays, *Proc. Roy Soc.* **A117**, 600 (1928); **A119**, 651 (1928); A. Reid, The diffraction of cathode rays by thin celluloid films, *Proc. Roy. Soc.* **A119**, 663 (1928).

Van der Ziel, A. *Solid State Physical Electronics.* Englewood Cliffs, N.J.: Prentice -Hall, 1976.

Yariv, A. *Introduction to Optical Electronics.* New York: Holt, Rinehart, and Winston, 1976.

For theory of dielectric waveguiding.

Yariv, A. *Quantum Electronics*, 2nd ed. New York: J. Wiley & Sons, 1975.

For additional information on semiconductor lasers:

(1) Panish, M. B. and H. C. Casey "Heterostructure Lasers" Academic Press, New York, 1978

(2) J. K. Butler, "Semiconductor Lasers." IEEE Press, New York, 1980

INDEX